Two Manuals Combined into One Book

Browning Machine Gun .50 HB M2

and

MK 19, 40-MM Grenade Machine Gun

Department of Defense

Pentagon
Publishing

*** C1, FM 23-65**

FIELD MANUAL
No. 23-65

HEADQUARTERS
DEPARTMENT OF THE ARMY
Washington, DC, 19 June 1991

BROWNING MACHINE GUN
CALIBER .50 HB, M2

CONTENTS

Page

DISTRIBUTION RESTRICTION: Approved for public release; distribution is unlimited.

---

*This publication supersedes FM 23-65, 19 May 1972, and TC 23-65-1, 19 September 1984.

* PREFACE

This manual provides technical information, training techniques, and guidance on the caliber .50 HB machine gun, M2. Unit leaders and designated gunners will use this information to successfully integrate the weapon into combat operations. They can instruct on the range or at concurrent training stations.

The material applies as is to both nuclear and conventional warfare.

Trainers must ensure that everyone observes safety procedures at all times. Commanders, trainers, and individual students must remember that safety is everyone's responsibility. Leaders will conduct all training as though each weapon were fully loaded. At no time during training will anyone allow the desire for speed or accuracy to override the requirement to follow safety procedures. Safe training is good training.

The proponent of this publication is US Army Infantry School. Send comments and recommendations on DA Form 2028 (Recommended Changes to Publications and Blank Forms) directly to the Commandant, US Army Infantry School, ATTN: ATSH-INB-O, Fort Benning, GA 31905-5594 or lusanoh@benning.army.mil.

Unless otherwise stated, the masculine gender refers to both men and women.

CHAPTER 1

INTRODUCTION

The procedures and methods used in Army machine gun marksmanship are based on the concept that soldiers must be skilled gunners who can effectively apply their firing skills in combat. The basic firing skills and exercises outlined in this manual must be a part of every unit's machine gun training program. The soldiers' proficiency depends on proper training and application of basic gunnery fundamental, which are taught in a progressive program to prepare gunners for combat.

1-1. TRAINING STRATEGY

Training strategy is the overall concept for integrating resources into a program to train individual and collective skills needed to perform a unit's wartime mission.

a. Training strategies for marksmanship are implemented in TRADOC institutions (NCOES, basic and advanced officer's courses) and in units. The overall training strategy is multifaceted and is inclusive of the specific strategies used in institution and unit programs. Also included are the supporting strategies that use resources such as publications, ranges, ammunition, training aids, devices, simulators, and simulations. These strategies focus on developing critical soldier skills, and on leader skills that are required for the intended outcome.

b. Two primary components compose the training strategies: initial training and sustainment training. Both may include individual and collective skills. Initial training is critical because a task that is taught correctly and learned well is retained longer. Well-trained skills can be more quickly regained and sustained if an interim of nonuse occurs. The more difficult and complex the task, the harder it is to sustain the skill. Personnel turnover is a main factor in decay of collective skills, since the loss of critical team members requires retraining to regain proficiency. If a long period elapses between initial and sustainment training sessions or training doctrine is altered, retraining maybe required.

c. The training strategy for caliber .50 MG marksmanship begins in selected resident training and continues in the unit. An example of this overall process is illustrated in Figure 1-1 and provides a concept of the flow of unit sustainment training. The soldiers graduating from selected resident training courses have been trained to maintain their MGs and to hit a variety of targets. They have learned range determination, target detection, application of marksmanship fundamentals, and other skills needed to engage a target. Task training during these courses may lead to qualification.

d. Training continues in units on the basic skills taught in combat arms. Additional skills, such as suppressive fire and supporting fire, are trained and then integrated into collective training exercises, which include platoon and squad live-fire STXs. (A unit-marksmanship training program is explained in Chapter 5.) The strategy for sustaining the basic marksmanship skills taught in combat arms is periodic preliminary instruction, followed by qualification range firing. However, a unit must set up a year-round program to sustain skills. Key elements include training of trainers and refresher training of nonfiring skills.

e. Additional skills trained in the unit include techniques for employment, suppressive fires, night fire, MOPP firing, and moving targets. Related soldier skills of camouflage, cover and concealment, maneuver, and preparation and selection of a fighting position are addressed in STP 21-24-SMCT, which must be integrated into tactical training.

f. In the unit, individual and leader proficiency of marksmanship tasks are integrated into collective training to include squad, section, and platoon drills and STXs. The collective tasks in these exercises, and how they are planned and conducted, are in the MTP and battle drill books for each organization. Based on the type organization, collective tasks are evaluated to standard and discussed during leader and trainer after action reviews. Objective evaluations of both individual and unit proficiency provide readiness indicators and future training requirements.

g. A critical step in the Army's overall marksmanship training strategy is to train the trainers and leaders first. Leader courses and unit publications develop officer and NCO proficiencies necessary to plan and conduct marksmanship training and to evaluate the effectiveness of unit marksmanship programs. Training support materials are provided by the proponent schools to include field manuals, training aids, devices, simulators, and programs that are doctrinal foundations and guidance for training the force.

h. Once the soldier understands the weapon and has demonstrated skill in zeroing, additional live-fire training and a target acquisition exercise at various ranges are conducted. Target types and scenarios of increasing difficulty must be mastered to develop proficiency.

i. Initial individual training culminates in the soldier's proficiency assessment, which is conducted on a transition/record fire range. This evaluation also provides an overview of unit proficiency and training effectiveness.

j. Unit training programs maintain the soldiers' proficiency level. The ultimate goal of a unit marksmanship program is to maintain well-trained gunners so a unit can survive and win on the battlefield. The trainer must realize that qualification is not an end, but a step toward reaching this combat requirement. (See Figure 1-1.)

(1) To reach this goal, the gunner must be able to position and use his weapon under the following combat conditions:

- Enemy personnel are seldom visible except when assaulting.
- Most combat fire must be directed at an area where the enemy has been detected or where he is suspected of being located but cannot be seen. Area targets consist of objects or outlines of men irregularly spaced along covered and concealed areas (ground folds, hedges, or borders of woods).
- Most combat targets can be detected by smoke, flash, dust, noise, or movement and are visible only for a moment.
- Some combat targets can be engaged by using nearby objects as reference points.
- The nature of the target and irregularities of terrain and vegetation may require a firer to use a variety of positions in addition to the prone or supported position to fire effectively on the target. In a defensive situation, the firer usually fires from a supported position.
- Most combat targets have a low contrast outline and are obscure. Therefore, choosing an aiming point in elevation is difficult.
- Time-stressed fire in combat can be divided into three types: a single, fleeing target that must be engaged quickly; distributed targets engaged within the time they remain available; and a surprise target that must be engaged at once with accurate, instinctive fire.

(2) The unit's program must provide fundamental training to sustain and improve the skills and proficiency the soldier has attained during his basic marksmanship training. Once basic skills have been mastered, these must be improved by conducting new or advanced individual and collective training. The program must develop collective firing skills by incorporating marksmanship into tactical exercises. This training must maintain the soldier's confidence in the weapon and his skills. A soldier's

survival may depend on his ability to defend himself or other members of the unit. Therefore, individual and collective firing skills must support the expected battlefield conditions and the unit's combat mission.

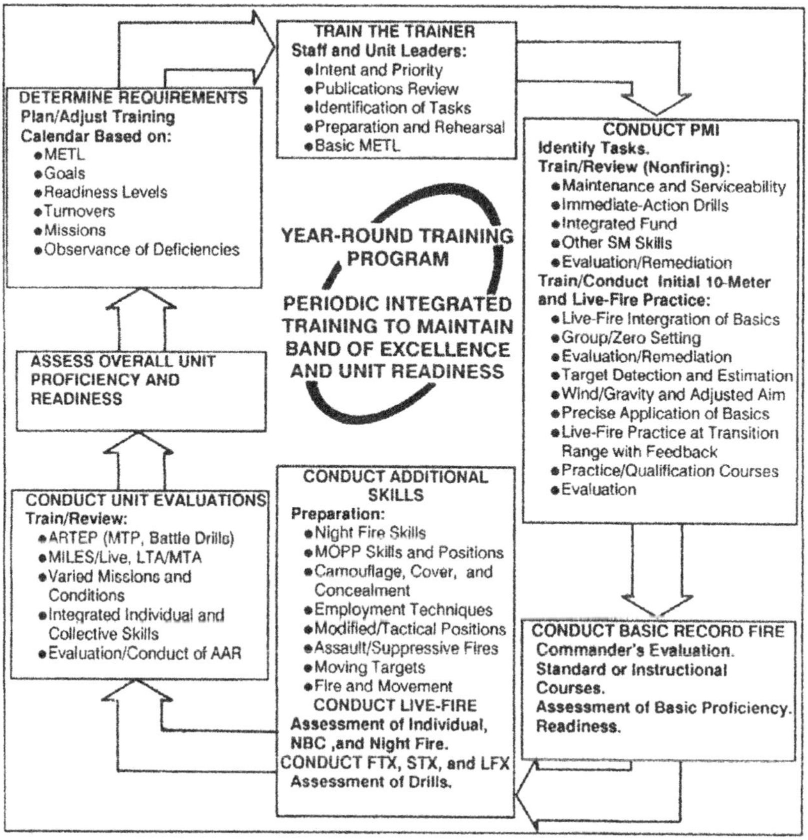

Figure 1-1. Unit marksmanship sustainment strategy.

(3) A unit's marksmanship program must be battlefield oriented. It must be based upon several individual combat tasks as well as organizational, operational, or contingency missions. It must have available resources such as ammunition, time, ranges, and qualified trainers. This manual provides the information a unit commander needs to develop an effective marksmanship program for his unit requirements.

(4) General marksmanship, training knowledge, and accurate firing are acquired skills that perish easily. Skill practice should be conducted for short periods throughout the year. Most units have a readiness requirement that all soldiers must zero their MGs within a certain time after unit assignment. Also, soldiers must confirm the zero of their assigned MGs before conducting a qualification firing.

1-2. DESCRIPTION

The Browning machine gun caliber .50 HB, M2 (Figure 1-2) is a belt-fed, recoil-operated, air-cooled, crew-served machine gun. The gun is capable of single shot, as well as automatic fire, and operates on the short recoil principle.

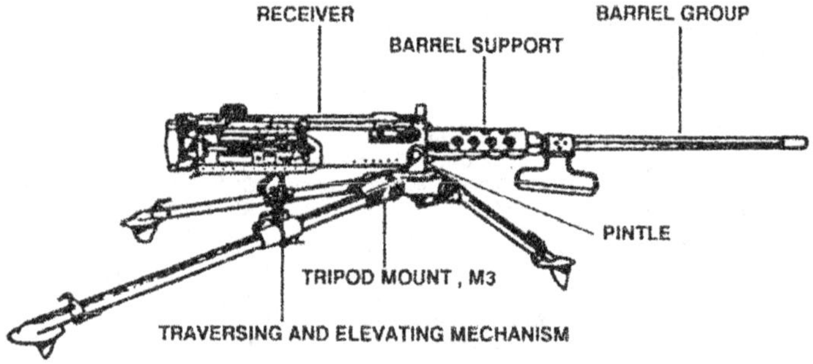

Figure 1-2. Browning machine gun.

a. The machine gun is capable of being fed from either the right or left by repositioning certain parts. The weapon has nonfixed headspace that must be set. Timing must also be adjusted to cause the gun to fire slightly out of battery to prevent damage to moving parts. The force for recoil operation is furnished by expanding powder gases, which are controlled by various springs, cams, and levers. Maximum surface of the barrel and receiver are exposed to permit air cooling. Perforations in the barrel support allow air to circulate around the breach end of the barrel and help in cooling the parts. A heavy barrel is used to retard early overheating.

b. The gun has a leaf-type rear sight (Figure 1-3), graduated in both yards and roils. The scale ranges from 100 to 2,600 in yards, and from 0 to 62 in mils. The windage knob-permits deflection changes to right or left of center. The front sight is a fixed blade type with cover (Figure 1-4).

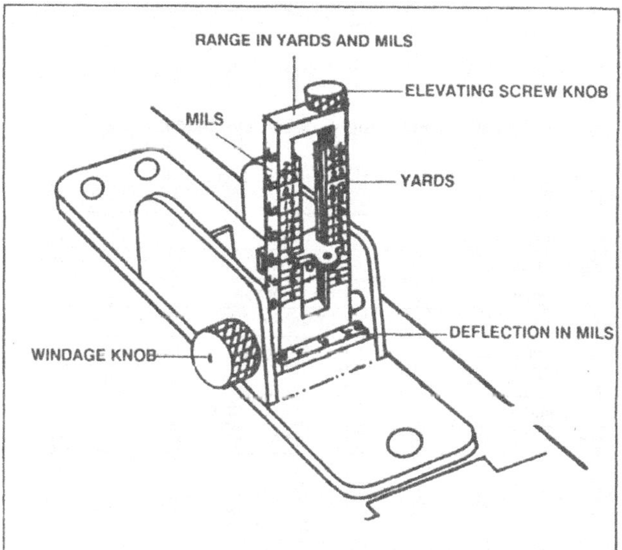

Figure 1-3. Leaf type rear sight.

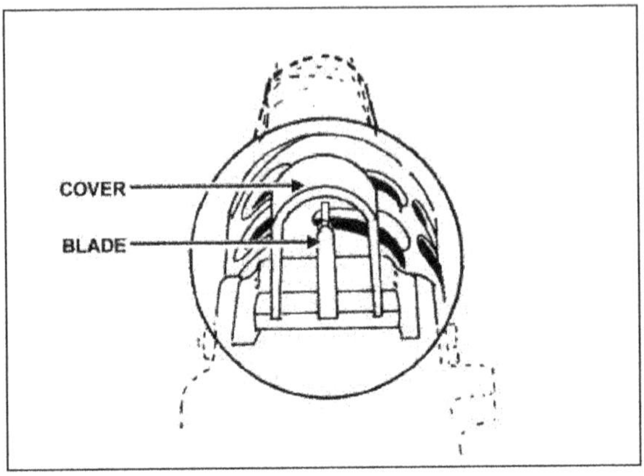

Figure 1-4. Front sight, cover, and blade.

c. Table 1-1 provides the general data on the caliber .50 MG.

Weight (approx)	84 pounds
Weight of barrel	24 pounds
Length of gun	65.13 inches
Length of barrel	45 inches
Length of rifling (approx)	41.88 inches
Number of lands and grooves	8
Twist, right-hand	one turn in 15 inches
Feed	link-belt
Operation	recoil
Cooling	air
Muzzle velocity (approx)	3,050 feet per second
Rate of fire (cyclic)	450 to 550 rounds per minute
Maximum range (approx)	7,440 yards or 6,764 meters
Maximum effective range (approx)	2,000 yards or 1,830 meters
• Area targets	1,830 meters
• Point targets, single shot	1,500 meters

Table 1-1. General data.

1-3. COMPONENTS

The major components of the caliber .50 MG and their purposes are shown in Table 1-2 and Figure 1-5.

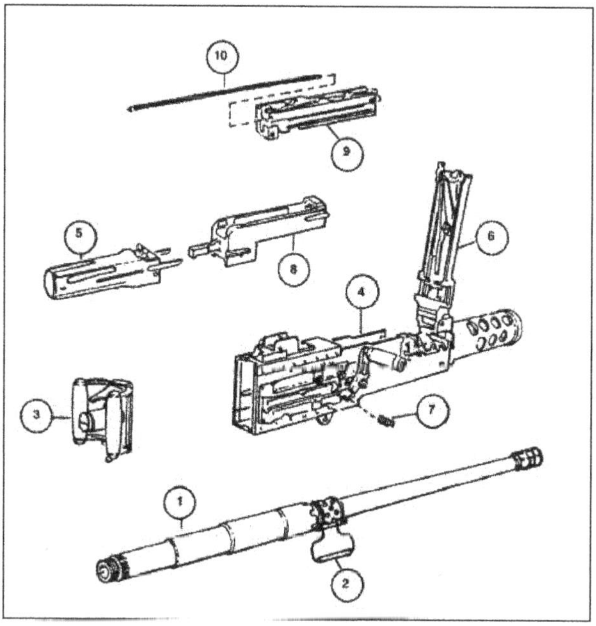

Figure 1-5. Components of the caliber .50 MG.

COMPONENTS	PURPOSES
l. Barrel Group	Houses cartridges for firing; directs projectile.
2. Carrier Assembly	Provides handle to carry barrel and to remove the barrel from the receiver.
3. Backplate Group	Houses the trigger, bolt latch release, buffer tube sleeve, and the left and right spade grips.
4. Receiver Group	Serves as a support for all major components; houses action of weapon, which controls functioning of weapon.
5. Bolt Group	Provides feeding, chambering, firing, and extracting, using the propellant gases and recoil spring for power.
6. Cover Group	Feeds linked belt ammunition; positions and holds cartridges in position for extracting, feeding, and chambering.
7. Bolt Stud	Provides a means to move the bolt to the rear with the retracting slide handle.
8. Barrel Extension Group	Secures the barrel to the recoiling parts.
9. Barrel Buffer Body	Assists in recoil and counterrecoil of the bolt group.
10. Driving Spring Rod Assembly	Drives the bolt forward when the bolt latch release is depressed.

Table 1-2. Components and their purposes.

1-4. GROUND MOUNTS

The two principal ground mounts used with the caliber .50 machine gun are the tripod mount, M3, and the antiaircraft mount, M63. The tripod mount, M3, is a ground mount designed for use against ground targets. The antiaircraft mount,

M63, is a ground mount principally designed for use against aerial targets. Its use against ground targets is limited because the mount tends to be unstable when the gun is fired at low angles.

a. **Tripod Mount, M3**. The M3 mount is the standard ground mount of the caliber .50 machine gun (Figure 1-6). It is a folding tripod with three, telescopic, tubular legs connected at the tripod head. Each leg ends in a metal shoe that can be stamped into the ground for greater stability. The two trail legs are joined together by the traversing bar. The traversing bar serves as a support for the traversing and elevating mechanism, which in turn supports the rear of the gun. The tripod head furnishes a front support for the mounted gun that is further supported by the short front leg. When the tripod is emplaced on flat terrain with all extensions closed, the adjustable front leg should form an angle of about 60 degrees with the ground. This places the gun on a low mount about 12 inches above the ground. To raise the tripod farther off the ground, extend the telescopic front and trail legs enough to keep the tripod level and maintain the stability of the mount.

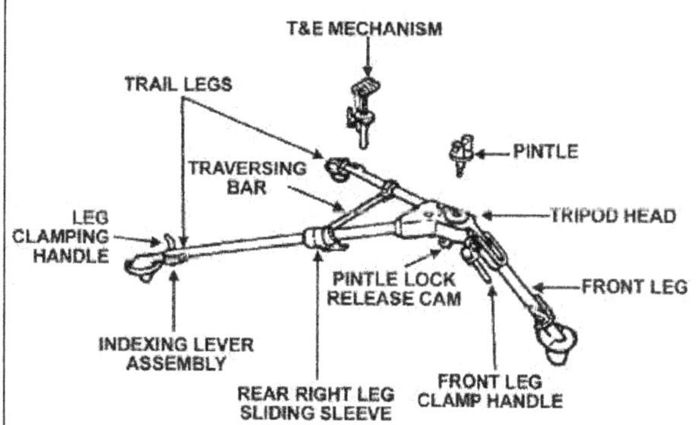

Figure 1-6. M3 tripod mount.

(1) To set the tripod trail legs--

(a) Unscrew the leg-clamping handle, press down on the indexing lever, and extend the leg to the desired length.

(b) Align the indexing lever stud with one of the holes in the tripod leg extension.

(c) Release the pressure on the indexing lever, allowing the stud to fit the desired hole. Tighten the leg-clamping handle.

(2) *To set the front leg of the tripod-*

(a) Turn the front leg clamp handle counterclockwise to loosen the front leg.

(b) Adjust the leg to the desired angle and tighten the front leg clamp.

(3) To secure the tripod legs, stamp the metal shoe on each tripod leg into the ground. Sandbag each leg to stabilize the M2 for firing.

b. **Antiaircraft Mount, M63.** The antiaircraft mount (Figure 1-7) is a four-legged, low silhouette, portable mount used for antiaircraft fire. Table 1-3 lists the general data pertaining to the M63.

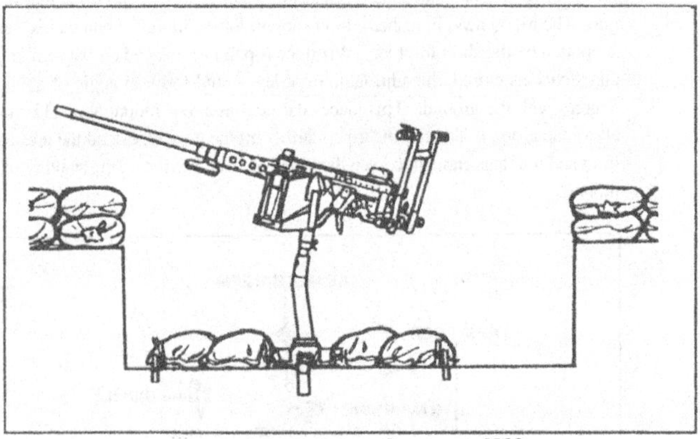

Figure 1-7. Antiaircraft mount, M63.

WEIGHTS:	
Overall	144 pounds
Four legs	24 pounds
Base assembly	54 pounds
Elevator assembly	12 pounds
Cradle assembly	44 pounds
Ammunition box-tray assembly	10 pounds
HEIGHT OVERALL	42 inches
LENGTH OF LEG	24 inches
DIAMETER OF BASE (WITH LEG ASSEMBLED)	52 inches
MAXIMUM ELEVATION	85 degrees
MAXIMUM DEPRESSION	29 degrees
MAXIMUM TRAVERSE	360 degrees

Table 1-3. M63 general data.

1-5. ACCESSORIES FOR GROUND MOUNTS

The following paragraph explains the functions of the traversing and elevating mechanism and pintle used in the mounting of the machine gun when used in the ground configuration.

a. **Traversing and Elevating Mechanism.** The T&E mechanism (Figure 1-8) is used to engage preselected target areas at night or during limited visibility conditions. Record direction and elevation readings from the traversing bar and T&E mechanism. Record all readings in mils.

(1) The traversing mechanism consists of a traversing bar, slide, and screw assembly.

(a) The traversing bar, graduated in 5-roil increments, fits between the trail legs of the tripod. The traversing slide and screw assembly are clamped in place on the traversing bar by the traversing slide lock lever. When the traversing slide is locked to the traversing bar, the traversing handwheel should be centered. The traversing slide is properly mounted when the lock lever is to the rear and the traversing handwheel is positioned to the left.

(b) To make changes in direction, loosen the traversing slide lock lever and move the slide along the traversing bar. This permits traverse of 400 mils left or right of the zero index in the center of the traversing bar. Readings on the traversing bar are taken from the left side of the traversing slide. For changes of 50 mils or less in deflection, turn the traversing handwheel of the screw assembly. This allows a traverse of 50 mils left or right of center. One click in the traversing handwheel signifies 1 mil change in direction.

(2) The elevating mechanism consists of an upper and lower elevating screw.

(a) It is connected to the gun by inserting the quick release pin assembly through the holes in the upper elevating screw yoke and the rear mounting lugs of the receiver. A scale, graduated in mils, is fitted to the upper screw to indicate elevation. This scale is marked to show 250 mils in depression and 100 mils in elevation from the zero setting.

(b) The elevating handwheel is graduated in l-mil increments up to 50 mils, and is fastened to the elevating screw by a screw lock. This synchronizes the handwheel graduations with those on the upper elevating screw. A spring-actuated index device produces a clicking sound when the handwheel is turned. Each click equals 1 mil change in elevation. The handwheel is turned clockwise to depress the barrel and counterclockwise to elevate.

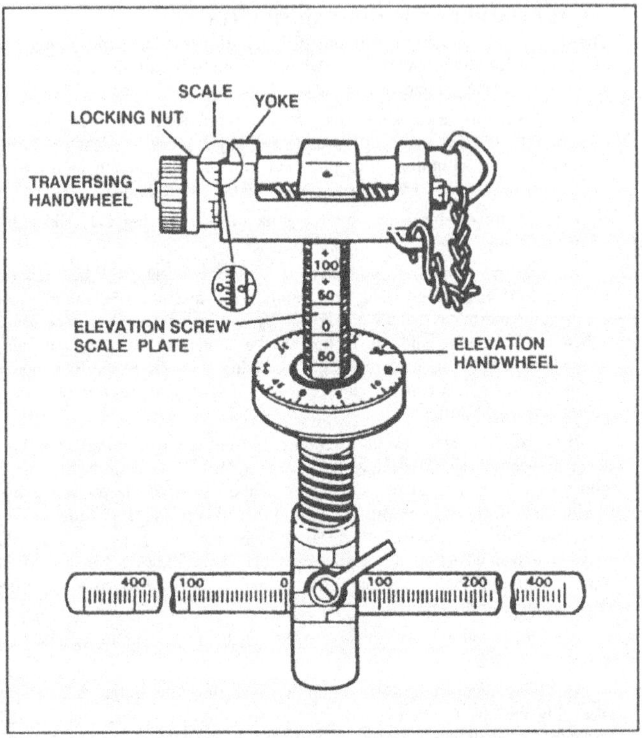

Figure 1-8. Traversing and elevating mechanism.

b. **Pintle.** The gun is connected to the tripod mount, M3, by a pintle (Figure 1-9, page 1-14). This pintle is semipermanently attached to the machine gun by a pintle bolt through the front mounting hole in the receiver. The tapered stem of the pintle seats in the tripod head. It is held secure by a pintle lock and spring. To release the pintle, raise the pintle lock, releasing the cam. The weight of the pintle and traversing and elevating mechanism are considered as part of the total weight of the tripod mount, M3 (44 pounds).

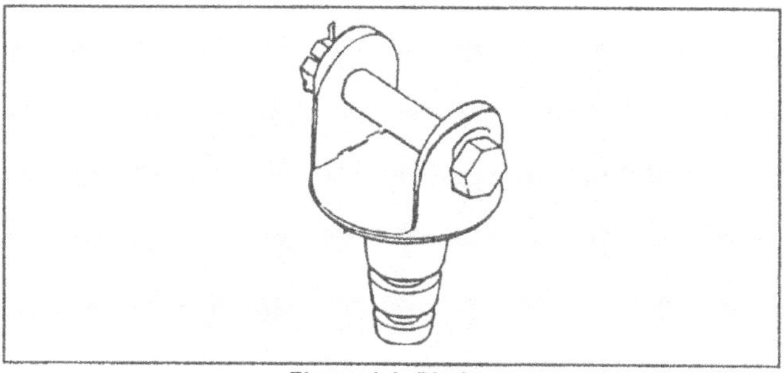

Figure 1-9. Pintle.

1-6. VEHICULAR MOUNTS

The four principal vehicular mounts used with the caliber .50 machine gun are the truck mount, M36; the pedestal truck mount, M31C and M24A2; the commander's cupola, MI13 armored personnel carrier; and the MK64 gun cradle.

a. **Truck Mount, M36.** This mount consists of a cradle with a roller carriage on a circular track (Figure 1-10). The cradle can be rotated in the pintle sleeve of the carriage and can be adjusted for elevation. The carriage is guided on the track by rollers. The track is secured to the vehicle by supports.

(1) To move the gun in elevation on the M36 mount, remove the cradle locking pin and place it in the carriage handle; grasp the spade grips and elevate or depress as desired. The gun is also moved in traverse by pressure on the spade grips.

(2) To move the gun on the track, raise the brake handle lever until it is retained by the brake detent plungers. The cradle may then be moved on the track by applying pressure on the carriage handle.

b. **Pedestal Truck Mount, M31C.** Pedestal mounts are component assemblies designed for installation on the 1/4-ton vehicles to support a machine gun mount. They are composed of a pintle socket, pintle clamping screw column, and braces (Figure 1-11).

c. **Armored Vehicle Cupola Mount.** A caliber .50 machine gun and mount are installed in the gun support on the commander's cupola of an MI 13 armored personnel carrier. The machine gun can be traversed 360 degrees, elevated 53 degrees, and depressed 21 degrees maximum (Figure 1-12).

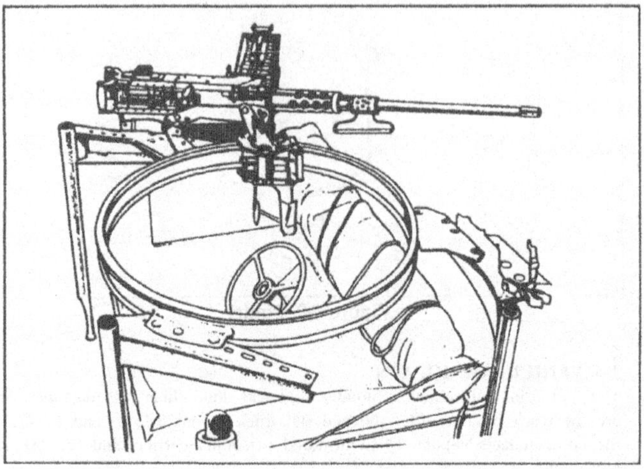
Figure 1-10. Truck mount, M36.

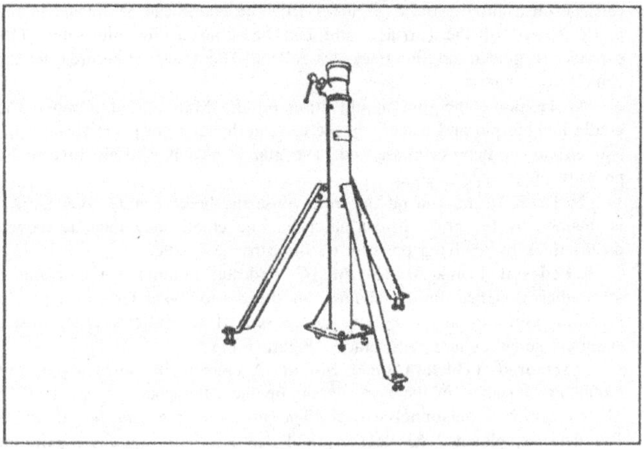
Figure 1-11. Pedestal truck mount, M31 C.

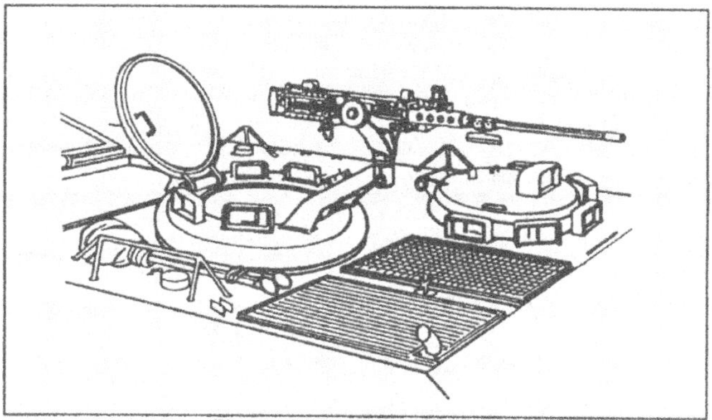

Figure 1-12. Cupola mount.

d. **MK64 Gun Cradle Mount.** This vehicle mount was primarily designed for the M2. However, because of its versatility, the MK64 will accept the MK 19 also (using the M2 mounting adapter assembly). The MK64 can be mounted on the following vehicles – M151 series, M966 HMMWV armament carrier, and the M113 series (Figure 1-13).

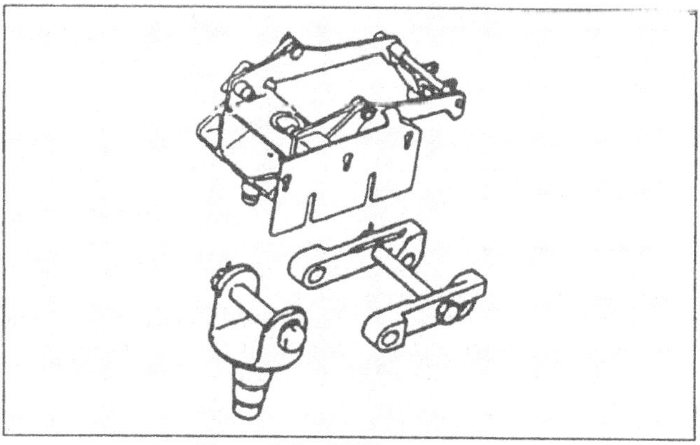

Figure 1-13. MK64 gun cradle.

1-7. AMMUNITION

This paragraph describes the ammunition used in caliber .50 machine guns (Figure 1-14). Soldiers should be able to recognize the types of ammunition available and know how to care for it. The caliber .50 cartridge consists of a cartridge case, primer, propelling charge, and the bullet. See TM 9-1300-200. The term bullet refers only to the small-arms projectile. There are eight types of ammunition issued for use in the caliber .50 machine gun. The tips of the various rounds are color-coded to indicate their type. The ammunition is linked with the M2 or M9 metallic links for use in the machine gun (Figure 1-15, page 1-18).

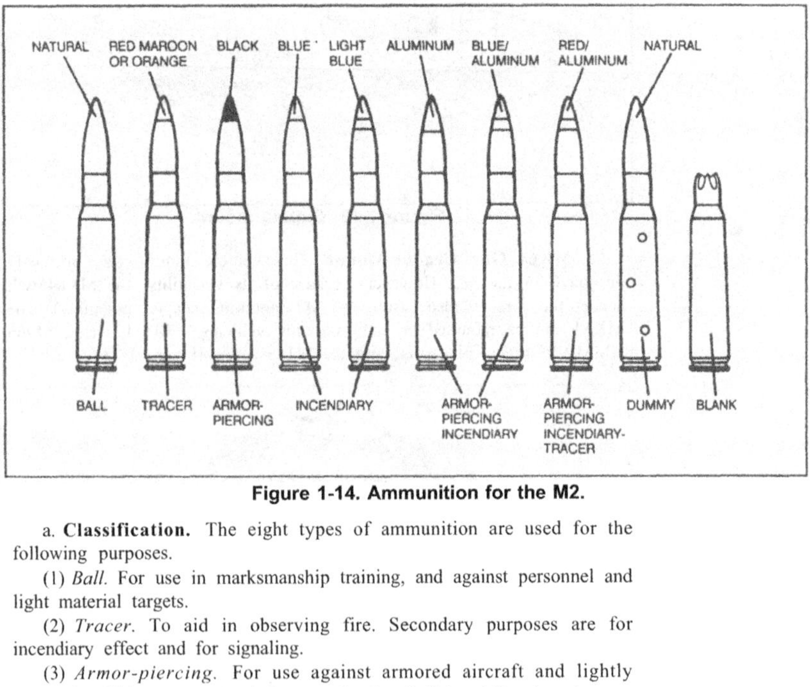

Figure 1-14. Ammunition for the M2.

a. **Classification.** The eight types of ammunition are used for the following purposes.

(1) *Ball.* For use in marksmanship training, and against personnel and light material targets.

(2) *Tracer.* To aid in observing fire. Secondary purposes are for incendiary effect and for signaling.

(3) *Armor-piercing.* For use against armored aircraft and lightly armored vehicles, concrete shelters, and other bullet-resisting targets.

(4) *Incendiary.* For incendiary effect, especially against aircraft.

(5) *Armor-piercing-incendiary.* For combined armor-piercing and incendiary effect.

(6) *Armor-piercing-incendiary-tracer.* For combined armor-piercing and incendiary effect, with the additional tracer feature.

(7) *Blank.* For simulated fire (contains no bullet).

(8) *High-pressure test.* For use only in proof firing of weapons and barrels.

(9) *Dummy.* For training (completely inert).

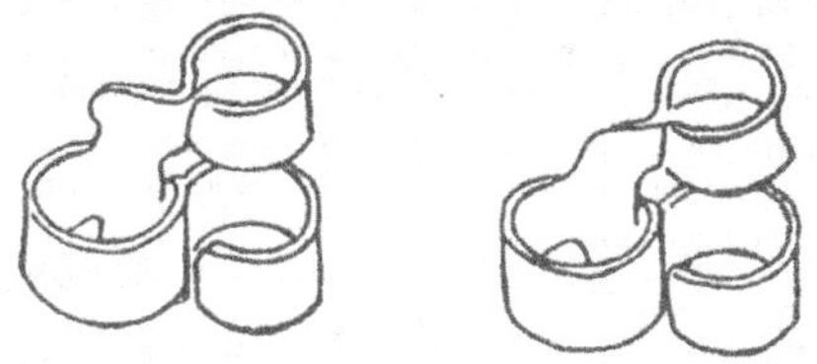

Figure 1-15. M2 and M9 links (closed loop).

b. **Ballistic Data.** The approximate maximum range and average muzzle velocity of some of the different types of caliber .50 ammunition authorized for use in the machine gun are noted in Table 1-4.

CARTRIDGE	MAXIMUM RANGE (meters)	METER TRACE	AVERAGE MUZZLE VELOCITY (feet per second)
Ball, M2	7,400	—	2,930
Tracer, M1 (with gilding metal jacket)	5,575	1,800	2,860
Tracer, M1 (with clad steel jacket)	5,450	1,800	3,030
Tracer, M17	5,450	2,450	3,030
Incendiary, M1	6,050	—	3,090
Armor-piercing, M2	7,400	—	2,930
Armor-piercing-incendiary,M8	6,470	—	3,050
Armor-piercing-incendiary-tracer, M20	6,470	*300-1,750	3,050

* This tracer is dim at near ranges but increases to bright as it moves farther from the gun.

Table 1-4. Ballistic data.

c. **Care, Handling, and Preservation**. Exercise care to prevent ammunition boxes from becoming broken or damaged. If they do, repair them immediately. Transfer all original markings to the new parts of the box. Do not open ammunition boxes until the ammunition is to be used. Ammunition removed from the airtight container, particularly in damp climates, is likely to corrode. Protect the ammunition from mud, sand, and water. If the ammunition gets wet or dirty, wipe it off at once with a clean, dry cloth. Wipe off light corrosion as soon as it is discovered. Turn in heavily corroded cartridges. Do not expose ammunition to the direct rays of the sun. If the powder is hot, excessive pressure may be developed when the weapon is fired. Do not oil or grease ammunition. Dust and other abrasives that collect on greasy ammunition are injurious to the operating parts of the gun. Moreover, oiled cartridges produce excessive chamber pressure. Do not fire dented cartridges, cartridges with loose bullets, or otherwise defective rounds.

d. **Storage**. Small-arms ammunition is not an explosive hazard, but under poor storage conditions it may become a fire hazard. Store ammunition of all classes away from radiators, hot water pipes, and other sources of heat. Whenever possible, store ammunition under cover. If it is necessary to leave ammunition in the open, keep it at least 6 inches off the ground and covered with a double thickness of tarpaulin. Place the tarpaulin so that it gives maximum protection and allows free circulation of air. Dig suitable trenches to prevent water from flowing under the ammunition pile.

e. **Miscellaneous Data**. Table 1-5 lists the maximum penetration in inches for an armor-piercing cartridge fired from the 45-inch barrel (muzzle velocity, 2,935 feet per second), which in some cases may enhance the leader's selection of targets to engage.

	INCHES AT:		
MATERIAL	**200 METERS**	**600 METERS**	**1,500 METERS**
Armor plate (homogeneous)	1.0	0.7	0.3
Armor plate (face-hardened)	0.9	0.5	0.2
Sand (100 pounds dry weight/cubic feet)	14.0	12.0	16.0
Clay (100 pounds dry weight/cubic feet)	28.0	27.0	21.0

Table 1-5. Maximum penetration for armor-piercing cartridge.

Table 1-6 lists the maximum penetration in inches for a ball cartridge fired from the 45-inch barrel (muzzle velocity, 2,935 feet per second):

	INCHES AT:		
MATERIAL	**200 METERS**	**600 METERS**	**1,500 METERS**
Sand (100 pounds dry weight/cubic feet)	14.0	12.0	6.0
Clay (100 pounds dry weight/cubic feet)	28.0	27.0	21.0
Concrete	2.0	1.0	1.0

Table 1-6. Maximum penetration for ball cartridge.

CHAPTER 2

MAINTENANCE

This chapter addresses the proper care of the machine gun to ensure its overall effectiveness and efficient functioning. The information includes the gunner's knowledge in disassembly and assembly, inspection, and maintenance procedures of the gun, its mount, the T&E, and its ammunition.

2-1. SAFETY

The paramount consideration while training with the machine gun is safety. It is imperative that the weapon be cleared properly before disassembly and inspection. (See Figure 2-1 for step-by-step procedures.)

1. Unlock the bolt-latch release and raise the cover.
2. Pull the bolt to the rear. Examine the chamber and the T-slot to ensure that they hold no rounds.
3. Place a wooden block inside the receiver between the bolt and the rear of the barrel.

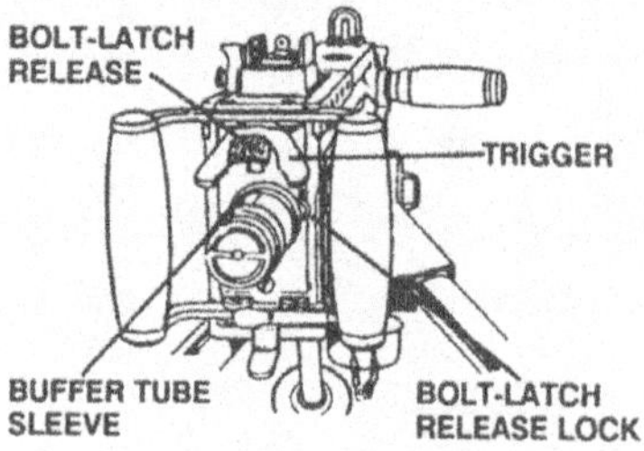

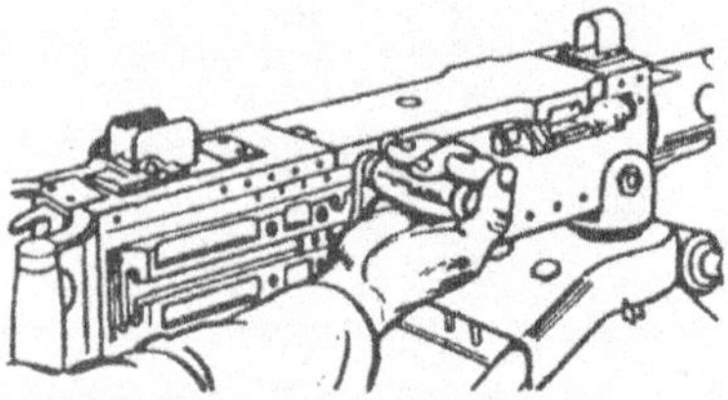

4. Insert a cleaning rod in the muzzle end of the barrel until the rod can be seen in the receiver.
5. Remove the rod. The gun is now clear.

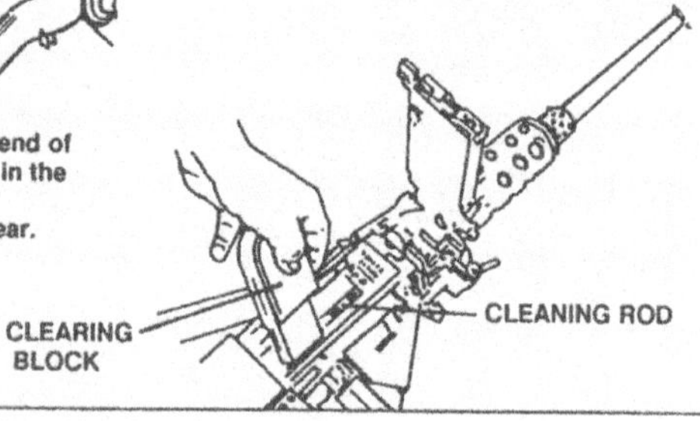

Figure 2-1. Clearing the weapon.

2-2. GENERAL DISASSEMBLY

The crew operating the MG must be fully familiar with its general disassembly, which consists of removing the major groups and assemblies for inspection and cleaning. The eight major groups that must be disassembled in the following order are barrel group, backplate group, driving spring rod assembly, bolt group, barrel extension group, barrel buffer body group, barrel buffer assembly, and receiver group (Figure 2-2).

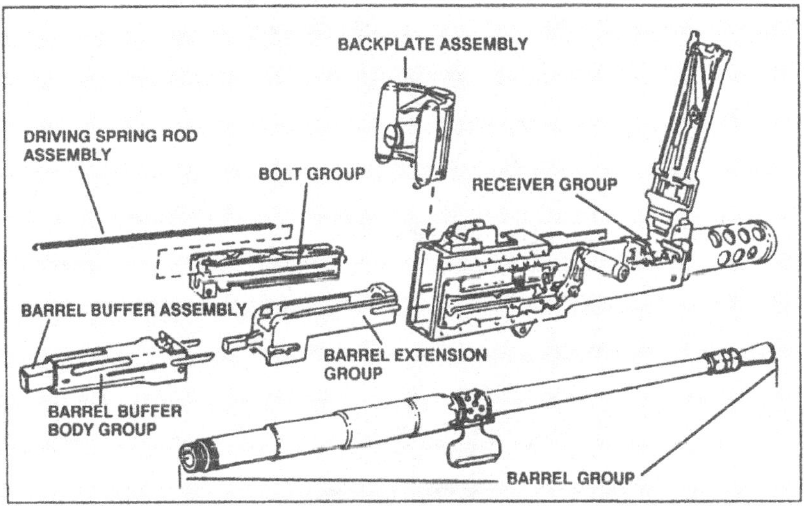

Figure 2-2. Major groups.

a. **Barrel Group.** Turn the cover latch and raise the cover group (Figure 2-3). Grasp the retracting slide handle with the right hand, palm up, and pull the recoiling parts to the rear until the lug on the barrel locking spring aligns with the 3/8-inch hole in the right sideplate of the receiver (just below the feedway exit). The barrel can be turned only when the lug is aligned with the 3/8-inch hole. Place the smallest loop of a caliber .50 link, or suitable spacer, between the trunnion block and the barrel extension (Figure 2-4). This holds the barrel locking spring lug aligned with the 3/8-inch hole in the right sideplate. Unscrew the barrel from the receiver (Figure 2-5, page 2-4). Be careful not to damage the threads or barrel locking notches when setting the barrel down. Pull back slightly on the retracting slide handle and remove the link or spacer from the receiver.

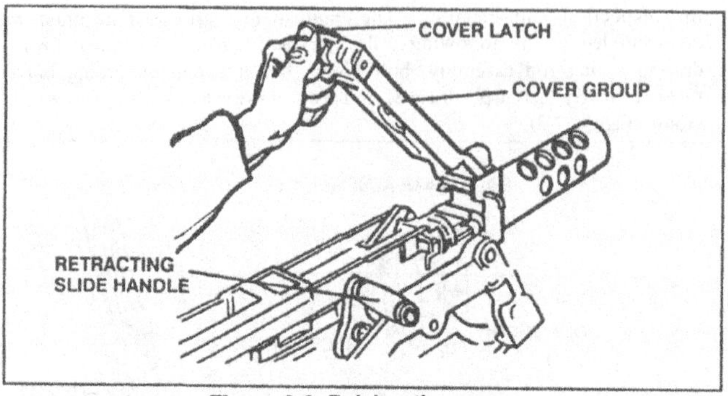

Figure 2-3. Raising the cover.

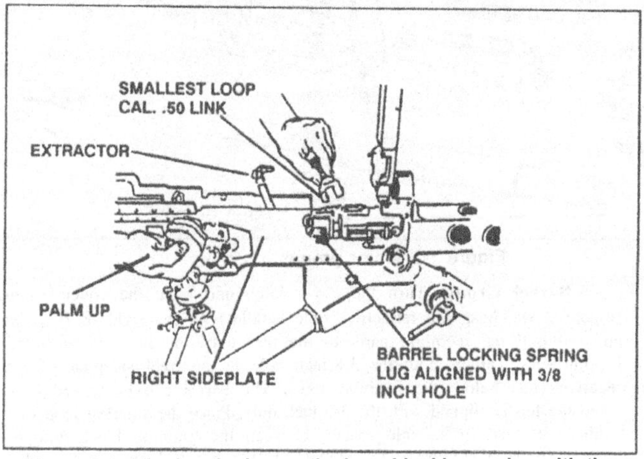

Figure 2-4. Aligning the lug on the barrel locking spring with the 3/8-inch hole in the right sideplate.

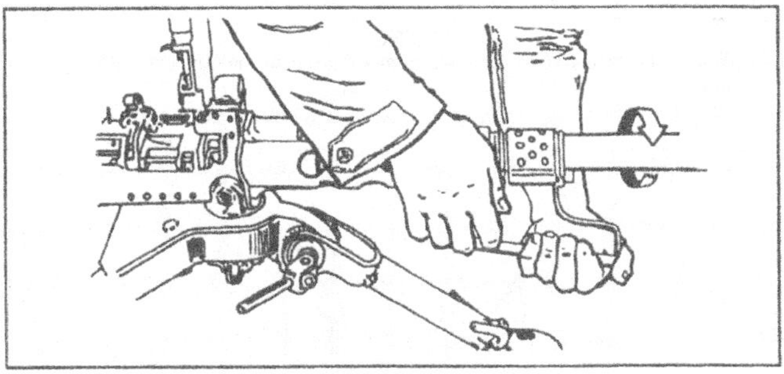

Figure 2-5. Removing the barrel.

b. **Backplate Group.** Ensure that the bolt latch release is up, free of the bolt latch release lock. If it is not, push down on the bolt latch release and turn the buffer tube sleeve to the right to free it (Figure 2-6). The bolt must be forward before the backplate is removed. If the bolt is to the rear, push down on the bolt latch release, place palm up on the retracting slide handle, and ease the bolt forward. The backplate latch lock and latch are below the buffer tube. Pull out on the lock and up on the latch; remove the backplate by lifting it straight up (Figure 2-7).

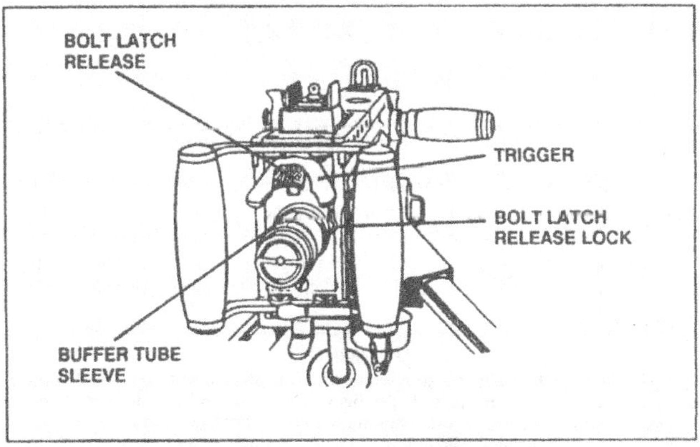

Figure 2-6. Releasing the bolt latch.

CAUTION

Care must be taken to prevent the bolt from slamming forward with the barrel removed.

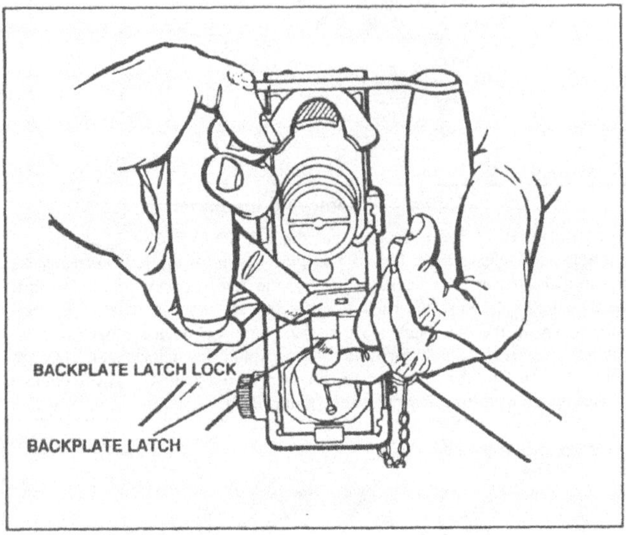

Figure 2-7. Removing the backplate.

c. **Driving Spring Rod Assembly.** The inner and outer driving springs and driving spring rod are located inside the receiver next to the right sideplate (Figure 2-8, page 2-6). Push in on the head of the driving spring rod and push to the left to remove the driving spring rod retaining pin from its seat in the right sideplate. Pull the driving spring assembly to the rear and out of the receiver.

WARNING

Never attempt to cock the gun while the backplate is off and the driving spring assembly is in place. If the backplate is off and the driving spring assembly is compressed, the retaining pin on the driving spring rod can slip from its seat in the sideplate and could cause serious injury to anyone behind the gun.

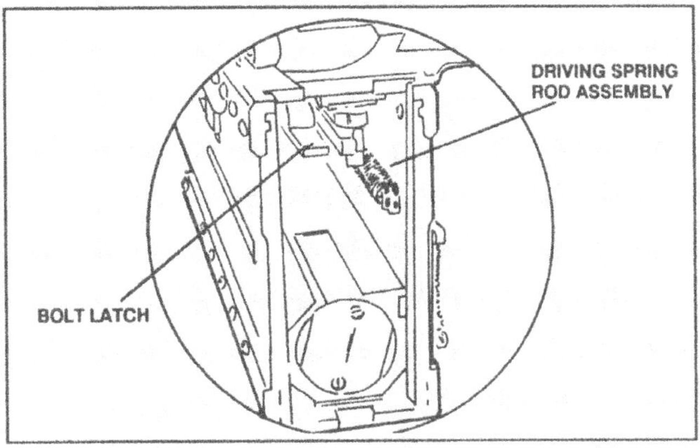

Figure 2-8. Removing the driving spring rod assembly.

d. **Bolt Stud.** Grasp the retracting slide handle and give it a quick jerk, freeing the bolt from the barrel extension. Align the collar of the bolt stud with the clearance hole in the bolt slot on the right sideplate, and remove the bolt stud (Figure 2-9). If the bolt is accidentally moved all the way to the rear, the bolt latch will engage in the bolt latch notches in the top of the bolt. If this occurs, raise the bolt latch (left of the trigger lever) and push the bolt forward to align the bolt stud with the clearance hole (Figure 2-10).

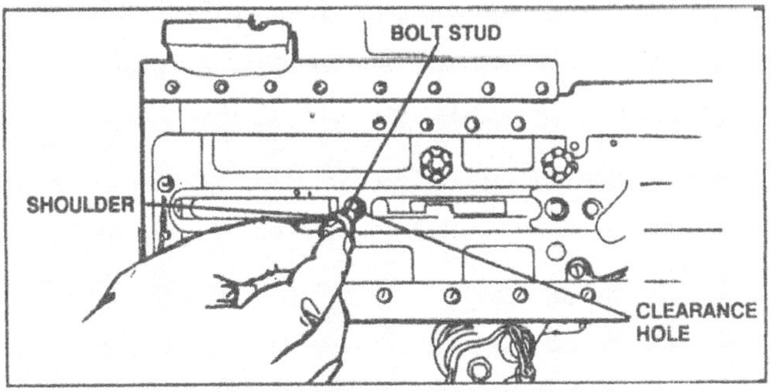

Figure 2-9. Removing the bolt stud.

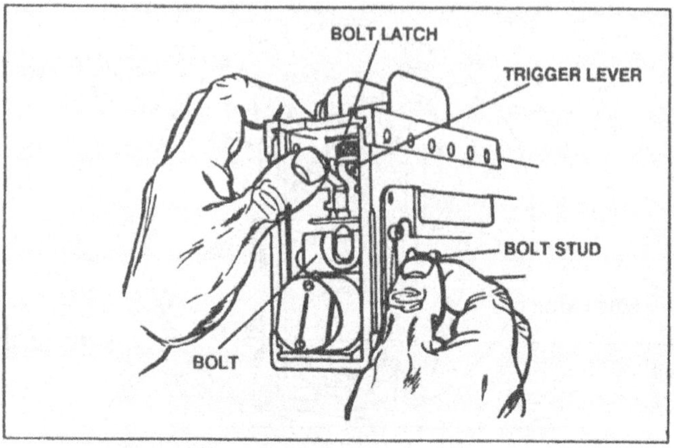

Figure 2-10. Freeing the bolt.

e. **Bolt Group.** After freeing the bolt, slide it to the rear and out of receiver (Figure 2-11). Place the bolt down on its right side (with the extractor arm up), so that the extractor will not fall from the bolt.

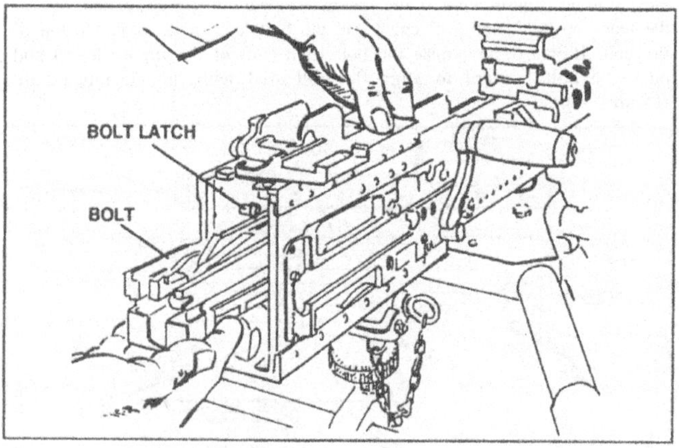

Figure 2-11. Removing the bolt from the receiver.

f. **Barrel Buffer Body Group and Barrel Extension Group.** Insert the drift of a combination tool, or other pointed instrument, through the hole in the lower rear corner of the right sideplate. Push in on the barrel buffer body lock. At the same time, place one hand in the receiver and push the barrel extension group and barrel buffer group to the rear (Figure 2-12). Remove the barrel buffer group and barrel extension group from the receiver. Separate the two groups by pushing forward on the tips of the accelerator (Figure 2-13).

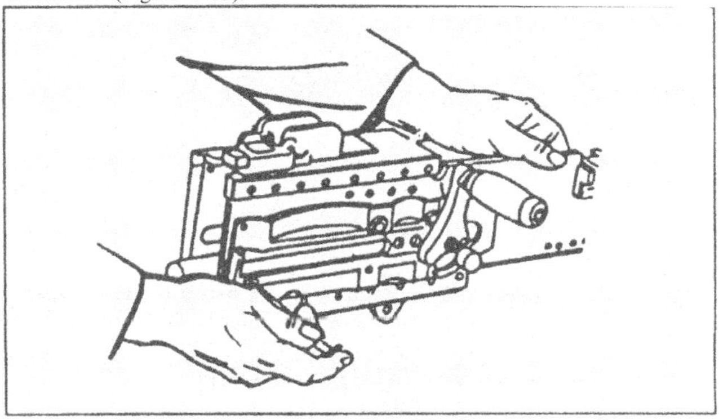

Figure 2-12. Removing barrel buffer group and barrel extension group.

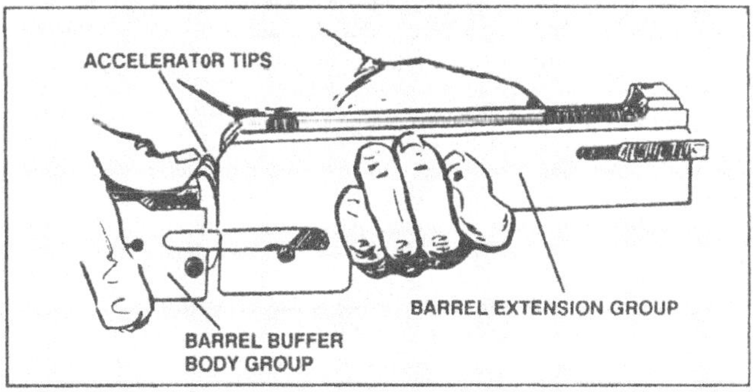

Figure 2-13. Separating the groups.

g. **Barrel Buffer Assembly.** Pull the barrel buffer assembly from the rear of the barrel buffer body group. The barrel buffer assembly will not be disassembled (Figure 2-14). This completes general disassembly.

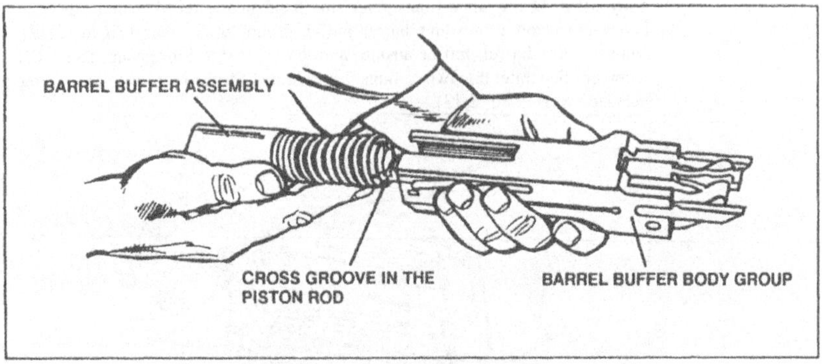

Figure 2-14. Separating the barrel buffer assembly from the barrel buffer body group.

2-3. CLEANING, INSPECTION, AND LUBRICATION

To ensure proper care of the MG, it is necessary to have a system of maintenance or an SOP for the frequency of cleaning. Each gun should be cleaned as soon after firing as possible and each time it is exposed to field conditions. Under combat conditions, the gun should be cleaned and oiled daily. Under extreme climatic and combat conditions, it maybe necessary to clean and lubricate more frequently. Under ideal conditions, where the gun is not used, and is stored in a clean, dry place, it may only be necessary to inspect, clean, and lubricate every 5 days. The gun should be disassembled, cleaned, and oiled in a clean, dry location. If possible, keep the gun covered with a gun cover, canvas, tarpaulin, or poncho when not in use.

a. **Routine Care and Cleaning.** Before firing (when the situation permits), take the following steps to ensure efficient functioning of the machine gun:

- Disassemble the gun into its major groups or assemblies.
- Clean the bore and chamber, and lightly oil them.
- Clean all metal parts thoroughly with CLP. See paragraph 2-3f for lubrication procedures.

b. **Care and Cleaning Under Unusual Conditions.** Extreme cold, hot, dry, and tropical climates affect the gun and its functioning. Care should be taken under these climatic conditions to ensure that the gun is cleaned daily with the prescribed lubricants and protected from the elements by some sort of cover if possible. Further information on care and cleaning of the gun under unusual climatic conditions can be found in TM 9-1005-213-10.

c. **Care and Cleaning of M3 Mount and Accessories.** The mount and accessories, such as the ammunition chest and spare parts, should also be kept clean and lubricated. Painted surfaces should be spot painted when necessary. Moving surfaces should be inspected and oiled with the prescribed lubricant. All external surfaces of the mount should be kept clean and lightly oiled. Be particularly careful that the pintle bushing is clean and lightly oiled, and that the pintle lock release cam is well-lubricated and free from grit. The sleeve lock indexing levers and telescopic legs should be clean and lubricated enough for ease in use. The mount should be cleaned and oiled with the same regularity and in the same manner as the gun.

d. **Maintenance and Inspection.** Units must establish guidelines and conduct regular maintenance and inspection to keep the machine gun and its mounts in operational conditions.

(1) *Gun maintenance.* The importance of a thorough knowledge of care, cleaning, and maintenance of the machine gun cannot be overemphasized because these actions determine whether or not the gun will function properly when needed. The bore and chamber must be properly maintained to preserve accuracy. Because of the close fit of working surfaces and the high speed at which the gun operates, the receiver and moving parts must be kept clean, correctly lubricated, and free from burrs, rust, dirt, or grease to ensure proper, efficient functioning.

(2) *Mount maintenance.* The care, cleaning, lubrication, and adjustment of the mounts used with the gun are no less important. The functioning of the gun and mount together determine overall effectiveness. All accessories and equipment used with the gun and mount, including ammunition, must also be properly maintained.

(3) *Inspection.* When inspected, the machine gun should be completely disassembled. Inspecting personnel should look for dirt, cracks, burrs, and rust.

e. **Inspection Checklist.** Table 2-1 is an inspection checklist to be used as a guide for crewmembers or inspecting personnel to ensure that the gun and equipment are properly maintained.

UNIT	INSPECTION
1. GUN.	
a. Barrel.	Inspect the bore and chamber for rust. See that they are clean and lightly oiled.
b. Moving parts.	See that they are clean and lightly oiled. Operate the retracting slide handle and bolt latch release several times to see that the parts function without excessive friction.
c. Headspace and timing.	Check with the gauges to ensure that headspace and timing are correct.
d. Rear sight and windage knob.	Ensure that the sight is in good condition, clean, free of grease or dirt, and lightly oiled. Elevation should be set at 1,000, windage zero, and the sight should be down.
2. MOUNT (M3, MK64, M36, or M4).	See that it is clean, lightly lubricated, and that all clamps are securely tightened. It should function properly and be complete.
3. SPARE PARTS AND TOOLS.	Inspect to see that they are clean and lightly oiled. See that spare parts kits are complete and in good condition. Replacement parts should be requisitioned and newly drawn parts examined.
4. T&E.	See that it is clean, lightly lubricated, and that both hand wheels work properly.
5. AMMUNITION.	See that ammunition is properly stored, and that boxes and ammunition are in good condition and not oiled.

Table 2-1. Inspection checklist.

f. **Lubrications.** Use cleaner, lubricant, preservative to clean the machine gun. As its name implies, it cleans, lubricates, and preserves all in one application.

(1) After cleaning the gun with CLP, wipe it dry and reapply a thin coating. Allow this thin coat to dry on the parts for a short time before reassembly. CLP deposits a thin coating on the metal which minimizes carbon buildup and prevents foreign material from sticking. It is this coating that provides the frictionless operation of the weapon parts, not liquid oil deposited on them. A gun treated with CLP will operate better and remain clean longer than one treated with any other cleaning material. Use of CLP will reduce maintenance costs and extend the life of the weapon.

(2) Rifle bore cleaner is a cleaning solvent which can be used to clean powder residue, carbon, and dirt from weapons. RBC does not preserve or lubricate a weapon. If you clean a weapon with RBC, dry the weapon and lubricate it with lubricating oil, semifluid (LSA); lubricating oil, special purpose (PL-S); or lubricating oil, general purpose (PL-M). The use of these oils will cause sand or grit to stick to the weapon. RBC and oil should be used only when CLP is not available.

2-4. MAINTENANCE PROCEDURES

There are certain actions that must be taken before, during, and after firing to properly maintain the gun. See Table 2-2 for a preoperation checklist.

PART	BEFORE	DURING FIRING OR TEMPORARY CESSATION	AFTER
Bore	Make sure it is clear and clean.		Clean and oil lightly.
Moving parts	Oil lightly and test for worn or broken parts. They should function without excessive friction.	Lubricate working parts. Observe the functioning of the gun to anticipate failures.	Inspect, clean, and oil lightly.
Headspace and timing	Check adjustment by gauge and correct if necessary.	Watch for bulged cases to prevent a ruptured (separated) case. If a separated case occurs, remove it and readjust headspace.	Check adjustment.
Rear sight and windage knob	See that the sight is clean and functions properly. Set sight at 1,000, windage zero.	Keep properly set.	Clean and oil; set at 1,000, windage zero.
Spare parts and tools	Clean and oil spare parts and tools. Check kits for completeness. Examine newly drawn parts.	Keep available.	Clean and oil. Check and replace damaged or missing parts.
Ammunition	Have an adequate supply; clean, correctly loaded, and in good condition. DO NOT OIL.	Keep correctly aligned with the feedway; check resupply. Protect from sun, moisture, and dirt. Watch for link stoppage.	Clean, store carefully, and replenish supply.

Table 2-2. Operational checklist.

2-5. MAINTENANCE UNDER NBC CONDITIONS

If contamination is anticipated, apply oil to all outer metal surfaces of the weapon. DO NOT OIL AMMUNITION. Keep the weapon covered as much as possible. If the weapon is contaminated, decontaminate it as prescribed by FM 3-5 and then clean and lubricate it.

2-6. GENERAL ASSEMBLY

To assemble the gun, replace the groups and assemblies in reverse order of their removal in disassembly.

a. **Barrel Buffer Assembly and Barrel Buffer Body Group.** Replace the barrel buffer assembly in the barrel buffer body group, with the key on the spring guide to the right. This key must fit in its slot in the right side of the barrel buffer body. Turn the barrel buffer tube until the screwdriver slot (in the rear of the tube) is vertical and the arrow is pointing to the right. The stud on the tube lock will now engage the serrations in the barrel buffer tube to keep the tube from turning. Push the barrel buffer assembly fully forward (Figures 2- 15 and 2-16).

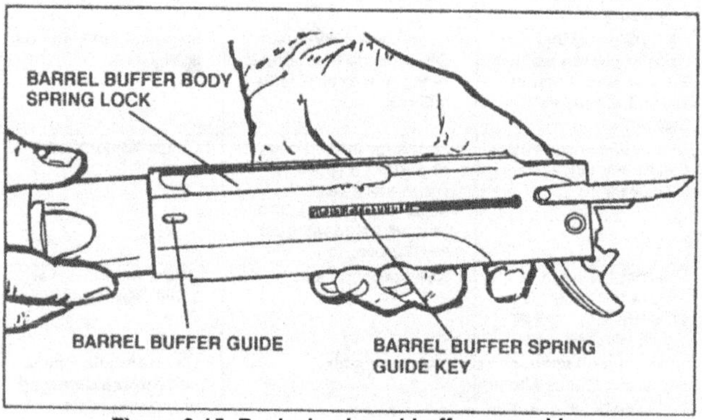

Figure 2-15. Replacing barrel buffer assembly.

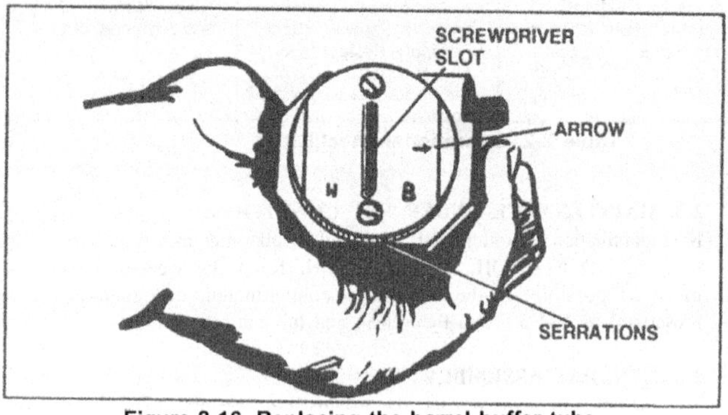

Figure 2-16. Replacing the barrel buffer tube.

b. **Barrel Buffer Group and Barrel Extension Group.** To join the two groups together, hold the barrel buffer group in the right hand, with the index finger supporting the accelerator. Join the notch on the shank of the barrel extension group with the cross-groove in the pistol rod of the barrel buffer assembly. At the same time, align the breech lock depressors with their guideways in the sides of the barrel extension, ensuring that the tips of the accelerator are against the rear end of the barrel extension (claws against the shank) (Figure 2-17). Push the groups together. As the accelerator rotates to the rear, press down on its tips to ensure positive locking of groups. Place the groups in the receiver, and push them forward until the barrel buffer body spring lock snaps into position. When the parts are properly locked in place, the barrel buffer tube should protrude about 1 1/8 inches from the rear of the barrel buffer body group.

c. **Bolt.** Place the bolt in the receiver, with the top of the cocking lever forward and the extractor down. The barrel extension, barrel buffer, and bolt groups may be assembled and returned to the receiver together (Figure 2-18).

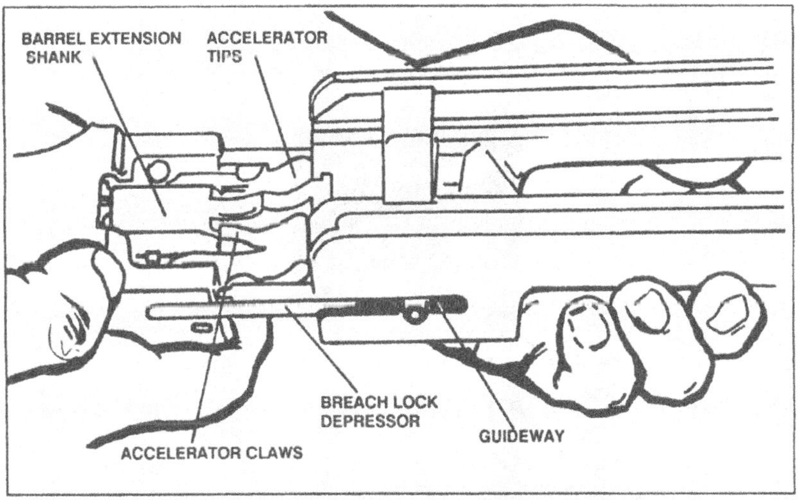

Figure 2-17. Joining the barrel extension group and barrel buffer group.

*** Figure 2-18. Returning the barrel extension, barrel buffer, and bolt groups together.**

*** CAUTION**

Before inserting the bolt group, ensure bolt switch mechanism is positioned on **L** (left-hand) feed for the weapon.

* d. **Bolt Stud**. Align the stud hole in the bolt with the clearance hole and replace the bolt stud, ensuring that the collar of the stud is inside the sideplate (Figure 2-19).

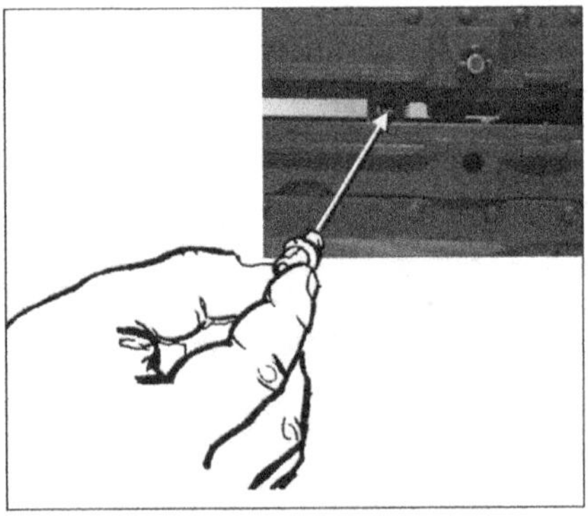

*** Figure 2-19. Replacing the bolt stud.**

* e. **Drive Spring Assembly**. Press up on the bolt latch and push the bolt all the way forward by pushing on the bolt stud only. Place the end of the drive spring rod in its hole in the rear of the bolt, and push forward on the drive spring assembly and the barrel buffer tube. Press in and to the right on the head of the drive spring rod and place the retaining pin in its seat in the right sideplate (Figure 2-20).

* **Figure 2-20. Inserting drive spring and drive spring rod (with drive spring rod inside drive spring).**

* NOTE: At this time, the barrel buffer tube should be completely inside the receiver. If not, the barrel buffer body spring is not properly seated.

* f. **Backplate Group**. Hold the backplate with the latch down and the trigger up; place the backplate guides in their guideways. Hold out on the latch lock and tap the backplate into position until the latch snaps into place (Figure 2-21). Release the latch lock and pull up on the backplate group to ensure it is firmly seated.

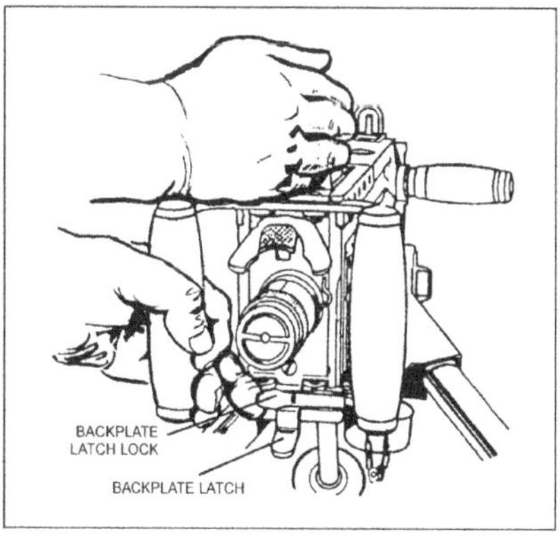

*** Figure 2-21. Replacing the backplate group.**

> *** CAUTION**
> Do not use the driving rod to drive the bolt forward from the rear position. This may damage the drive spring group and cause a stoppage.

* g. **Barrel**. Pull the retracting slide handle to the rear until the lug on the barrel locking spring is visible through the 3/8-inch hole in the right sideplate. Place the smallest loop of a caliber .50 link, or suitable spacer, between the trunnion block and the barrel extension. Screw the barrel all the way into the barrel extension; then unscrew the barrel two notches. Remove the link and close the cover. This completes general assembly.

2-7. FUNCTION CHECK

A function check must be performed as soon as the weapon is assembled to ensure that the weapon has been assembled correctly. The following procedures should be taken to check the function of the weapon.

a. Place the weapon in the single-shot mode.

b. Open the cover and lock the bolt to the rear (bolt should stay to rear while in the single-shot mode).

c. Hold the retractor handles, press the bolt latch release, and ride the bolt forward.

d. Press down on the trigger; weapon should fire. (Check T-slot to ensure that firing pin does protrude.)

e. Place the weapon in the automatic-fire mode

f. Pull the retractor slide handle to the rear and hold. (Bolt should not lock to rear.)

g. Release the pressure on the slide handles and ride the bolt forward

h. Make sure firing pin does not protrude

i. Press trigger; weapon should fire

j. Make sure firing pin does protrude.

NOTE: Before firing a newly assembled weapon, first set the headspace and timing. Chapter 3 describes these procedures.

* CHAPTER 3

OPERATION AND FUNCTIONING

This chapter explains the operation of the MG. It discusses the loading, unloading, and clearing procedures, and the cycle of functioning of the weapon. When training the cycle of functioning using dummy ammunition, it is imperative that all safety procedures be followed.

* 3-1. OPERATION

The overall operation of the MG includes how to load, unload, and clear the weapon. During the weapon's operation, it is mandatory that all ammunition be free of dirt and corrosion, that the ammunition be properly linked, and that the double-linked end be at the top of the ammunition can.

* 3-2. LOADING PROCEDURES

Trainers must ensure that the weapon functions correctly and that proper headspace and timing have been set before loading. When loading in either mode, the ammunition is fed into the MG in the same manner (Figure 3-1). Ensure that the bolt is forward and the cover is closed. Insert the double-loop end of the ammunition belt into the feedway until the first round is engaged by the belt-holding pawl. Figure 3-2 shows the correct position of the bolt latch in the single-shot or automatic mode.

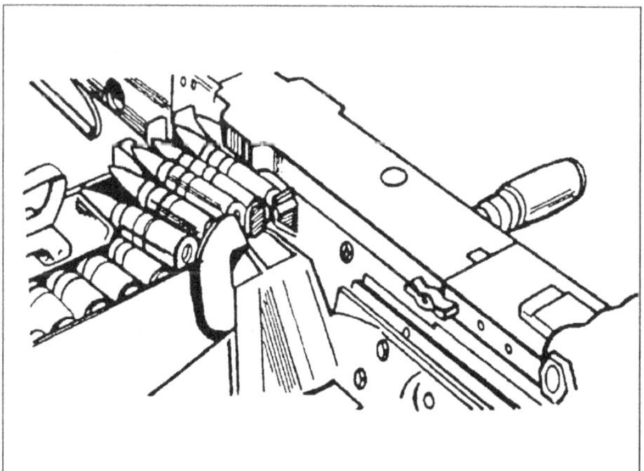

Figure 3-1. Inserting ammunition.

a. **Single-Shot Mode**. When engaging targets at ranges greater than 1,100 meters, using the single-shot mode (firing one round at a time) allows the gunner to deliver well-aimed fire on the target. To load in the single-shot mode--

(1) Keep the bolt-latch release unlocked in the up position and release it manually for each round.

(2) Jerk the retracting slide handle to the rear and lock it in position. Return the retracting slide handle to the forward position and then release the bolt by pressing the bolt latch release. The gun is now half-loaded.

(3) To complete loading, jerk the retracting slide handle to the rear and lock it in position. Return the retracting slide handle to the forward position. Press the bolt latch release. When the bolt goes forward for the second time, the gun is loaded.

b. **Automatic Mode**. To load in the automatic mode--

(1) Lock the bolt-latch release down with the bolt- latch release lock.

(2) Jerk the retracting slide handle to the rear and release it. The gun is now half-loaded.

(3) To complete loading, jerk the retracting slide handle to the rear a second time and release it. When the bolt goes forward for the second time, the gun is loaded.

Figure 3-2. Firing modes.

* 3-3. UNLOADING PROCEDURES

To unload the MG, the gunner must first ensure that the weapon is in the single-shot mode. The cover is then lifted and the assistant gunner removes the ammunition belt from the feedway. The bolt is then locked to the rear. If a round is chambered, it will release, unfired, when the bolt locks to the rear. Once the bolt is locked to the rear, the chamber and

T-slot are examined to ensure that they are not holding rounds. In darkness, this must be done by feeling the areas. After the examination has been done (during training), a wooden block is inserted in the receiver between the bolt and the rear of the barrel, extending above and below the receiver about one inch. Then a cleaning rod is inserted in the muzzle end of the barrel and pushed through the bore until it can be seen in the receiver. Remove the rod, the gun is now clear (Figure 3-3).

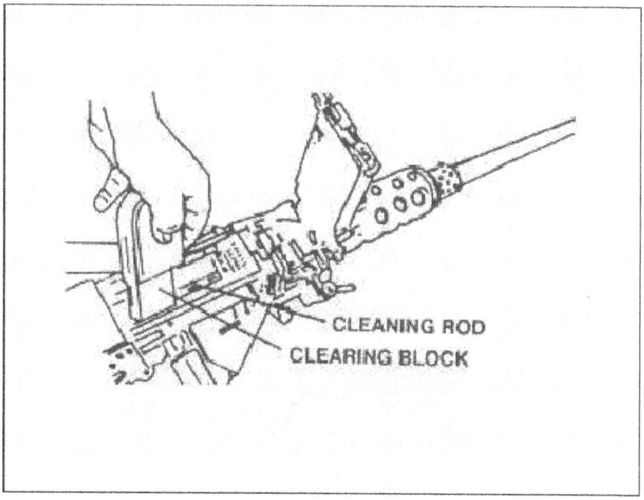

Figure 3-3. The clearing block.

* 3-4. CYCLE OF FUNCTIONING

The cycle of functioning is broken down into basic steps: feeding, chambering, locking, firing, unlocking, extracting, ejecting, and cocking. Some of these steps may occur at the same time.

a. **Feeding**. Feeding is the act of placing a cartridge in the receiver, approximately in back of the barrel, ready for chambering. When the bolt is fully forward and the top is closed, the ammunition belt is held in the feedway by the belt-holding pawl (Figure 3-4).

(1) As the bolt is moved to the rear, the belted ammunition is moved over and then held in a stationary position by the belt-holding pawl. At the same time, the belt-feed pawl rides up and over the link, holding the first round in place. When the bolt is all the way to the rear, the belt-feed slide moves out far enough to allow the belt-feed pawl spring to force the pawl up between the first and second rounds (Figure 3-5).

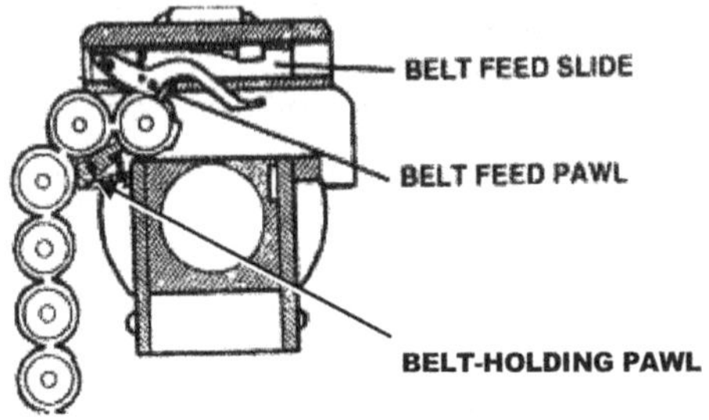

Figure 3-4. Feeding--step 1.

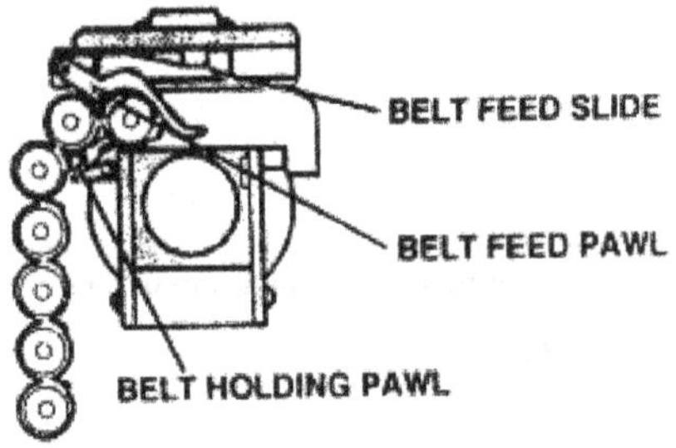

Figure 3-5. Feeding--step 2.

(2) As the bolt moves forward, the belt-feed slide is moved back into the receiver, pulling with it the next linked cartridge. When the bolt reaches the fully forward position, the belt-holding pawl will snap into place behind the second linked cartridge (Figure 3-7), holding it in place. The extractor will then grasp the rim of the first cartridge, preparing to release it from the belt on the next rearward motion (Figure 3-8).

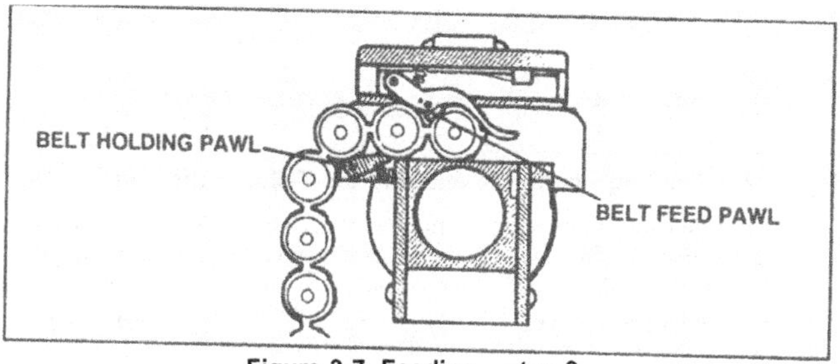

Figure 3-7. Feeding – step 3.

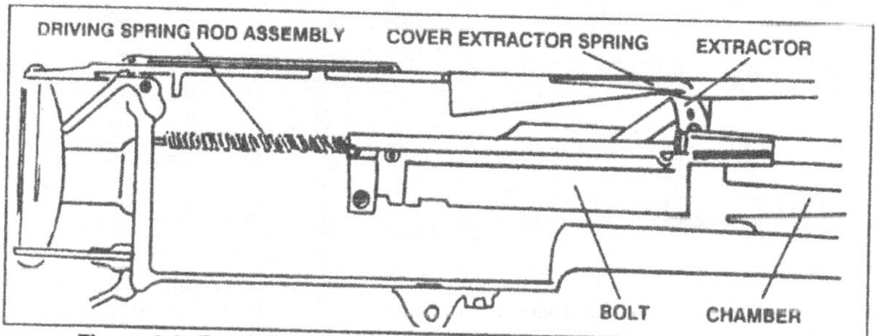

Figure 3-8. Feeding – withdrawing the first round from the feedway.

(3) As the bolt then moves to the rear, the extractor will pull the cartridge with it, releasing it from the belt. As it moves to the rear, the extractor is forced down by the extractor cam, causing the cartridge to be moved into the T-slot in the bolt face, preparing the cartridge to be chambered (Figure 3-9, page 3-6). It is connected under the extractor switch on the side of the receiver until it is repositioned by the forward movement of the bolt, and pressure of the cover extractor spring forces it over the next round.

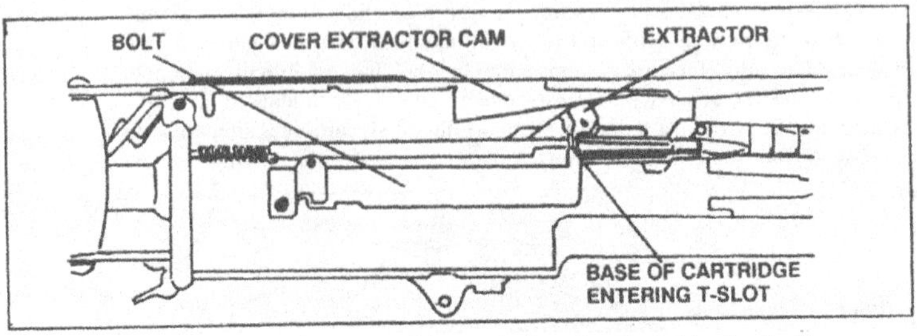

Figure 3-9. Feeding - cartridge entering the T-slot in the bolt.

b. **Cambering.** Cambering is placing the cartridge into the chamber of the weapon. During this cycle, the bolt moves forward, carrying the cartridge in the T-slot in a direct route to the chamber of the weapon. At the same time, the extractor rides up the extractor cam and when the bolt is fully forward, the extractor grasps the next linked cartridge (Figure 3-10).

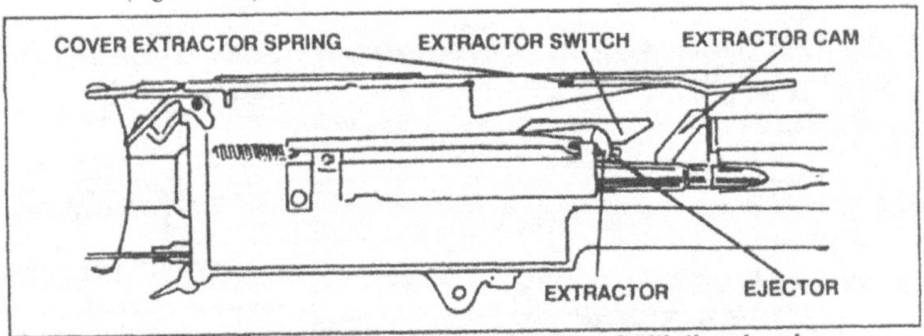

Figure 3-10. Cambering – new round aligned with the chamber.

c. **Locking.** The bolt is locked to the barrel and barrel extension.

(1) Initially, the bolt is forced forward in counter-recoil by the energy stored in the driving spring assembly and the compressed buffer disks. At the start of counter-recoil, the barrel buffer body tube lock keeps the accelerator tips from bounding up too soon and catching in the breech lock recess in the bolt. After the bolt travels forward about 5 inches, the lower rear projection of the bolt strikes the tips of the accelerator, turning the accelerator forward. This unlocks the barrel extension from the barrel buffer body group and releases the barrel buffer spring. The barrel buffer spring expands, forcing the piston rod forward.

(2) Since the cross groove in the piston rod engages the notch on the barrel extension shank, the barrel extension and barrel are also forced forward by the action of the barrel buffer spring. Some of the forward motion of the bolt is transmitted to the barrel extension through the accelerator. As the accelerator rotates forward, the front of the accelerator speeds up the barrel extension; at the same time, the accelerator tips slow down the bolt.

(3) Locking begins 1 1/8 inches before the recoiling groups (bolt, barrel extension, and barrel) are fully forward. The breech lock in the barrel extension rides up the breech lock cam in the bottom of the receiver into the breech lock recess in the bottom of the bolt, locking the recoiling groups together. The recoiling groups are completely locked together three-fourths of an inch before the groups are fully forward (Figure 3-11).

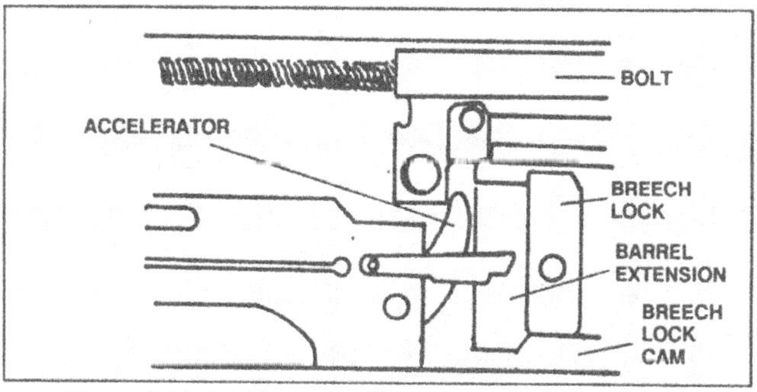

Figure 3-11. Locking – recoiling groups locked together.

d. **Firing.** The firing pin is released, igniting the primer of the cartridge.

(1) As the trigger impressed down, it pivots on the trigger pin, so that the trigger cam on the inside of the backplate engages and raises the rear end of the trigger lever. This in turn pivots on the trigger lever pin assembly, causing the front end of the trigger lever to press down on the top of the sear stud. The sear is forced down until the hooked notch of the firing pin extension is disengaged from the sear notch. The firing pin and firing pin extension are driven forward by the firing pin spring; the striker of the firing pin hits the primer of the cartridge, firing the round (Figures 3-12 and 3-13, page 3-8).

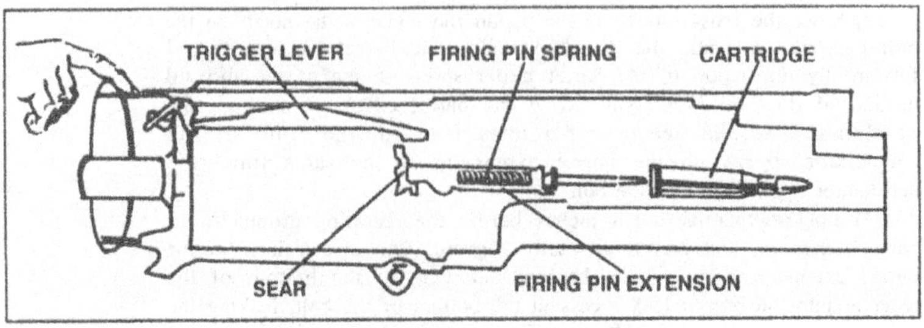

Figure 3-12. Firing – ready to fire.

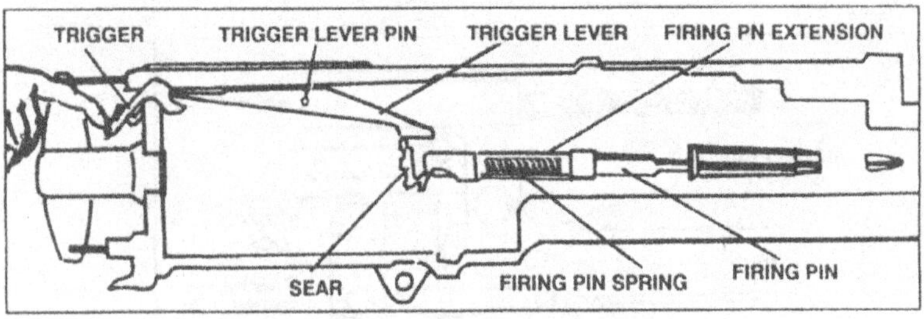

Figure 3-13. Firing – round ignited.

(2) For automatic firing, the bolt-latch release must be locked or held depressed, so that the bolt latch will not engage the notches in top of the bolt, holding the bolt to the rear as in single-shot firing. The trigger is pressed and held down. Each time the bolt travels forward in counter-recoil, the trigger lever depresses the sear, releasing the firing pin extension assembly and the firing pin. This automatically fires the next round when the forward movement of the recoiling groups is nearly completed. The gun should fire about one-sixteenth of an inch before the recoiling groups are fully forward. Only the first round should be fired with the parts fully forward. The gun fires automatically as long as the trigger and bolt latch are held down and ammunition is fed into the gun.

e. **Unlocking.** The bolt is unlocked from the barrel and barrel extension.

(1) At the instant of firing, the bolt is locked to the barrel extension and against the rear end of the barrel by the breech lock, which is on top

of the breech lock cam and in the breech lock recess in the bottom of the bolt. When the cartridge explodes, the bullet travels out of the barrel; the force of recoil drives the recoiling groups rearward. During the first three-fourths of an inch, the recoiling groups are locked together. As this movement takes place, the breech lock is moved off the breech lock cam stop, allowing the breech lock depressors (acting on the breech lock pin) to force the breech lock down, out of its recess from the bottom of the bolt. At the end of the first three-fourths of an inch of recoil, the bolt is unlocked, free to move to the rear independent of the barrel and barrel extension.

(2) As the recoiling groups move to the rear, the barrel extension causes the tips of the accelerator to rotate rearward. The accelerator tips strike the lower rear projection of the bolt, accelerating the movement of the bolt to the rear. The barrel and barrel extension continue to travel to the rear an additional three-eighths of an inch, or an approximate total distance of 1 1/8 inches until they are stopped by the barrel buffer assembly (Figure 3-14).

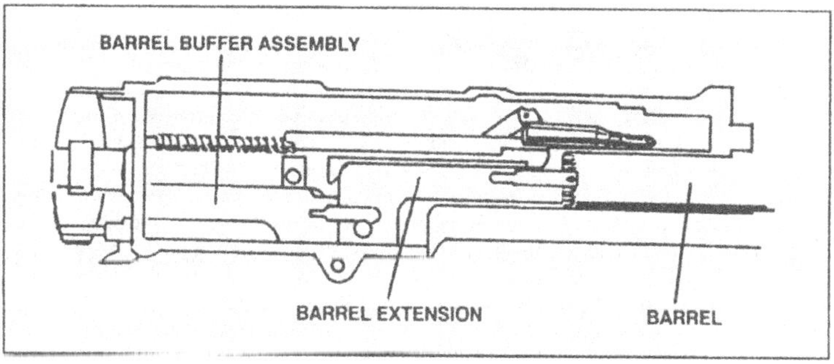

Figure 3-14. Unlocking – barrel and barrel extension stopped by the barrel buffer assembly.

(3) During the recoil of 1 1/8 inches, the barrel buffer spring is compressed by the barrel extension shank, since the notch on the shank is engaged in the cross groove in the piston rod head. The spring is locked in the compressed position by the claws of the accelerator, which engage the shoulders of the barrel extension shank. After its initial travel of three-fourths of an inch, the bolt travels an additional 6 3/8 inches to the rear, after it is unlocked from the barrel and barrel extension, for a total of

7 1/8 inches. During this movement, the driving springs are compressed. The rearward movement of the bolt is stopped as the bolt strikes the buffer plate. Part of the recoil energy of the bolt is stored by the driving spring rod assembly, and part is absorbed by the buffer disks in the backplate (Figure 3-15).

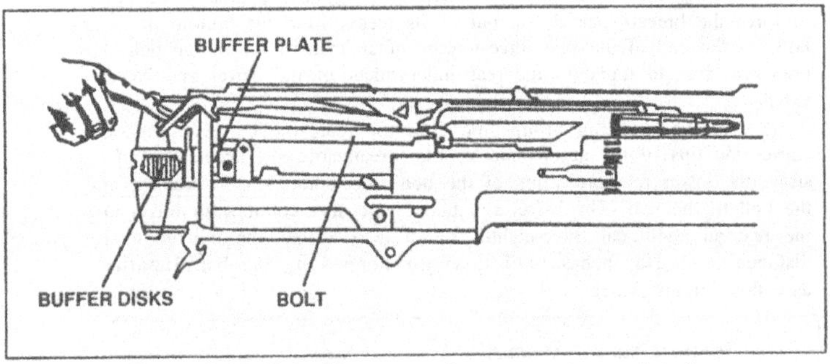

Figure 3-15. Unlocking – recoil movement completed.

f. **Extracting.** The empty cartridge case is pulled from the chamber.

(1) The empty case, held by the T-slot, has been expanded by the force of the explosion; therefore, it fits snugly in the chamber. If the case is withdrawn from the chamber too rapidly, it may be torn. To prevent this, and to ensure slow initial extraction of the case, the top forward edge of the breech lock and the forward edge of the lock recess in the bolt are beveled. As the breech lock is unlocked, the initial movement of the bolt away from the barrel and barrel extension is gradual.

(2) The slope of the locking faces facilitates locking and unlocking and prevents sticking. The leverage of the accelerator tips on the bolt speeds extraction after it is started by kicking the bolt to the rear to extract the empty case from the chamber.

g. **Ejecting.** The empty cartridge case is expelled from the receiver.

(1) As the bolt starts its forward movement (counter-recoil), the extractor lug rides below the extractor switch, forcing the extractor assembly farther down until the round is in the center of the T-slot of the bolt.

(2) The round, still gripped by the extractor, ejects the empty case from the T-slot. The last empty case of an ammunition belt is pushed out by the ejector.

h. **Cocking.** The firing pin is withdrawn into the cocked position.

(1) When the recoiling groups are fully forward, the top of the cocking lever rests on the rear half of the V-slot in the top plate bracket. As the bolt moves to the rear, the top of the cocking lever is forced forward. The lower end pivots to the rear on the cocking lever pin. The rounded nose of the cocking lever, which fits through the slot in the firing pin extension, forces the extension to the rear, compressing the firing pin spring against the sear stop pin (accelerator stop). As the firing pin extension is pressed to the rear, the hooked notch of the extension rides over the sear notch, forcing the sear down. The sear spring forces the sear back up after the hooked notch of the firing pin extension has entered the sear notch.

(2) The pressure of the sear and firing pin springs holds the two notches locked together. There is a slight overtravel of the firing pin extension in its movement to the rear to ensure proper engagement with the sear. As the bolt starts forward, the overtravel is taken up and completed when the cocking lever enters the V-slot of the top plate bracket, and is caromed toward the rear; pressure on the cocking lever is relieved as the bolt starts forward.

3-5. LEFT-HAND FEED

By repositioning some of the components, the MG is capable of alternate feed. Ammunition can be fed into the weapon from the right or left side of the receiver; however the Army uses only left-hand feed. (See Table 3-1).

PART	POSITION
Belt-feed lever, shoulder headless pin, and spring.	Upper (rear) hole; lug of feed lever is on left side of cover.
Belt-feed slide.	Feed pawl is on left side of cover.
Belt-feed pawl arm.	Arm toward latch end of cover, pointing right.
Cover latch shaft lever	Left side of cover.
Cartridge stops and link stripper.	Right-hand rear cartridge stop assembly and front cartridge stop on right side of feedway.
Retracting slide with handle.	Right sideplate.
Belt-holding pawl.	Left side of feedway.
Bolt switch.	Cam grove in line with "L" on bolt.

Table 3-1. Position of parts for left-hand feed.

DANGER

Clear the gun of ammunition before setting headspace.

3-6. HEADSPACE AND TIMING

Headspace is the distance between the face of the bolt and the base of the cartridge case, fully seated in the chamber. Timing is the adjustment of the gun so that firing takes place when the recoiling parts are in the correct position for firing. Because the cartridge is held by the T-slot of the bolt, headspace with the MG is measured as the distance between the rear of the barrel and the face of the bolt. This occurs when the recoiling parts are forward and there is positive contact between the breech lock recess in the bolt and the lock in the barrel extensions. Periodic calibration checks should be made of the gauge by direct support personnel at least annually.

WARNING

Firing a weapon that has improperly set headspace and timing could result in damage to the machine gun, or injury to the gunner. Damage may also occur in the trunnion block, base of the barrel, or face of the bolt. This warning applies whether the gun is firing service ammunition or M1E1 blanks. (The weapon has improper early timing when two rounds are fired – and firing stops.)

a. **Gauges.** The headspace and timing gauge consists of a headspace gauge and two timing gauges (Figure 3-16). These gauges provide an accurate means of checking the adjustment of headspace and timing.

NOTE: The headspace and timing gauge should be kept with the gun at all times.

Figure 3-16. Headspace and timing gauge.

b. **Headspace.** Check and set headspace before firing, after assembling the gun, and after replacing the barrel or receiver group. Use the following procedures to set headspace.

(1) Raise the cover all the way up. Grasp the retracting slide handle (Figure 3-17). Using the retracting slide handle, retract the bolt until the barrel-locking-spring lug is centered in the 3/8-inch hole on the right side of the receiver (Figure 3-18).

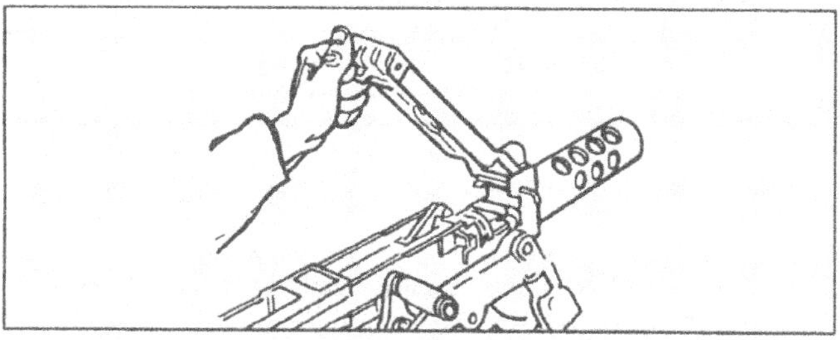

Figure 3-17. Raising the cover.

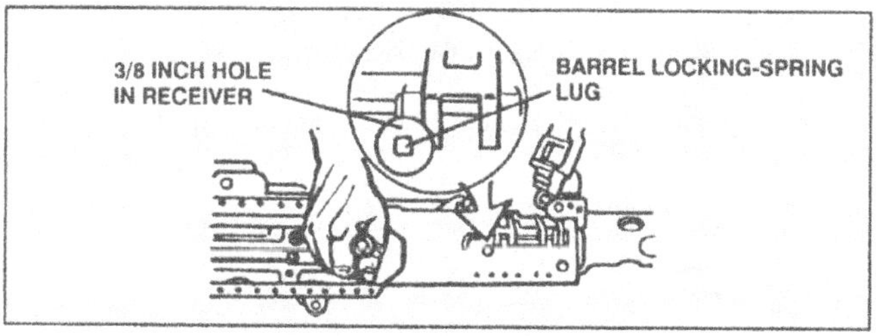

Figure 3-18. Retracting the bolt.

(2) Hold the bolt in this position and screw the barrel fully into the barrel extension (Figure 3-19).

Figure 3-19. Screwing in the barrel.

WARNING

When resetting the headspace and timing of a gun that has been fired, use an asbestos mitt to avoid burns.

(3) With the bolt still retracted, unscrew the barrel two notches (clicks). Release the retracting slide handle and allow the bolt to go forward.

NOTE: At this point, check the barrel for rotation. Attempt to turn the barrel in either direction. The barrel should not turn. If the barrel does turn, stop here and check barrel notches and the barrel-locking spring for damage.

(4) Pull the bolt to the rear with the retracting slide handle and hold. This cocks the weapon. Otherwise, the headspace gauge will not fit.

(5) Holding the retracting slide handle, release the bolt, and slowly return the bolt fully forward (Figure 3-20). Do not press the trigger or let the bolt slam forward.

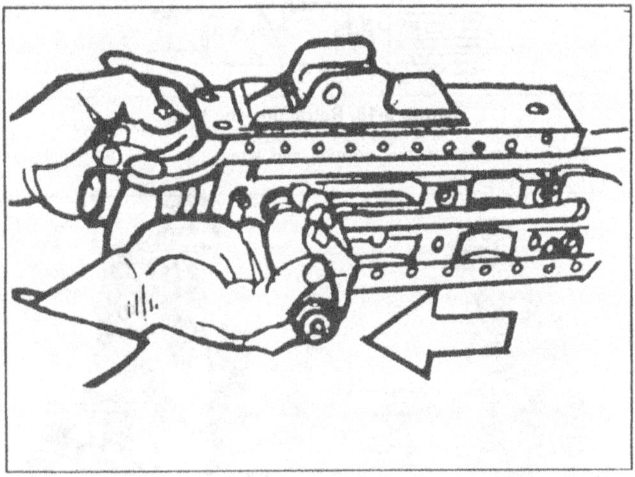

Figure 3-20. Releasing the bolt.

(6) Retract the retracting slide handle and separate the barrel extension from the trunnion block by a 1/16-inch gap (Figure 3-21).

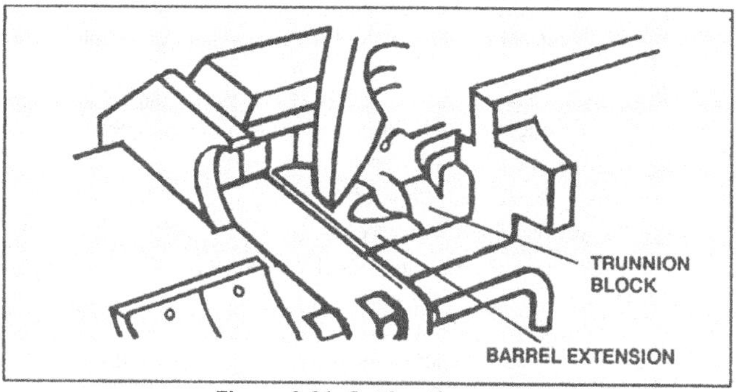

Figure 3-21. Setting the gap.

(7) Raise the extractor out of the way to clear the top of the T-slot and try both ends of the go/no go gauge as shown in Figure 3-22.

NOTE: Insert the go end of the gauge between the face of the bolt and the end of the barrel all the way up to the ring.

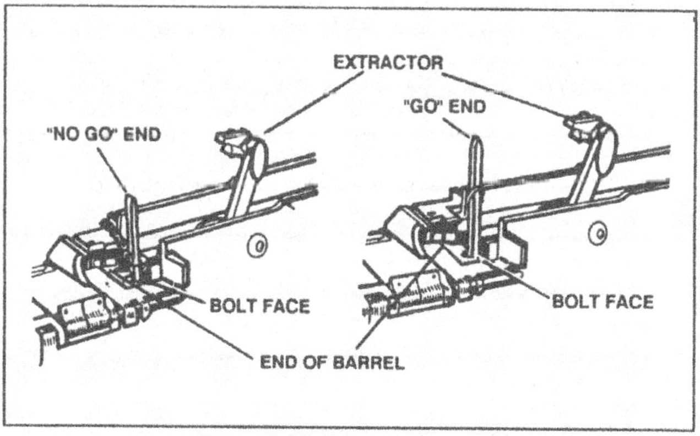

Figure 3-22. Using the gauge.

(8) If the go end of the gauge enters the T-slot freely to the center ring of the gauge, and the no-go end will not enter, headspace is correct. Remove gauge. Headspace setting is now complete. If the go end of the gauge will not enter the T-slot freely, headspace is too tight. Continue as follows.

(9) Retract the bolt so you can see the barrel-locking lug in the center of the 3/8th-inch alignment hole on the right side of the receiver.

(10) Unscrew the barrel one notch (click).

(11) Return the bolt fully forward.

(12) Recheck headspace (step 9).

(13) Repeat steps 10 through 13 until the go gauge fits but the no-go gauge does not fit.

NOTE: You should not have to unscrew the barrel more than five notches (clicks) beyond the first setting of two clicks. If this condition does occur, turn in the machine gun to your unit armorer for inspection.

(14) If the no-go end of the gauge enters the T-slot, headspace is too loose. Adjust it using the same procedures as above, screwing the barrel into the barrel extension rather than out.

(15) Repeat steps 10 through 13, one click at a time, until the no-go gauge does not fit but the go gauge does fit.

c. **Timing.** Timing is the adjustment of the weapon so that firing takes place when the recoiling parts are between .020 and .116 inch out of battery to prevent contact between the front end of the barrel extension and the trunnion block. Use the following procedures to set timing.

WARNING

Make sure the gun is clear of ammunition before starting.

(1) Check headspace first as previously described.

(2) Pull the bolt to the rear with the retracting slide handle and then ease bolt fully forward to cock the machine gun.

(3) Grasp the retracting slide handle and retract the bolt just enough (1/16 inch) to insert the no-fire gauge between the barrel extension and the trunnion block. Release the retracting slide handle (Figure 3-23).

(4) Depress the trigger. Gun should not fire.

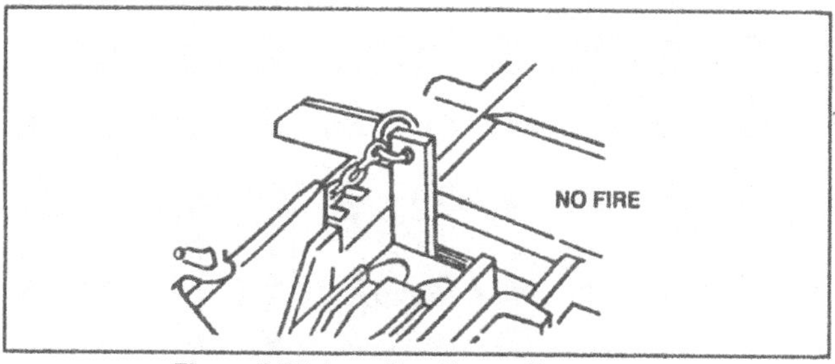

Figure 3-23. Inserting the no-fire gauge.

NOTE: Insert timing gauge with bevel against barrel notches.

NOTE: If the gun does not fire, go to step 5. If the gun does fire, you have early timing. Go on to steps 7 through 14.

(5) Grasp the retracting slide handle and retract the bolt just enough to remove the no-fire gauge and insert the fire gauge in the same place (Figure 3-24). Release the retracting slide handle.

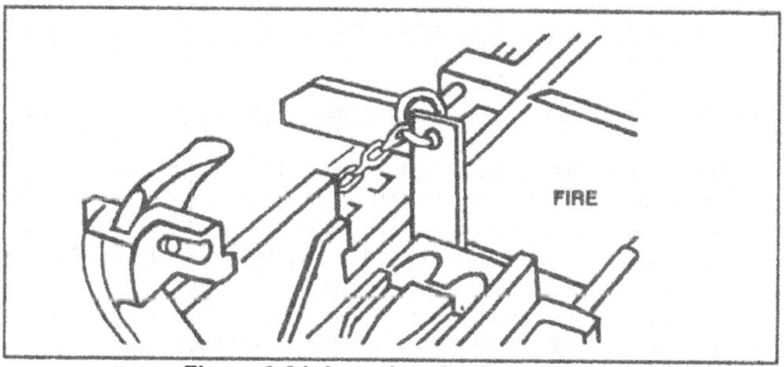

Figure 3-24. Inserting the fire gauge.

(6) Depress the trigger. Gun should fire. If it does, timing adjustment is now complete.

NOTE: If the gun does not fire, you have late timing. Go to steps 7 through 14.

(7) Remove the gauge, cock the gun, and return the bolt forward

(8) Insert the fire gauge.

(9) Remove the backplate.

(10) Screw the timing adjustment nut all the way down until it touches the trigger lever (Figure 3-25).

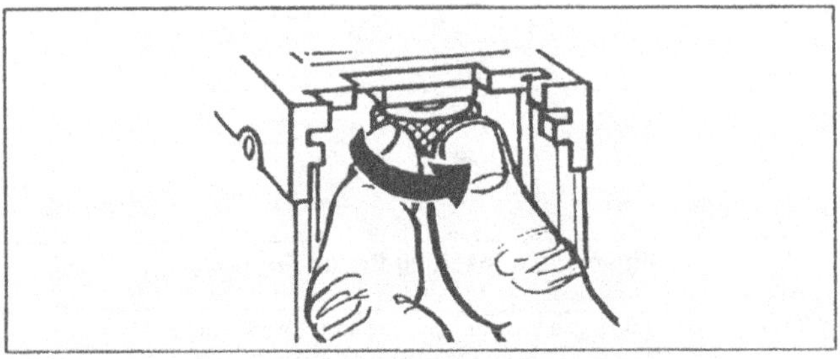

Figure 3-25. Adjusting the timing nut.

WARNING

Never cock the gun with the backplate off.

(11) Try to fire the MG by pushing up on the rear of the trigger lever. Gun should not fire.

(12) Screw the timing adjustment nut up (to the right) one click at a time. Push up on the trigger lever after each click. Keep doing this until the gun fires.

(13) Turn the timing adjustment nut no more than two more clicks up (to the right).

(14) Remove the gauge, replace the backplate, and pull the bolt to the rear to cock the machine gun. Ease the bolt forward with the charging handle. Do not allow the bolt to slam forward.

(15) Recheck the timing with the fire/no-fire gauge twice after the backplate is installed to ensure that the setting is correct.

d. **Field Expedient Methods.** When a go/no-go gauge is not available, you can still set the headspace and timing using field expedient methods. However, this method should be used only in combat.

(1) *To set headspace:*

(a) Raise the cover and retract the bolt in the normal manner until the barrel-locking-spring lug is centered in the 3/8-inch hole on the right side of the receiver.

(b) Hold the bolt in this position and screw the barrel fully into the barrel extension; then unscrew the barrel two clicks or notches.

(2) *To set timing:*

(a) Use a dog tag or a dime as a fire gauge.

(b) Use a nickel and a dime or four dog tags as a no-fire gauge.

(c) Set the timing using the normal procedure.

(3) *To check for correct settings:*

(a) Attempt to fire the weapon. If it fires sluggishly, clear the weapon then unscrew the barrel one more notch.

(b) Recheck the rate of fire. Repeat the procedures in paragraph (1); however, do not exceed two more clicks.

(c) Do not unscrew the barrel more than one notch between test firings.

CHAPTER 4

PERFORMANCE PROBLEMS AND DESTRUCTION

This chapter discusses the malfunctions, stoppages, immediate actions, remedial actions, and destruction procedures of the MG. All corrective actions must comply with safety procedures.

4-1. MALFUNCTIONS

A malfunction is any failure of the gun to function satisfactorily. Examples of malfunctions are:

a. **Failure to Function Freely.** Sluggish operation is usually due to human failure to eliminate excessive friction caused by dirt, lack of proper lubrication, burred parts, incorrect headspace adjustment, or incorrect timing.

b. **Uncontrolled Automatic Fire.** Uncontrolled automatic fire (runaway gun) is when fire continues even when the trigger or trigger control mechanism is released. If the cause is present before the gun is fired, the gun will start to fire when the recoiling groups move into battery the second time. If the defect occurs during firing, the gun will continue firing when the trigger control mechanism is released. A runaway gun may be caused –

- By a bent trigger lever, forward end of the trigger lever sprung downward.
- By burred beveled contacting surfaces of the trigger lever and sear.
- By a jammed or broken sideplate trigger.

To stop the uncontrolled automatic fire:

(1) Keep the gun laid on target.

(2) Twist the belt, causing the gun to jam.

(3) Caution, do not unlatch the cover.

(4) Wait 5 minutes to guard against cook off.

(5) Clear weapon, replace broken, worn, or burred parts. Check the sideplate trigger and trigger control mechanism, when applicable.

4-2. STOPPAGES

A stoppage is any interruption in the cycle of operation caused by the faulty action of the gun or ammunition. Stoppages are classified as follows:

a. **Failure to Feed.** Prevents the round from being properly positioned in the receiver group.

b. **Failure to Chamber.** Prevents the complete chambering of the round.

c. **Failure to Lock.** Prevents the breech lock from correctly entering its recess in the bolt.

d. **Failure to Fire.** Prevents the ignition of the round.

e. **Failure to Unlock.** Prevents the breech lock from moving out of its recess in the bolt.

f. **Failure to Extract.** Prevents the extraction of the expended cartridge from the chamber.

g. **Failure to Eject.** Prevents the ejection of the expended cartridge from the receiver.

h. **Failure to Cock.** Prevents the firing pin extension from being engaged with the sear.

Table 4-1 lists the causes of most stoppages.

NATURE OF STOPPAGE	USUAL CAUSES	OTHER CAUSES
Failure to feed.	Defective ammunition belt. Defective feed mechanism. Defective extractor.	Improperly loaded belt. Short round
Failure to chamber.	Broken part or obstruction in T-slot or chamber. Separated (ruptured) case.	Thick or thin rim, bulged round, protruding primer.
Failure to lock	Incorrect headspace.	Broken parts. Battered breach lock. Battered breach lock cam. Faulty breech lock cam adjustment.
Failure to fire.	Defective parts in firing mechanism. Defective ammunition. Incorrect timing.	
Failure to unlock.	Broken parts in receiver.	Worn or faulty breech lock cam, or faulty adjustment.
Failure to extract.	Dirty chamber.	Defective cartridge case.
Failure to eject.	Defective ejector.	Burred T-slot.
Failure to cock.	Broken sear. Worn sear notch. Weak sear spring. Worn hooked notch on firing pin extension.	Broken cocking lever.

Table 4-1. Stoppages and their causes.

4-3. IMMEDIATE ACTION

The first thing to do when a stoppage occurs is to apply immediate action, which is the prompt action taken by the firer to reduce a stoppage.

Immediate action is performed by the gunner; however, every crewmember must be trained to apply immediate action. The following procedures will assist in reducing most stoppages without analyzing their cause in detail.

WARNING

Failure to follow procedures may result in damage to weapon and injury to personnel.

a. If gun fails to fire, take the following action:

(1) Wait 5 seconds; a hangfire maybe causing the misfire. In the next 5 seconds, pull the bolt to the rear (check for ejection and feeding of belt), release it, re-lay on the target, and attempt to fire. If the bolt-latch release and trigger are depressed at the same time, the bolt goes forward and the weapon should fire automatically.

(2) If the gun again fails to fire, wait 5 seconds, pull the bolt to the rear (engage with bolt latch if applicable), and return the retracting slide handle to its forward position. Open the cover and remove the belted ammunition. Inspect the gun to determine the cause of stoppage.

b. A hangfire or cookoff can cause injury to personnel or damage to the weapon. To avoid these, the gunner must take the following precautions:

(1) Always keep the round locked in the chamber the first 5 seconds after a misfire occurs. This prevents an explosion outside of the gun in event of a hangfire.

(2) If the barrel is hot, the round must be extracted within the next 5 seconds to prevent a cookoff. When more than 150 rounds have been fired in a 2-minute period, the barrel is hot enough to produce a cookoff.

(3) If the barrel is hot and the round cannot be extracted within the 10 seconds, it must remain locked in the chamber for at least 5 minutes, to guard against a cookoff.

(4) Keep the gun cover closed during the waiting periods.

4-4. REMEDIAL ACTION

When immediate action does not correct the malfunction, the quickest way to resume firing is to replace the defective part.

a. **Removal of a Cartridge from the T-Slot.** If the cartridge does not fall out, hold the bolt to the rear, and with the extractor raised, use a screwdriver to push the cartridge out the bottom of the receiver.

b. **Removal of a Ruptured Cartridge.** A ruptured (separated) cartridge case may be removed with a cleaning rod or ruptured cartridge extractor. When using the ruptured cartridge extractor, raise the cover and pull the bolt to the rear. Place the extractor in the T-slot of the bolt in the same manners as that of a cartridge, so that it is held in line with the bore by the ejector of the extractor assembly of the gun. With the extractor aligned with the bore and held firmly in the T-slot, let the bolt go forward into the ruptured case, and the shoulders will spring out in front of the case. Pull the bolt to the rear and remove the ruptured case and extractor (Figures 4-1 and 4-2).

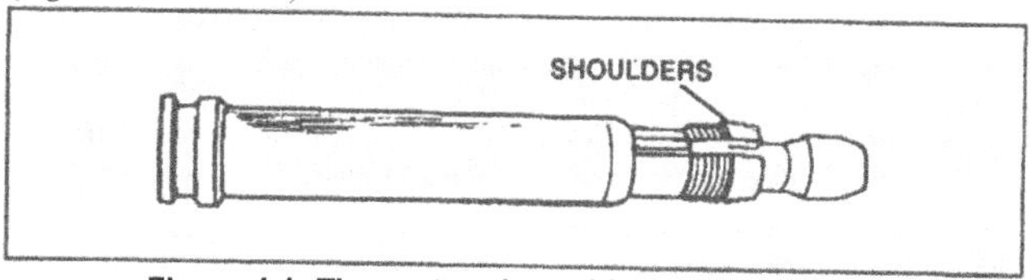

Figure 4-1. The ruptured cartridge case extractor.

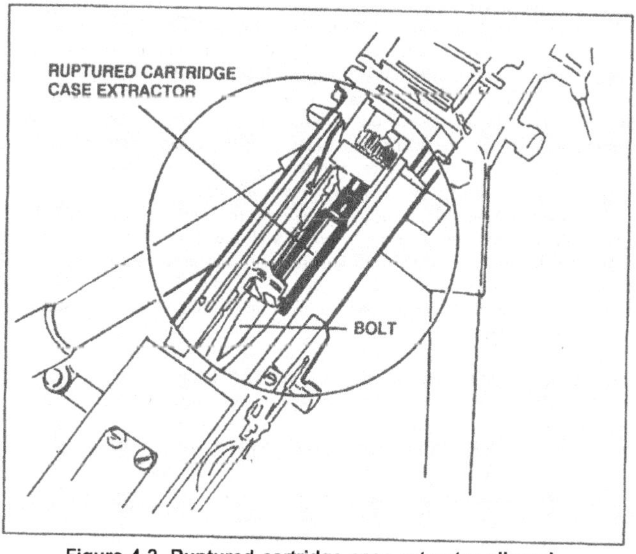

Figure 4-2. Ruptured cartridge case extractor aligned with the chamber.

4-5. DESTRUCTION PROCEDURES

The decision to destroy the gun to prevent its capture and use by the enemy is a command decision. It will be ordered and carried out only on authority delegated by the major unit commander.

- Destroy the machine gun and mount only when they are subject to capture or abandonment. Destruction must be as complete as circumstances permit.
- Lacking time for complete destruction, destroy the parts essential to operation of the gun, beginning with those parts most difficult for the enemy to duplicate.
- Destroy the same parts of each gun to prevent the reconstruction of a complete gun from several damaged guns.

The following methods may be used to destroy specific components of the MG.

a. **Gun.** Field strip the gun. Use the barrel as a sledge. Raise the cover and smash the cover forward and down toward the barrel support. Smash the backplate group. Remove the firing pin from the bolt; place the striker in the hole in the face of the bolt and bend it until broken. Remove the barrel buffer tube lock assembly from the barrel buffer body group and bend and deform it. Smash and bend the breech lock depressors. Place the barrel extension in the rear of the receiver with the barrel extension shank protruding; knock off the shank by striking it with the barrel from the side. Deform and crack the receiver by striking it with the barrel at the sideplate corners nearest the feedway. Smash the extractor.

b. **Tripod Mount, M3.** Leave the pintle on the tripod by removing the pintle bolt from the gun. Use the barrel as a sledge. Strike the sides of the pintle and deform it. Fold the trail legs and turn the mount over. Stand on the folded trail legs and knock off the pintle latch (pintle lock release cam). Smash the elevating mechanism with the barrel. If possible, smash the rear legs to prevent unfolding.

c. **Antiaircraft Mount, M63.** Remove the sideplate trigger control mechanism from its container or the gun, and deform it by using the barrel. Lock the cradle and yoke assembly in the horizontal position and beat the trigger frame assembly and cradle until they are bent down along the elevator assembly. Strike the elevator from the side with the barrel until it is bent so that the elevator will not rotate in the base.

d. **Spare Parts.** Destroy the bolt, barrel extension, firing pins, and barrel buffer groups. Break or deform all other parts.

e. **Ammunition.** When time permits and material is available, destroy ammunition by burning. Unpack all ammunition from boxes or cartons, stack the ammunition in a heap and, using flammable material available, ignite and take cover immediately.

f. **Burning.** To destroy the gun by burning, place a thermite grenade in the receiver on the bolt (with the cover resting on the grenade) and fire the grenade. (This method may require the use of more than one grenade.) Remove the backplate group, place a thermite grenade in the rear of the receiver and fire the grenade.

g. **Disposal.** Bury in suitable holes, or dump parts into streams, mud, snow, sumps, latrines, or scatter the parts over a wide area.

CHAPTER 5

MARKSMANSHIP TRAINING

The procedures, exercises, and techniques for implementing a proper marksmanship program are based on the concept that all machine gunners must understand common firing principles, be proficient marksmen, and be confident in applying these skills in combat. This depends on their understanding of the machine gun and their application of marksmanship fundamentals. Proficiency is accomplished through practice that is supervised by qualified trainers and through objective performance amassments by the unit leaders. The structure of this chapter is in four sections: planning, fundamentals, basic marksmanship, and advanced gunnery. All advanced exercises are conducted under conditions that are as much like tactical conditions as possible.
(See Appendix B for training aids and devices.)

Section I. PLANNING

The planning of MG training is no different from other marksmanship training. Guidelines are provided to assist the trainers in understanding, preparing, and ensuring that all training is conducted to standard. This section addresses the objectives, the responsibilities of the commanders, and the phases of training. It also introduces the training devices that assist in training, and designates when remedial and sustainment training should be conducted.

5-1. OBJECTIVES

The objectives of the MG training program are to guide the trainers, leaders, and gunners through a sequence of training to standard that produces a gunner who is able to maintain the gun and effectively engage targets in combat.

5-2. COMMANDER'S RESPONSIBILITIES

The responsibilities of the commander are to ensure the instructors and their assistants are thoroughly trained and rehearsed in the planning, knowledge, and presentation of all MG training. He will ensure that safety is emphasized and never overlooked during training. Serviceable weapons are a must for good training and commanders must make sure they are used properly. He also ensures sufficient time is scheduled.

5-3. PHASES OF TRAINING

The initial training strategy is divided into three phases of marksmanship instruction — the fundamental phase, the basic marksmanship phase, and the advanced gunner phase. Gunner safety is continuously stressed during all phases of training.

a. In the fundamental phase, the gunner learns the necessary common skills, such as dry-fire exercises, preparation of positions, manipulation of the T&E, range determination, and sighting and aiming. The gunner must master these skills before he is allowed to progress to the basic marksmanship phase.

b. During the basic marksmanship phase, the gunner learns the basics of loading, zeroing, laying, and engaging of single and multiple targets from the tripod and vehicle mode.

c. The advanced gunnery phase trains the gunner in engaging moving targets, night firing, NBC firing, mounted firing, and firing using fire commands. The gunner will be placed under the stress and strain of simulated combat conditions.

5-4. SUSTAINMENT TRAINING

Once individuals and units have trained to a required level of proficiency, leaders must structure collective and individual training plans to conduct critical task training at the frequency necessary for the sustainment training strategy. Mission training plans and individual training evaluation programs help achieve and sustain collective/individual proficiency. Sustainment training prevents skill degradation. Army units must be prepared to accomplish their wartime missions at any time — they cannot rely on infrequent peaking to the appropriate level.

5-5. REMEDIAL TRAINING

Remedial training will be conducted in any of the phases of training where the gunner does not meet the standard. The trainer must be instantly aware of any gunner that seems to be having trouble. Once the problem has been identified, the gunner should be retrained as soon as possible so that he will maintain the same level of proficiency as the other gunners.

Section II. FUNDAMENTALS

The fundamentals are necessary basic skills that a gunner must be trained in before he can be expected to effectively engage targets. Personnel conducting marksmanship training must fully understand the fundamentals and be well rehearsed in applying them. The basics in MG training are assuming a proper firing position, sighting, aiming, determining range, and manipulating the T&E mechanism.

5-6. FIRING POSITIONS

Before a gunner can hit targets, he must learn to get behind the weapon in a position that allows him to be comfortable, affords him protection, and enhances mission accomplishment.

a. The tripod firing positions are prone, sitting, and standing. They are assumed in the following manner.

(1) The prone position is used when firing from the tripod that is set in a low position. It is assumed by lying on the ground directly behind the gun. The gunner then spreads his legs a comfortable distance apart with his toes turned outward. His left elbow rests on the ground, and his left hand grasps the elevating handwheel of the T&E. His right hand lightly grasps the right spade grip with his right thumb in a position to press the trigger. The position of his body can then be adjusted to position his firing eye in alignment with the sights of the weapon (Figure 5-1).

Figure 5-1. Prone position with tripod mount.

(2) The sitting position can be used when the tripod is set in a high or low position. The gunner sits directly behind the gun between the legs of the tripod. He may extend his legs under the tripod or cross them, depending on his physique. The gunner then places both elbows on the inside of his thighs to get the best support. He grasps the elevating

handwheel of the T&E with the left hand, and lightly grasps the right spade grip with his right hand. He must ensure that the right thumb is in position to press the trigger (Figure 5-2).

Figure 5-2. Sitting position with tripod mount.

(3) The standing position is used when the gunner is firing from a fighting position. This position is assumed by standing directly behind the gun with the feet spread a comfortable distance apart. The gunner grasps the elevating handwheel of the T&E with the left hand. He lightly grasps the right spade grip with the right hand, ensuring that the right thumb is in a position to press the trigger. Adjustment of the body is allowed in order to align the firing eye with the sights on the weapon (Figure 5-3).

Figure 5-3. Standing position with tripod mount.

b. The vehicular firing position for the MG is standing. It is assumed by constructing a solid platform to stand on, using sandbags or ammunition boxes; or, in the case of the M113 APC, using the commander's seat. The gunner must then ensure that his platform is high enough to place the spade grips of the gun about chest high. He grasps the spade grips with both hands and places both thumbs in a position to press the trigger. The gunner holds the gun tightly to his chest for stabilization; his elbows should be locked tightly to his sides. He sights over the weapon and adjusts his position by flexing his knees and leaning forward to absorb any recoil (Figure 5-4).

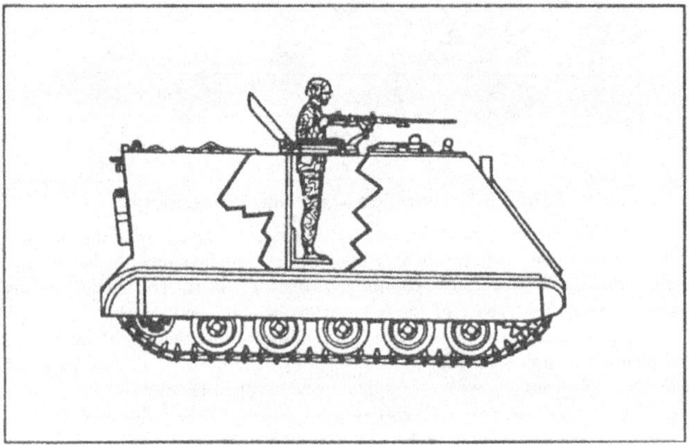

Figure 5-4. Standing position using cupola mount.

c. The antiaircraft firing position uses a standing position when firing from the M63 mount. To assume the position, the gunner stands with his feet spread comfortably apart with his shoulders squarely behind the gun (Figure 5-5, page 5-6). When the gunner is engaging aerial targets, he grasps the upper extension handles with both hands. When engaging low-level aircraft or ground targets, he grasps the lower extension handles with both hands.

NOTE: The kneeling position may be used; it has the advantage of presenting a lower profile of the gunner and also aligns the gunner's eye closer to the axis of the barrel.

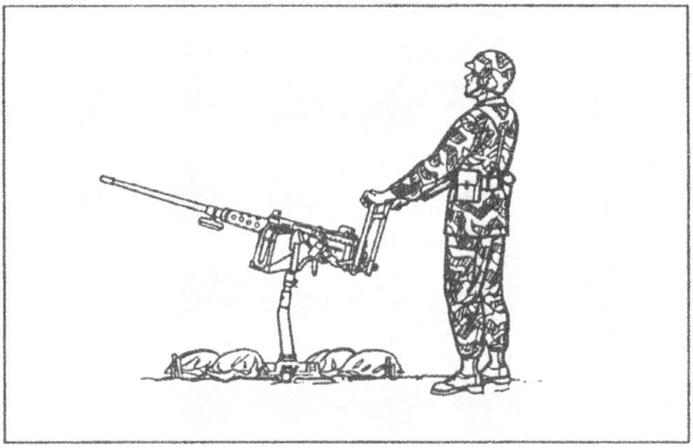

Figure 5-5. Antiaircraft firing position.

5-7. DRY-FIRE TRAINING

Dry-fire training is designed to teach the gunner the essentials of MG gunnery, including safety. Dry-fire training also includes sighting, aiming, sight setting, laying, manipulating the gun, manipulating the T&E mechanism, and determining the range. Thorough, carefully supervised training of these essentials is necessary to conserve time and ammunition during live fire. Practical exercises should be used to determine gunners' proficiency. Mastery of these skills is a must before the gunner is allowed to move on to the next phase of training. Practice is a must to achieve mastery.

a. **Sighting and Aiming.** Sighting is the ability of the gunner to use correct sight alignment and correct sight picture to engage targets.

(1) The first step in proper sighting is finding a natural, comfortable spot where the gunner is able to see the front sight blade through the rear peep sight. It is important the gunner understands that the spot he chooses to sight from must be constant throughout his firing.

(2) The second step in sighting is to move the weapon until the top center of the front sight blade is exactly in the center of the rear sight peep hole. The gunner can achieve this by drawing imaginary lines that bisect in the center of the rear peephole and then placing the top of the front ight blade center of them. This is correct sight alignment (Figure 5-6).

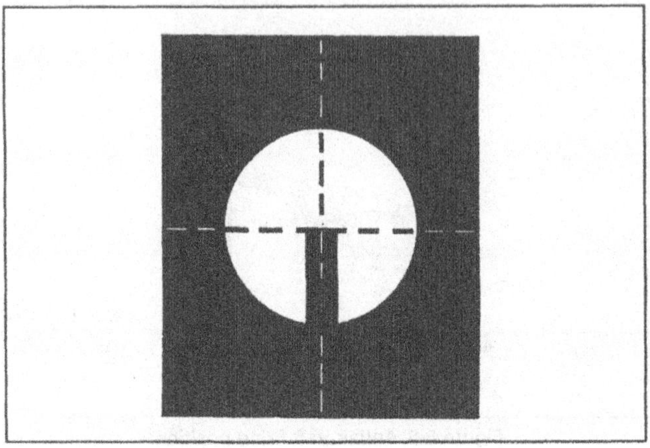

Figure 5-6. Correct sight alignment.

(3) The third step is to establish correct sight picture. To perform this task, the gunner adjusts the weapon until the top center of the front sight blade is bottom center of the intended target. Correct sight picture is a combination of sight alignment and placement of point of aim (Figure 5-7).

Figure 5-7. Correct sight picture.

b. **Range Setting and Laying.** Range setting and laying the gun are important elements in marksmanship training. It is this training that prepares the gunner to accurately and rapidly place fire on his target in combat. To properly set ranges, the gunner must be trained in rear sight operation.

(1) Setting ranges on the rear sight is a simple but important task. The gunner just has to rotate the elevating screw knob in a clockwise manner to move the peep sight up or counterclockwise to move it down. The range scale on the left is graduated in mils and the scale on the right in yards. The gunner must align the hairline index of the peep sight with the scale index line at the desired range as quickly and accurately as possible.

(2) Laying is placing the barrel of the weapon on a direct line with the target, using the sights. This must also be done as quickly and accurately as possible.

(3) The range setting and laying exercises are designed to require the gunner to practice and the instructor to evaluate both correct sight alignments and correct placements of aiming points. The exercise starts with the gunner in the sitting firing position with rear sight down. The coach will announce a range and general aiming point. The gunner will then repeat the range and direction of target. The coach will then announce, "Begin." The gunner will then raise and set his sights, and align the weapon on the aim point. When the gunner completes this task, he will announce "Up." The coach will then get behind the weapon and check the range setting and aim point and critique the gunner on his findings. This exercise will be continued until speed and accuracy is obtained.

c. **Traversing and Elevating Mechanism.** Manipulation of the T&E mechanism (Figure 5-8) is another key factor in effectively engaging targets. The gunner is taught how to instinctively manipulate the T&E mechanism to shift from one target to another. The gunners are trained to use the traversing handwheel, the traversing slide lock, and the elevating handwheel.

(1) The traversing mechanism consists of a traversing handwheel, locking nut, scale, and yoke. The T&E mechanism is attached to the traversing bar of the M3 bipod.

(a) The traversing bar is graduated in 5-mil increments and fits between the trail legs of the M3 tripod. The traversing slide and screw assembly are clamped in place on the traversing bar by the traversing slide lock lever. When the traversing slide is locked to the traversing bar, the traversing handwheel should be centered. The traversing slide is properly mounted when the lock lever is to the rear and the traversing handwheel is positioned to the left.

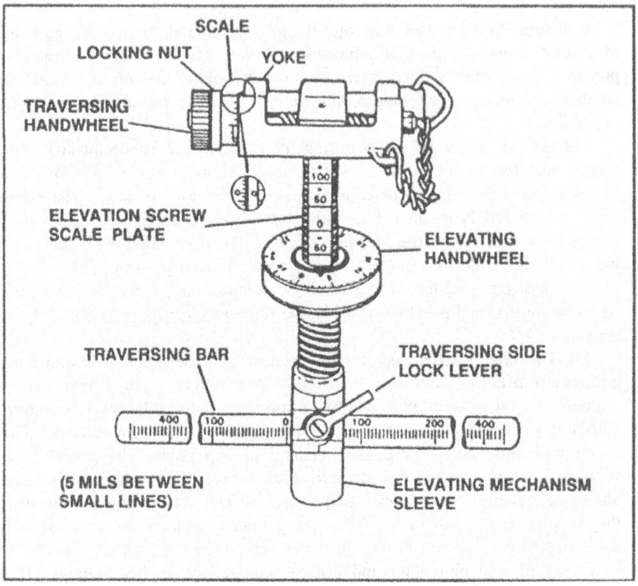

Figure 5-8. Traversing and elevating mechanism.

(b) To make changes in direction, loosen the traversing slide lock lever and move the slide along the traversing bar. This permits traverse of 400 mils left or right of the zero index in the center of the traversing bar. Readings on the traversing bar are taken from the left side of the traversing slide. For changes of 50 mils or less in deflection, turn the traversing handwheel. This allows a traverse of 50 mils left or right of center. One click in the traversing handwheel signifies 1 mil change in direction.

(2) The elevating mechanism consists of an upper and lower elevating screw, which is connected to the gun by inserting the quick release pin assembly through the holes in the upper elevating screw yoke and the rear mounting lugs of the receiver. A scale, graduated in mils, is fitted to the upper screw to indicate elevation. This scale is marked to show (-) minus 250 mils in depression and (+) plus 100 mils in elevation from the zero setting.

(3) The elevating handwheel is graduated in 1 mil increments from 0 to 50 mils and is fastened to the elevating screw by a screw lock. This synchronizes the handwheel graduations with those on the upper elevating screw. A spring-actuated index device produces a clicking sound when the handwheel is turned. Each click equals 1 mil change in elevation. The handwheel is turned clockwise to depress the barrel and counterclockwise to elevate.

(4) Direction and elevation readings constitute the data necessary to engage preselected target areas during limited visibility. These readings are measured by and recorded from the traversing bar and the T&E mechanism. To obtain accurate readings, the T&E must be first zeroed with all measurements recorded in mils.

(a) To zero the traversing handwheel, the gunner must first hold the T&E so that the traversing handwheel is on his left as he looks at it. He then turns the handwheel toward himself until it stops, loosens the locking nut slightly, and aligns the zero on the scale with the zero on the elevating screw yoke. Once the zeros are aligned, he tightens the locking nut. He must then turn the handwheel two complete turns away from the body and stop. The scale should again be on the zero. If this procedure is done at night, the gunner will turn 50 clicks away from him.

(b) To zero the elevating handwheel, the gunner must first turn the handwheel up or down until the handwheel is level with the line directly under the zero on the elevating screw plate scale, and the elevating handwheel indicator is pointing to the zero on the top of the handwheel. The elevating mechanism sleeve is then rotated up until it is stopped by the handwheel. The gunner then rotates the sleeve down until it stops, making sure he counted each complete rotation. He then divides the number of rotations by two, rotates the sleeve back up that number, and stops. The T&E mechanism is now ready to be attached to the tripod.

(c) To obtain and record direction readings, the gunner sets the sight on the proper range to hit the target, loosens the traversing slide lock lever, and slides the T&E mechanism along the traversing bar until the weapon is sighted on the aiming point of the target. The T&E mechanism is then locked down by tightening the traversing slide lock lever. All readings are taken from the left side of the sleeve mechanism. If the left side of the sleeve is not exactly on one of the 5-mil tick marks, the gunner must slide the sleeve to the next smaller tick mark to align it exactly. The traversing handwheel is then used to move the weapon back on point of aim. The direction is now ready to be recorded. The reading is taken from the number on the traversing bar and the direction from the direction of the barrel of the weapon. If the sleeve mechanism is on the right side of the zero on the traversing bar, then the reading is left; if it is on the left side of the zero, then it is a right reading. The width of a target may also be measured and recorded by first moving the traversing handwheel until the sights are aligned with the right or left side of the target. The clicks required to do this measure the width.

NOTE: Before repositioning the weapon for another target, the gunner must realign the handwheel.

(d) To obtain an elevation reading, the gunner must first ensure that the sights are aligned and at the desired aim point of the target. The elevation reading is made up of two portions, a major reading and a minor reading. The major reading is taken from the elevating screw plate scale. The scale is graduated in 50-mil increments and ranges from a minus (-) 250 mils to a plus (+) 100 mils with a zero between them. There is an index line below each number and a plus or minus sign above each number, with the exception of the zero. The zero does not have a plus or minus sign. To obtain the elevation reading, the gunner should lower his head until his eyes are level with the elevating handwheel. The major reading is the first number with a plus or minus sign, with its index line just visible above the elevating handwheel. The minor reading is taken from the top surface of the elevating handwheel. It is graduated in 1-mil increments for a total of 50 mils. The handwheel is also equipped with an indicator that points to each number on the handwheel as it is turned. Once the gunner has the major reading from the screw plate scale, he will then get the minor reading by looking at the number at which the indicator is pointing. Both portions of the elevation reading are recorded by placing a slash (/) mark between the two portions.

(e) An elevation reading is valid only on one T&E mechanism. If the same data is placed on another T&E mechanism using the same weapon, the data may be inaccurate. The number of threads exposed on the T&E must remain the same both when obtaining and recording data. If the number of exposed threads is changed in any manner, the firing will be off target. For example, when a gun is freed to engage targets in the secondary sector, the data will be correct if the gunner ensures that the same amount of threads is exposed when he returns to his primary sector of fire.

(f) To ensure that the data is correct, the gunner should fire and adjust his weapon.

(5) The T&E manipulation exercise gives the gunner practice and the instructor a tool to evaluate the gunner's progress (Figure 5-9, page 5-12). The exercise is conducted in two stages. Both stages require the coach to give directions and the gunner to respond. In the first stage, the coach

positions himself about 10 paces to the front of the gun. He then directs the gunner to manipulate the weapon in certain directions. He indicates the direction by the use of hand signals. The gunner responds by manipulating the T&E mechanism with his left hand. The gunner must keep his eyes on the coach; at no time during this exercise is he permitted to look at the T&E mechanism. The coach must be very attentive during the first stage because the gunner will be manipulating using the elevating handwheel and the traversing handwheel. The second stage is done in the same way except the gunner must make bold changes in elevation and deflection. The exercise continues until the instructor is satisfied that the gunner can manipulate the weapon by T&E without looking at the device. This exercise can also be conducted using the basic MG target. The gunner will be shifted from one selected target to another. The coach must observe all movements of the gunner during this training.

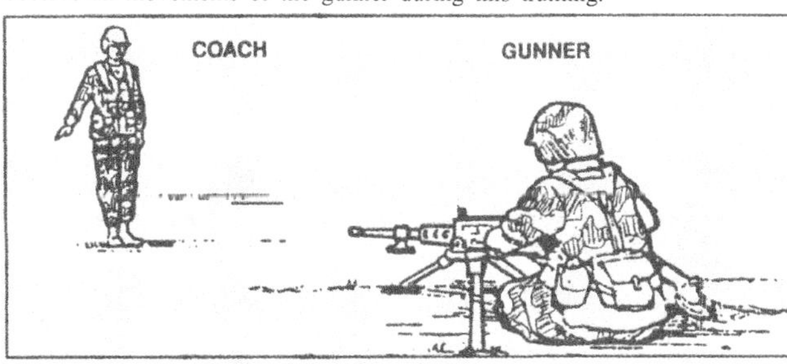

Figure 5-9. Manipulation exercise.

5-8. RANGE DETERMINATION

Range determination is the process of estimating the distance to a target from a gunner's position. The ability of the gunner to get the range to, sight on, and destroy a target is the realism of combat. Under combat conditions, ranges are seldom known in advance; therefore the effectiveness of fire depends largely upon the accuracy and speed of the gunner in determining range. Some methods of determining range are estimating by eye (Table 5-1, page 5-15), firing the gun, measuring range from a map or aerial photograph, stepping off the distance, or securing information from other units. Ranges are determined to the nearest 100 meters for machine gun firing. In combat, the most commonly used methods are estimating by eye and firing the gun. There is also a method used for measuring lateral distance.

a. The two techniques of eye estimation are the 100-meter unit of measure method and the appearance of objects method.

(1) When using the 100-meter unit of measure method, the gunner must be able to visualize what 100 meters looks like on the ground. With this distance in mind, the gunner can mentally determine the number of 100-meter units between his position and the target. The accuracy of this method is limited to 500 meters or less, and it requires constant practice (Figure 5-10).

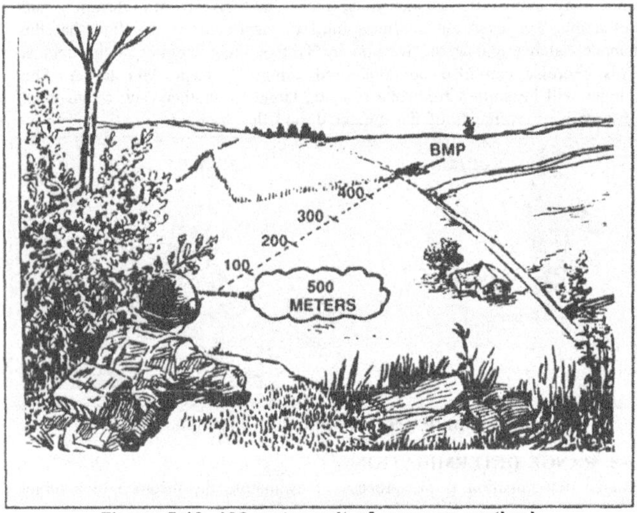

Figure 5-10. 100-meter unit of measure method, less than 500 meters.

(2) For targets that appear to be more than 500 meters, the gunner must modify this technique. The gunner selects what he thinks is the halfway point between the target and his position. He then mentally counts the number of 100-meter units to the halfway point and doubles it. This method of range determination is not accurate beyond 1,000 meters (Figure 5-11, page 5-14).

(3) Some terrain affects the appearance of 100-meter units of measure. When the terrain slopes upward toward the target, 100 meters appears longer than on level terrain. It appears shorter on downward

sloping terrain. The gunner must consider these two factors when using the 100-meter unit of measure method.

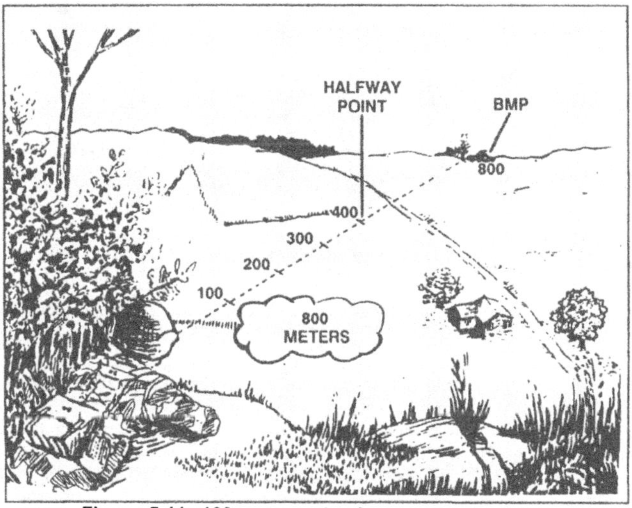

Figure 5-11. 100-meter unit of measure method, more than 500 meters.

(4) The appearance of objects method may be used if the gunner is unable to use the 100-meter unit of measure method because of terrain. To use this method, the gunner learns through practice how familiar objects look at various known ranges. This can be achieved by studying the appearance of a man standing 100 meters away. The gunner must then fix the appearance of the man firmly in his mind to include the size and details of his uniform and equipment. Next, he studies the same man at the same distance in the kneeling and prone positions. This procedure is used at 200, 300, 400, and 500 meters. By comparing the appearance of the man at these known ranges, he can establish a series of mental images that will help him determine range on unfamiliar terrain out to 500 meters. This training could also be conducted to familiarize the gunner with other objects, such as weapons and vehicles, at various ranges.

FACTOR	APPEARS NEARER (Range is underestimated when)–	APPEARS MORE DISTANT (Range is overestimated when)–
Target visibility.	Most of the target is visible and offers a clear outline.	Only a small part of the target is seen or target is small in relation to its surroundings.
Terrain, or position of the observer.	Looking across a depression, most of which is hidden from view. Looking down from high ground. Looking down a straight, open road or along a railroad track. Looking over uniform surfaces such as water, snow, desert, or grain fields.	Looking across a depression, all of which is visible. Looking from low toward high ground. When vision is narrowly confined as in streets, draws, or forest trails.
Light and atmosphere	In bright light or the sun is shining from behind the observer. The target is in sharp contrast with the background, or is silhouetted by reason of size,shape, or color,or is seen in the clear atmosphere of high altitudes.	In poor light such as dawn and dusk, in rain, snow, fog, or when the sun is in the observer's eyes. The target blends into the background or terrain.

Table 5-1. Factors affecting range estimation by eye.

b. Firing the gun is another method of determining range. In this method, the gunner opens fire on the target at the estimated range and moves the center of the beaten zone into the center base of the target by means of the T&E handwheels. He resets the sight so the new line of aim is at the center base of the target and notes the range setting on the rear sight. This range setting may apply only to this gun. When the ground in the vicinity of the target does not permit observation of the strike of the rounds, or when surprise fire on the target is desired, fire is adjusted on a point that offers observation and is known to be the same range as the target. The gunner then lays his gun on the target when ordered. When moving into position occupied by other units, range cards prepared by those units can furnish valuable range information on targets, suspected targets, and various terrain features. When the tactical situation and time permit, range may be determined by pacing off the distance.

c. Lateral distance measure is a method that the gunner may use to determine the distance from one target to another from left to right or right to left. When the gun is mounted on the M3 tripod, width can be measured by aiming on a point and manipulating the traversing handwheel, counting the clicks from one point to another point of aim. Each click equals one meter at 1,000 meters or one-half meter at 500 meters. This method is accurate but time-consuming. The finger measurement method is not a method of range determination; it is a method of measuring the lateral distance (in fingers or mils) between two points. To measure the distance in fingers between a reference point and a target, extend the arm with palm outward, the fingers cupped, and elbow locked. Close one eye, raise the index finger, and sight along its edge, placing the edge of the finger along the flank of the target or reference point (Figure 5-12). The remaining space is then filled in by raising fingers until the space is covered. The measurement is then stated as being one or more fingers or so many mils, depending on the number of fingers used (Figure 5-13).

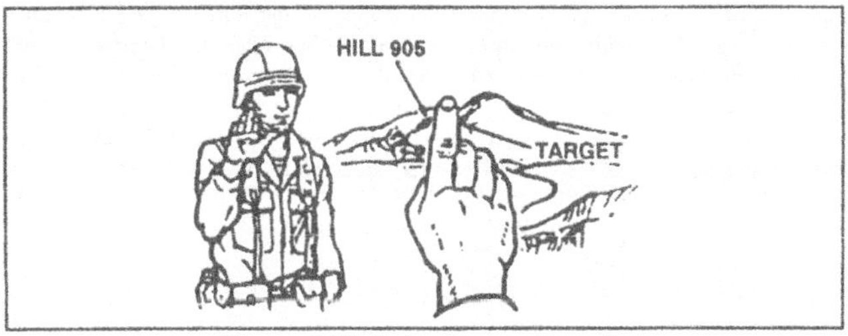

Figure 5-12. Index finger aligned.

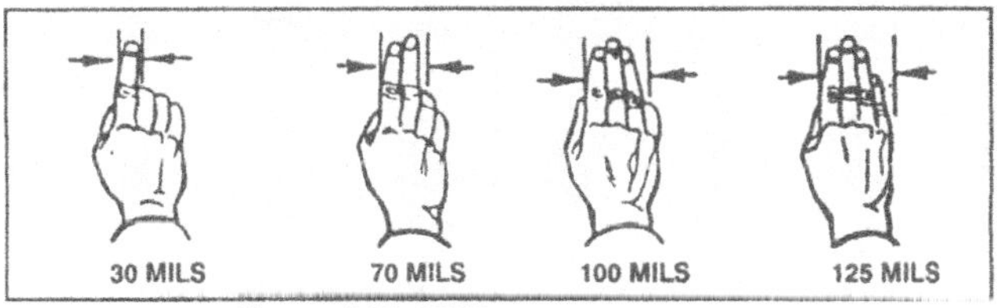

Figure 5-13. Mil/finger relationships.

5-9. OBSERVATION AND ADJUSTMENT OF FIRE

The purpose of observation and adjustment of fire practice is to teach the adjustment of fire by observing the strike of the bullets and the flight of the tracers, or by frequent re-laying on the target using sights.

a. Observation is used when firing on the 10-meter range because the impact of the round is visible on the target. When firing at greater distances, the strike of the round on the ground may cause dust to rise that is visible to the gunner; however, during wet weather, the strike cannot always be seen. In this case, use tracer ammunition that allows the gunner or crew to note the strike of the burst in relation to the target.

b. Adjustments on the target can be made using the mil relation; that is, one click of traversing or elevating handwheel moves the strike of the round one-half inch on the target at 10 meters. When firing on field targets, adjustment is made by moving the burst into the target. One click of traverse will move the strike of the round one-half meter at 500 meters, or one meter at 1,000 meters (Figure 5-14). However, the distance one click of elevation will move the strike of the round depends on the range to the target and the slope of the ground. The gunner determines the number of mils necessary to move the center of the strike into the target, and he manipulates the gun the required number of mils. This does not require the use of sights. For example, should the gunner fire on a target at 500 meters and observe the strike 10 meters to the right of the target and short about 50 meters, he would traverse the gun to the left 20 clicks (mils) and add one or more clicks (mils), depending on the slope.

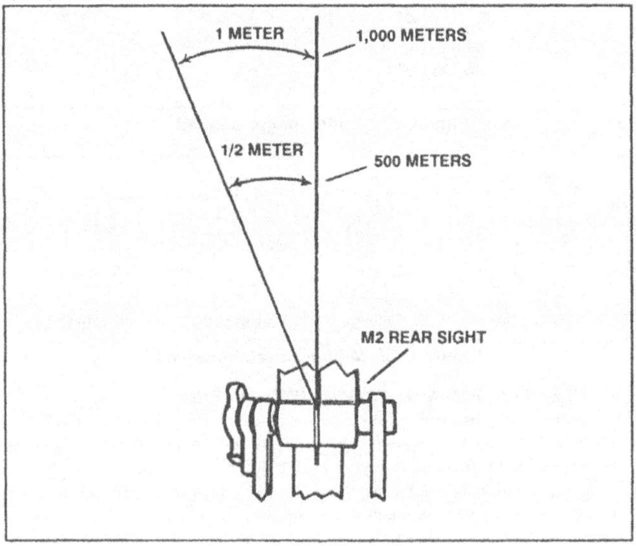

Figure 5-14. Mil relation.

c. The gunner may use the adjusted aiming point method to adjust the fire. In this method, the gunner must use his sights. He selects an aiming point that will place the next burst on target. For example, when the gunner fires on a target at 500 meters and estimates that the rounds impacted 20 meters short and 10 meters to the right, he would rapidly select an aiming point about 20 meters beyond the target and 10 meters to the left of it and lay on that aiming point and fire (Figure 5-15).

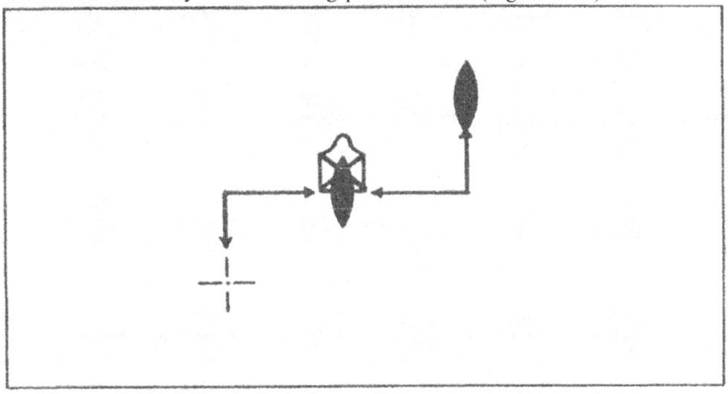

Figure 5-15. Adjusted aiming point method of fire adjustment.

5-10. FIRE COMMANDS

Fire commands are technical instructions issued by a leader to enable the unit or crew to accomplish a desired fire mission. Fire commands have been standardized for infantry direct fire weapons, and they follow the same sequence. There are two types – initial fire commands, issued to engage a target; and subsequent fire commands, which are issued to adjust fire, change the rate of fire, interrupt fire, shift fire to a new target, or to terminate the alert. A correct fire command is one that is as brief as clarity permits and yet includes all the elements necessary for the accomplishment of the fire mission. It is given in the proper sequence, transmitted clearly at a rate that permits receipt and application of instructions without confusion.

a. **Elements of the Initial Fire Command.** There are six essential elements of the initial fire command for the machine gun, which are given or implied by using one or more of the methods of control. During training, the gun crew repeats each element as it is given. This is done to avoid confusion and to train the crew to think and act in the proper sequence. The six elements are the alert, direction, description, range, method of fire, and the command to open fire.

(1) *Alert.* This element brings the crew to a state of readiness to receive further instructions. Once alerted, the gunner ensures the gun is loaded. The assistant gunner continuously checks with the leader for orders or instructions and passes them on to the gunner. The oral alert is announced as FIRE MISSION. At this command, the gunners are alerted that a target has been detected and fire may be delivered upon it. When the leader announces the alert, such as FIRE MISSION, both gun crews react to the alert. If only a certain gun is to engage, the leader announces NUMBER 1 (or 2). The other crew follows the fire mission, loads, and lays on the target to take up the fire, if required.

(2) *Direction.* This element indicates the general direction to the target and may be given in one or a combination of the following ways:

(a) The leader gives the direction orally to the target in relation to the position of the gun(s). For example, FRONT, RIGHT FRONT, LEFT FRONT.

(b) The leader can designate a small or obscure target by pointing with his arm and hand or aiming the machine gun. When pointing with his arm and hand, a man standing behind him should be able to look over his shoulder and sight along his arm and index finger to locate the target. When a gun has been aimed at a target, a soldier looking through the sights should be able to see the target.

(c) Tracer ammunition is a quick and sure method of designating a target that is not clearly visible. When using this method, the leader should first give the general direction to direct the gun crew's attention to the desired area. To minimize the loss of surprise when using tracer ammunition, the leader does not fire until he has given all the elements of the fire command except the command to fire. The leader may use his individual weapon or fire one or more bursts from the machine gun. The firing of the tracer(s) then becomes the last element of the fire command and is the signal to open fire. For example:

FIRE MISSION.
FRONT.
BUNKER.
WATCH MY TRACER(S).
SLOW (or SINGLE SHOT).

The leader fires his individual weapon or a machine gun at the enemy bunker, then his gun crew(s) opens fire.

(d) Another method of designating obscure targets is by using easily recognizable reference points. Prominent terrain features and man-made objects make good reference points. All leaders and members of the crew(s) must be familiar with the terrain features and the terminology used to describe them. The general direction to the reference point should be given. For example:

FIRE MISSION, NUMBER 2.
FRONT.
REFERENCE: LONE PINE TREE.
TARGET: TRUCK.

Sometimes a target must be designated by using successive reference points. For example:

FIRE MISSION, NUMBER 1.
RIGHT FRONT.
REFERENCE: RED-ROOFED HOUSE, LEFT TO HAYSTACK, LEFT TO BARN.
TARGET: MACHINE GUN.

Finger measurements can be used to direct the gun crew's attention to the right or left of reference points. For example:

FIRE MISSION.
LEFT FRONT.
REFERENCE: CROSSROAD. RIGHT FOUR FINGERS
TARGET: LINE OF TROOPS.

When the guns are mounted on tripods, lateral distance from reference points can be accurately announced. When gunners are firing the tripod-mounted gun, lateral distance is assumed to be in mils unless otherwise indicated, so the word "mils" is not necessary. For example:

FIRE MISSION.
FRONT.
REFERENCE: KNOCKED-OUT TANK. LEFT FOUR ZERO.
TARGET: COLUMN OF TROOPS.

(3) *Description.* The target description is used to create a picture of the target in the minds of the gun crew. The gun crew must know the type of target they are to engage to properly apply their fire. The leader should describe it briefly but accurately. For example:

Dismounted enemy personnel .... TROOPS
Automatic weapons .... MACHINE GUN
Armored vehicles .... TANK
Artillery or antitank weapon .... ANTITANK
Airplanes or helicopters .... AIRCRAFT

If the target is obvious, no description is necessary. Finger measurements or mil measurements can be used to designate the width of a linear target when the flanks cannot be pinpointed.

(4) *Range.* The range to the target is given so the gun crew knows how far to look for the target and immediately knows what range setting to place on the rear sight. Range is determined and announced in meters. Since the meter is the standard unit of range measurement, the word "meters" is not announced. With machine guns, the range is determined and announced in even hundreds and thousands. For example: THREE HUNDRED, ONE THOUSAND, ONE ONE HUNDRED. This element may be omitted when the gunners can obviously determine the range; however, it is desirable in some situations to announce the range.

(5) *Method of fire.* This element includes manipulation and rate of fire.

(a) Manipulation is used to prescribe the class of fire with respect to the gun. It is announced FIXED, TRAVERSE, SEARCH, TRAVERSE AND SEARCH, SWINGING TRAVERSE, or FREE GUN.

(b) To control the rate of fire, the gunner may use single shot, slow, rapid, or cyclic.

- Single shot. Place the gun in the single-shot mode and engage the target with aimed shots. The MG is accurate out to 1,500 meters.
- Slow fire. Slow fire consists of less than 40 rounds per minute, in bursts of five to seven rounds, fired at 10- to 15-second intervals.
- Rapid fire. Rapid fire consists of more than 40 rounds per minute, fired in bursts of five to seven rounds, at 5- to 10-second intervals.
- Cyclic fire. Cyclic fire is when the weapon fires 450 to 550 rounds per minute.

(6) *Command to open fire.* If surprise fire is not desired, the command FIRE is given without pause. It is often important that machine gun fire be withheld for maximum effect of surprise fire. To ensure this, the leader may preface the command to commence firing with the words AT MY COMMAND or AT MY SIGNAL. When the gunner(s) is ready to engage the target, he reports UP to the assistant gunner(s) who signals READY to the leader. For example:

FIRE MISSION.
FRONT.
TROOPS.
AT MY COMMAND. (Pause until crew members are ready and fire is desired.)
FIRE (or appropriate arm-and-hand signal).

When the leader makes a mistake in the initial fire command, he corrects it by announcing CORRECTION, and then gives the corrected element(s). For example:

FIRE MISSION.
FRONT.
TROOPS.
FIVE HUNDRED.
CORRECTION.
SIX HUNDRED.
TRAVERSE.
AT MY COMMAND.

When the leader makes an error in the subsequent fire command, he may correct it by announcing CORRECTION, and then repeating the entire subsequent fire command. For example:

LEFT FIVE, DROP ONE.
CORRECTION
LEFT FIVE, DROP ONE ZERO.

b. **Subsequent Fire Commands.** If the gunner fails to adjust his fire on the target, the leader must promptly correct him by announcing or signaling the desired changes. When changes are given, the gunner makes the required corrections and continues to engage the target without further command. When firing under the control of a leader, the assistant gunner checks with the leader for instructions, which he passes on to the gunner. Changes in the rate of fire are given orally and by arm-and-hand signals. To interrupt firing, the leader announces CEASE FIRE or gives a signal to cease fire. The gun crew(s) remains on the alert and firing can be resumed on the same target by announcing FIRE. To terminate the alert, the leader announces CEASE FIRE, END OF MISSION.

5-11. CREW EXERCISES

The purpose of crew exercise is to develop precision, speed, skill, and teamwork in examining equipment, placing the gun into action, and taking it out of action. In crew exercise, precision must be stressed. Once that is attained, speed, skill, and teamwork will follow. Duties are rotated during crew exercise to allow each member of the gun crew to become familiar with all the duties. During crew exercise, all oral or visual signals are repeated. When the fire command is completed, the gunner will give the assistant gunner an UP. The assistant gunner will extend his hand and arm into the air in the direction of the leader (to indicate READY) and announce, UP. With the M3 mount, the crew must consist of at least four men, including the leader. There is no designated crew in the TOE for a dismounted caliber .50 MG. The following paragraphs are only suggestions for the breakdown of equipment and member designation that may be established by the commander.

a. **Crew Equipment.** In addition to individual arms and equipment, crew members carry the following equipment for the tripod-mounted machine gun:

CREW MEMBER	SUGGESTED MINIMUM EQUIPMENT
Squad or crew leader.	Binoculars, compass, one box of ammunition.
No. 1 Assistant gunner.	Tripod.
No. 2 Gunner.	Receiver, T&E mechanism attached, and headspace and timing gauge.
No. 3 Ammunition bearer.	Barrel, barrel cover, and box of ammunition.

b. **Form for Crew Exercise.** The crew leader commands, FORM FOR CREW DRILL.

(1) *Positions with equipment.* The crew forms in column, facing the crew leader with five paces between men (Figure 5-16, page 5-24). They are in the following order: assistant gunner, gunner, and ammunition bearer. When the crew members reach their correct positions, they assume the prone position with equipment arranged as follows:

(a) No. 1: Tripod to his left, trail legs to the rear, front leg uppermost.

(b) No. 2: Receiver across his front, backplate to the right, retracting slide handle uppermost.

(c) No. 3: Barrel to his right, muzzle to the rear, ammunition box to his left front with latch to the right (latch to the front for the new box).

(d) Other members, if present: Ammunition boxes in front, one foot apart, latches to the right (front).

(e) Crew leader: Ammunition box to his right as he faces the crew, latch to the right (front).

(2) *Rotation of duties.* Duties are rotated to ensure that each member learns and is capable of performing the duties of the other members.

(a) The command to rotate all personnel is, FALL OUT LEADER. At this command, each member of the crew rises, moves forward, and assumes a new duty. The crew leader becomes the ammunition bearer. The assistant gunner moves forward and becomes the crew leader. The gunner moves forward and becomes the assistant gunner. The ammunition bearer moves forward and becomes the gunner.

(b) If the leader is not changed, the command, FALL OUT ASSISTANT GUNNER, is given. At this command, the crew members rise, the gunner becomes the assistant gunner, the ammunition bearer becomes the gunner, and the assistant gunner becomes the ammunition bearer. When the crew members have assumed their new position, they call out their new duties in order, ASSISTANT GUNNER, GUNNER, AMMUNITION BEARER.

NOTE: An additional crew exercise, which the crew maybe required to practice, is the setting of headspace and timing. These procedures are outlined in paragraph 3-6.

Figure 5-16. Crew formed in column.

c. Inspection of Equipment Before Firing. When the crew is formed with equipment, the command is, INSPECT EQUIPMENT BEFORE FIRING. At this command, the crew proceeds as follows:

(1) The assistant gunner inspects the tripod M3 mount to ensure that:

(a) The indexing levers and clamps on the front and trail legs function properly, and the legs are in the short (low) position.

(b) The front leg and trail legs are closely folded, and the front leg clamp is hand tight.

(c) The sleeve lock latch and pintle lock release cam are in working order, and the pintle lock release cam is down.

(d) The pintle bushing is free from dirt and burrs.

(2) The gunner inspects the receiver group to ensure that:

(a) The barrel support and breech bearing are free of dirt.

(b) The gun pintle is free of dirt.

(c) The feed mechanism and bolt switch are properly assembled to feed from left (soldier raises cover for proper inspection).

(d) The striker projects through the aperture in the face of bolt (soldier closes the cover).

(e) The rear sight is set at 1,000 yards (900 meters) windage zero.

(f) The T&E mechanism is securely attached to the receiver.

(g) The traversing handwheel is centered.

(h) The elevating screws are equally exposed (about 2 inches) above and below elevating handwheel.

(i) The backplate is latched and locked in place.

(j) The bolt latch release is locked in the down position by the bolt latch release lock.

(3) The ammunition bearer inspects the barrel and ammunition box to ensure that:

(a) The barrel is clear.

(b) The barrel carrier assembly is securely attached to the barrel.

(c) The barrel threads are free of dirt.

(d) The metallic links are clean (soldier opens ammunition box).

(e) The belt is properly loaded and placed in box with the double-looped end up.

(f) Dummy ammunition is used during crew exercise, and no live ammunition is present.

(g) The box is closed and latched.

(4) When the ammunition bearer completes his inspection, he moves to the gunner's position with the barrel in his right hand and ammunition box in his left hand. With the aid of the gunner, he screws the barrel into the barrel extension (Figure 5-17, page 5-26). The headspace and timing adjustment is made. The ammunition bearer remains on the left and on line with the gunner.

(5) The crew leader examines his ammunition as described in paragraph (3).

(6) At the completion of the inspection, a report is rendered as follows:

(a) The ammunition bearer reports: AMMUNITION CORRECT (or any deficiencies).

(b) The gunner reports: GUN AND AMMUNITION CORRECT (or any deficiencies).

(c) The assistant gunner reports: ALL CORRECT (or any deficiencies).

Figure 5-17. Screwing the barrel into the barrel extension.

d. **Placement of the Gun Into Action.** To place the gun into action, the crew leader commands and signals, GUN TO BE MOUNTED HERE (pointing to the position where the gun is to be mounted), FRONT (pointing in the direction of fire), ACTION (vigorously pumping his fist in the direction of the designated gun position).

(1) At the command or signal ACTION, the assistant gunner grasps the left trail leg near the center with his left hand. Springing to his feet and grasping the tripod head with his right hand, he lifts the tripod across the front of his body with the front leg up, and carries the tripod to the desired location. Upon arrival at the position, he places the trail leg pointing upward. Steadying the tripod with his left hand on the front leg, he loosens the front leg clamp with his right hand (Figure 5-18), positions the front leg with his left hand, and tightens the front leg clamp with his right hand. With his right hand on the tripod head, he slides his left hand down on the left trail leg and with a snapping motion, pulls the left leg (to the left), engaging the sleeve latch (Figure 5-19). He then aligns the tripod for direction, drops the mount to the ground, stamps the right and left trail shoes with his right or left foot, and assumes the prone position behind the mount.

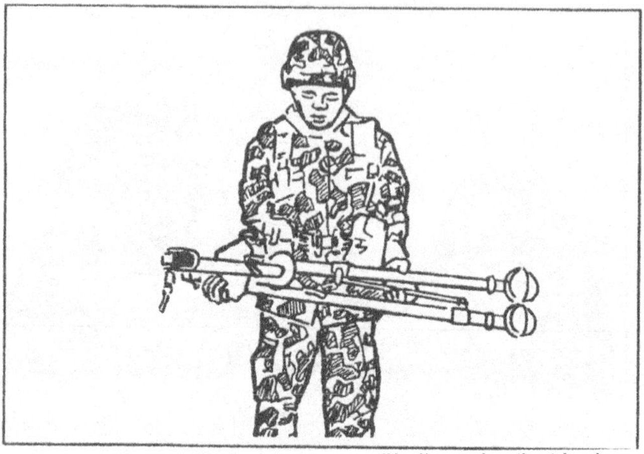

Figure 5-18. Assistant gunner-(No.1) opening the tripod.

Figure 5-19. Emplacing the mount.

(2) The gunner and ammunition bearer move together. When the tripod is nearly mounted, the gunner and ammunition bearer spring to their feet. The gunner places both hands on the spade grips, the ammunition bearer grasp the ammunition box in his left hand and the gun carrying handle with this right hand. Both men lift the gun from the ground and move rapidly to the gun position. Then the ammunition bearer places the ammunition box on line and in front of the leader's ammunition box. The gunner, assisted by the ammunition bearer, inserts the pintle into the pintle bushing. The ammunition bearer stamps the front shoe into the ground with his left foot (Figure 5-20), releases the carrying handle, lowers his right hand, and lifts the gun pintle lock release cam. When the gun pintle is fully seated, he presses down the pintle lock release cam with his right hand, turns to his left, and returns to his original position.

Figure 5-20. Mounting the gun.

(3) The assistant gunner holds the left spade grip with his left hand. With his right hand, he lowers the T&E mechanism to the traversing bar, ensuring that the traversing handwheel is to the left and the lock lever is to the rear. He then assumes a semiprone position to the left of the gun with his feet to the rear and his head on line with the feedway. He unlatches and raises the cover of the ammunition box, removes the ammunition belt,

inserts the double-looped end into the feedway, and taps the feedway with his right hand to ensure it is closed.

(4) With his right hand palm-up, the gunner grasps the retracting slide handle, vigorously jerks it to the rear, and releases it. The gun is half-loaded (Figure 5-21). He then assumes a prone position directly behind the gun with his legs spread and heels down. His right hand lightly grasps the right spade grip with the thumb in position to depress the trigger. His left hand is on the elevating handwheel (palm down) with his thumb near the traversing slide lock lever. He rests on his left elbow with his head as close as possible to the rear sight (Figure 5-22, page 5-30). He then gives the assistant gunner an UP. The assistant gunner announces UP and extends his hand and arm into the air in the direction of the leader.

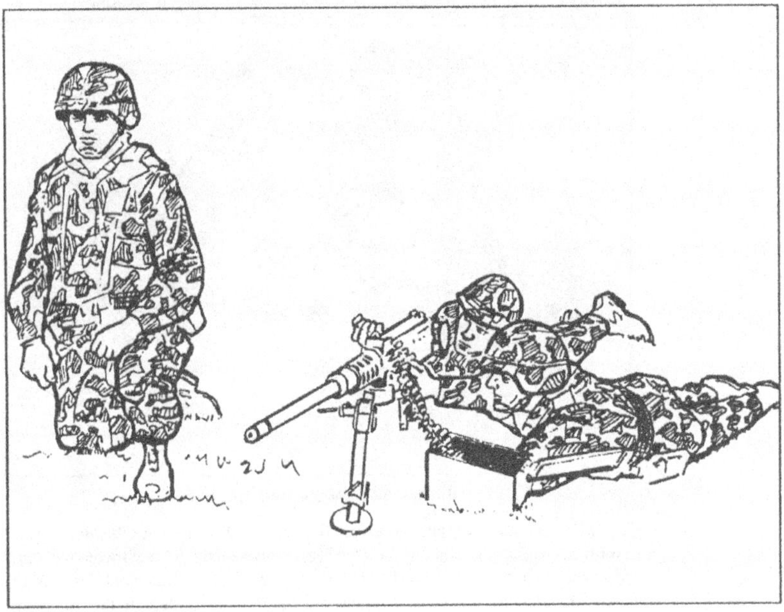

Figure 5-21. Gun mounted, gunner half-loading the gun.

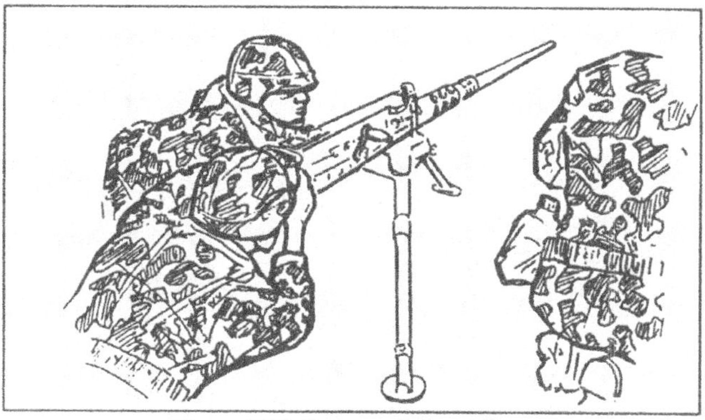

Figure 5-22. Crew ready for action.

e. **Removal of the Gun From Action.** To take the gun out of action, the command is OUT OF ACTION.

(1) The gunner raises the cover with his left hand, the assistant gunner lifts the ammunition out of the feedway, replaces the ammunition in the box, and closes and latches the box. The gunner closes the cover, pulls the retracting slide handle to the rear with right hand (palm up) and releases it, presses the trigger with his right thumb, and loosens the traversing slide lock with his left hand. He rises to his feet, grasping both spade grips. At this time, the ammunition bearer will arrive at the gun position. With his left hand, the ammunition bearer grasps the ammunition box and places his right foot on the front leg shoe. With his right hand, he reaches down and lifts up the pintle lock release cam and grasps the carrying handle (Figure 5-23).

(2) The gunner and ammunition bearer lift the gun from the tripod, turn right, and carry the gun back to their original position. They set the gun down with the muzzle to the left and the retracting slide handle up. The gunner pulls the retracting slide handle to the rear and aligns the lug on the barrel locking spring with the 3/8-inch hole in the right side of the receiver. The ammunition bearer unscrews the barrel from the receiver, picks up the ammunition box and barrel, moves five paces to the rear, places the barrel to his right with the muzzle to the rear and the ammunition box to his left, and assumes the prone position.

Figure 5-23. Crew taking the gun out of action.

(3) The gunner rotates the bolt latch release lock to the right and releases the bolt latch. He checks his sights to ensure they are properly set.

(4) The assistant gunner assumes a kneeling position behind the tripod and releases the sleeve lock latch with his right hand. He rises to his feet, grasps the front leg with his left hand, and rotates the tripod to a vertical position on the trail legs. With his right hand, he loosens the front leg clamp, folds down the front leg, then tightens the clamp. With his right hand on the tripod head, he rotates the tripod on the right trail leg and releases the sleeve latch. Then he folds the left trail leg against the right with his left hand (Figure 5-24, page 5-32). Holding the tripod head with his right hand, trail legs with his left, he lifts the tripod across the front of his body with the front leg up. He turns to the right and returns to his original position. At this time, the crew leader picks up his ammunition box and faces the crew. The assistant gunner places the tripod on the ground, assumes a prone position to the right of the tripod, and announces UP.

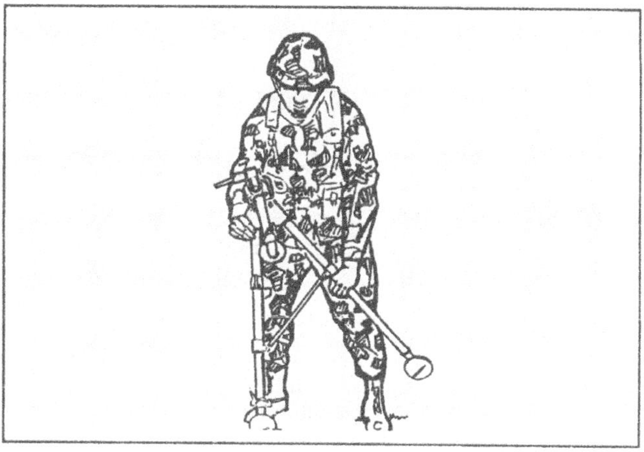

Figure 5-24. Folding the trail legs.

f. **Duties of The Crew.** To hand-carry the gun and equipment, the command is SECURE EQUIPMENT, FOLLOW ME. At this command,–

(1) The crew leader carries his ammunition box in his right hand.

(2) The assistant gunner carries the tripod over either shoulder.

(3) The gunner carries the receiver on either shoulder with the spade grips to the front.

(4) The ammunition bearer carries his ammunition box in his left hand and the barrel in his right hand with the muzzle to the rear (or on his left shoulder with the muzzle to the front).

g. Relocation of Tripod-Mounted Gun. When the gun is mounted on the tripod, it can be moved for short distances by dragging or by a two- or three-man carry. (In the latter, the men should move in step to make carrying easier.)

(1) *Dragging.* The gun is dragged when there is limited cover, or when the situation requires the gun to be moved in this manner. The gunner and assistant gunner drag the mounted gun to the desired position (Figure 5-25).

(2) *Two-man carry* With the gunner on the right and assistant gunner on the left, each grasps the front leg with his forward hand and a trail leg with the other hand, just above the traversing bar (Figure 5-26).

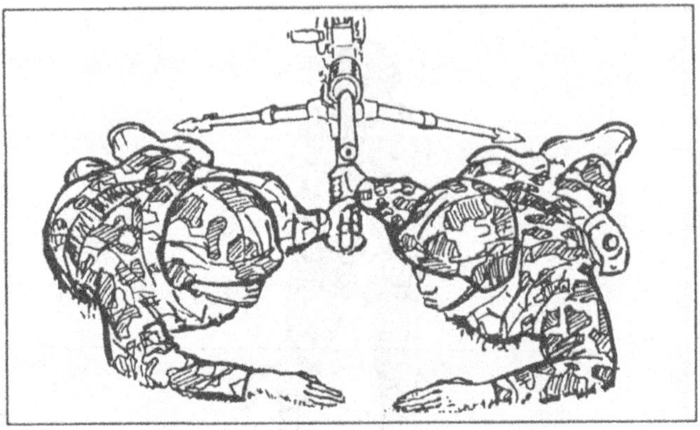

Figure 5-25. Dragging the gun into position.

Figure 5-26. Two-man carry.

(3) *Three-man carry*. When the barrel is hot, the gunner gets behind the tripod with a trail leg in each hand. With the assistant gunner on the left and the ammunition bearer on the right, each grasps the carrying handle. In addition, the assistant gunner carries the ammunition in his left hand (Figure 5-27). When the barrel is cool, the ammunition bearer and the assistant gunner each grasp the front leg (Figure 5-28).

Figure 5-27. Three-man carry (hot barrel).

Figure 5-28. Three-man carry (cola barrel).

NOTE: Carrying the gun by the barrel may cause damage to the barrel support and the barrel extension.

h. **Movement of the Gun to Other Mounts.** With the mount prepared to receive the gun, the cradle of the mount is placed in a horizontal position. To move the gun to the mount, the gunner carries the right spade grip in his left hand and a box of ammunition in his right. The assistant gunner grasps the carrying handle with his left hand and a box of ammunition in his right hand. When they get to the mount, the gunner and assistant gunner place their ammunition boxes near the mount. The gunner removes the rear mounting (gun-locking) pin with his right hand. The assistant gunner removes the front mounting (gun-locking) pin with his right hand. They place the gun on the mount. The gunner aligns the holes in the rear mounting lugs of the receiver with the rear mounting bracket and inserts the rear mounting pin. The assistant gunner aligns the front mounting hole in the front of the receiver with the front mounting bracket and inserts the front mounting pin. (For use of the sideplate trigger with the M63 mount, see TM 9-1005-213-10.)

5-12. MACHINE GUN FUNDAMENTAL SKILLS TEST

The machine gun fundamental skills test should be held periodically to ensure that proficiency with the MG is maintained by all crewmen. It consists of 10 fundamental skills; all tasks are hands-on (Figure 5-29). The test should be given prior to range firing on a go/no-go basis.

TASK	GO/NO-GO	
1. Clear the MG.	✓	
2. Disassemble the MG.	✓	
3. Assemble the MG.	✓	
4. Do a function check.	✓	
5. Maintain the MG.	✓	
6. Set the headspace on the MG.	✓	
7. Set the timing on the MG.	✓	
8. Load the MG.	✓	
9. Perform immediate action on the MG.	✓	
10. Unload the MG.	✓	

Figure 5-29. The MG fundamental skills test.

Section III. BASIC MARKSMANSHIP

This phase of training is designed to allow the gunner to apply the fundamentals of marksmanship that he learned earlier. During this training, the gunner is introduced to the basic machine gun target (see Appendix C), procedures for both the 10-meter and field fire ranges, and how to acquire targets. He also fires practice and qualification.

5-13. CONCEPT OF ZEROING/TARGETING

The concept of zero is very simple; it is no more than the best way to adjust the sights of the weapon so the point of aim of the sights and the point of impact of the rounds are the same at any given range. A properly zeroed M2 gives the gunner the highest probability of hit for most combat targets with the least adjustment to the point of aim. There are three methods of zeroing/targeting used with the .50 caliber MG.

a. Ten-meter zero is the basic and the most common method of zeroing the M2 MG. Once zeroed on a 10-meter range using the standard machine gun target, the weapon is ready for field fire. As other weapons, the sight on the M2 must also be set at an initial start point (Figure 5-30). The initial sight setting for field zero is basically the same; except the range setting during field zero will depend on the range to the target, and it is always 1,000 yards for 10 meters.

(1) *Set elevation.* Raise the rear sight by lifting straight up until it snaps into the upright position. Adjust the range scale to 1,000 yards by rotating the elevation screw knob in the necessary direction. (Clockwise moves the scale up; counterclockwise moves the scale down.)

(2) *Set windage.* Rotate the windage knob until the zero index mark on the base rear sight is aligned with the index mark on the top of the receiver. (Clockwise moves the windage scale to the left; counterclockwise moves it to the right.)

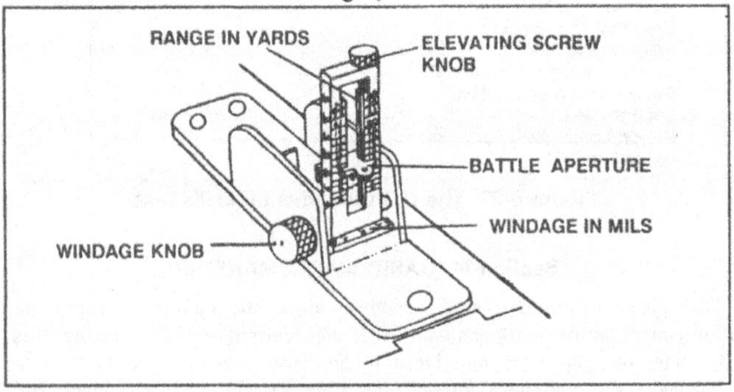

Figure 5-30. Rear sight setting.

(3) *Obtain proper sight picture.* Obtain the proper sight picture by looking through the zero aperture and centering the front sight blade in it. Once the sight alignment is obtained, place this combination on the center base of the selected target (Figure 5-31).

Figure 5-31. Proper sight picture.

b. Field zeroing/targeting is an expedient method of obtaining a zero when a 10-meter zero cannot be conducted. When preparing to field zero, make sure the M2 is mounted securely on the M3 tripod, make sure the T&E is working properly, and finally, know the distance to your zero target. The only difference in initial sight setting for field zero is range setting on the scale. The gunner must also remember that the range scale on the M2 is indicated in yards. Therefore, in order to get as close to the target as possible, you may have to convert the meters to the target into yards so you can set the range on the rear sight. Conversion of meters to yards is accomplished by multiplying the number of meters by 1.094. For example, 600 meters x 1.094 = 656.4 yards; the gunner would set his range scale at 650.

c. The AN/TVS 5 is an effective night fire aid. By using this device, the gunner can observe the area and detect and engage any suitable target. But, as usual, the device is only as good as its zero; the zeroing procedure requires practice to become proficient.

(1) *Mount the mounting bracket.* To mount the bracket, the gunner must ensure that the rear sight is in the down position. He then releases the catch on the left side of the top cover assembly and raises the cover to the upright position. He pushes the mounting bracket over the breech of the machine gun and slides the bracket rearward until it stops (Figure 5-32).

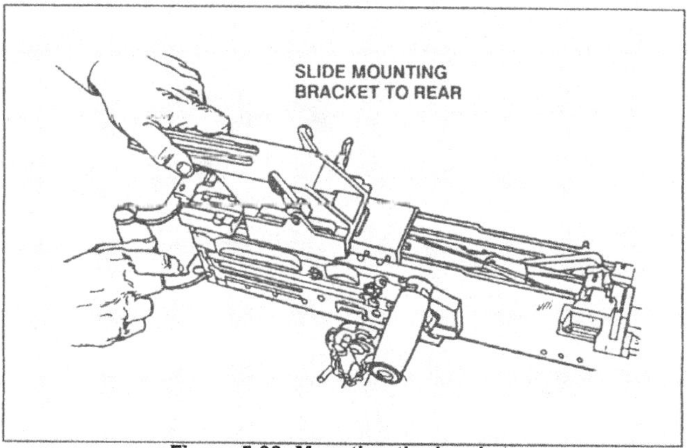

Figure 5-32. Mounting the bracket.

He pushes the three locking cams to secure the bracket (Figure 5-33) and closes the top cover assembly.

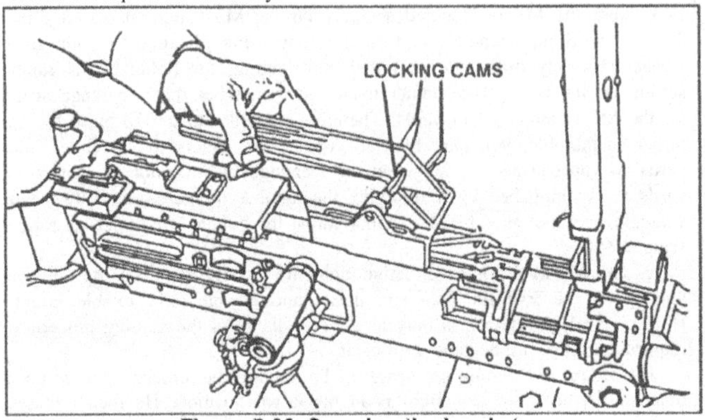

Figure 5-33. Securing the bracket.

(2) *Install the sight on the mount.* To install the sight, the gunner must align the scribe lines on the sight with the scribe lines on the bracket. He places the sight in the groove at the top of the bracket and tightens the lever screw to secure the sight to the bracket (Figure 5-34).

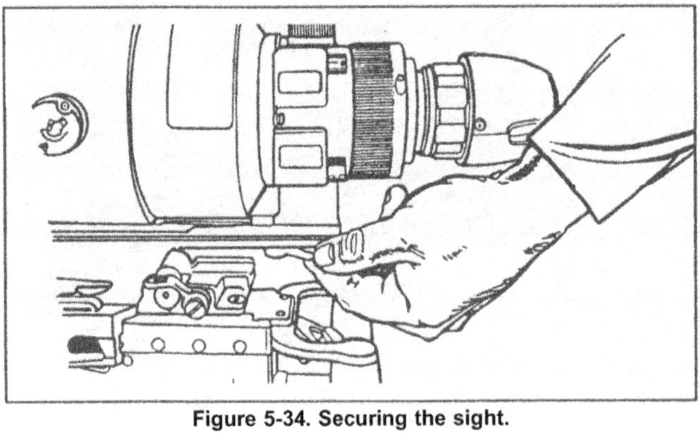

Figure 5-34. Securing the sight.

(3) *Zero the AN/TVS-5.* A gunner must zero the MG with its NVD on it before he can effectively engage targets with the AN/TVS-5. The zeroing of this device to the MG must be done at 50 meters. An M16A1, A2, or Canadian bull zero target may be used. The point of impact for either target must be 9.8 centimeters below the target aiming point. To accomplish this task, place the reticle aiming point on the target aiming point (Figure 5-35).

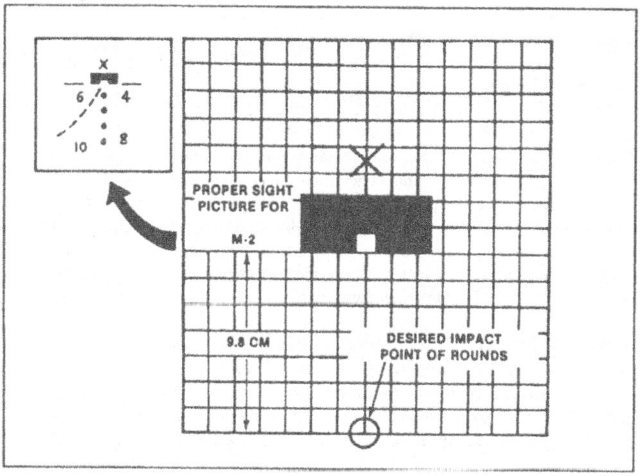

Figure 5-35. Desired impact point for the AN/TVS-5.

(a) Fire enough single-shot rounds to obtain a good shot group. Locate the center of the shot group. Adjust the AN/TVS 5 azimuth and elevation knobs until the shot group impacts 9.8 centimeters below the target aiming point. Move the weapon so that the reticle aiming point is again on the target aiming point. Repeat this process until the desired point of impact is obtained.

(b) When adjusting, move the azimuth or elevation adjustment actuator one click to move the strike of the round .5 inch at 50 meters. One click of adjustment will move the reticle about one square on the Canadian bull-type target.

NOTE: The lens cover with the peephole may be required to prevent scope washout from the muzzle flash.

(4) *Dismount the sight.* The gunner dismounts the sight from the bracket first by loosening the lever screws. Then he lifts the sight from the bracket. He must then dismount the bracket in the reverse mounting procedure.

5-14. TEN-METER FIRING EXERCISE

The purpose of 10-meter firing is to develop skills in the delivery of initial burst on target. When conducted properly, it will train gunners in the basic skills, such as 10-meter zero, controlled bursts, traverse, and traverse and search firing techniques. (See Appendix C for details on setup and conduct of firing.)

5-15. TRANSITION DAY FIRING EXERCISE

Transition day firing of the M2 machine gun will teach the gunner some techniques of fire that he may encounter in combat situations. The gunner will field zero his weapon and engage point and area targets from the tripod-mounted firing position. Within this training, the gunner will be required to apply all the fundamentals of gunnery learned in preparatory gunnery training and 10-meter firing. (See Appendix C for details on the setup and conduct of transition day fire.)

5-16. NBC FIRING

Since NBC plays an important part in our preparation for war on the modern battlefield, it is important that each soldier is prepared to accomplish the mission even if the area is contaminated and he must wear protective gear. (See Appendix C for details on setup and conduct of fire.)

5-17. NIGHT FIRE EXERCISE

The night fire exercise gives the soldiers the practical application of engaging targets using the AN/TVS-5 at night or during limited visibility. (See Appendix C for details on setup and conduct of fire.)

Section IV. ADVANCED GUNNERY

After firing 10-meter, day, NBC, and night, gunners need practice in applying what they have learned. They also need experience in engaging targets that depict realistic enemy formations. Advanced gunner exercises provide this experience in mounted, mounted NBC, and predetermined firing exercises.

5-18. OBJECTIVES

The objectives of the advanced gunnery phase are to prepare the gunners for combat. During this phase, training should be directed toward –

- Tracking and leading.
- Mounted target engagement techniques.
- Mounted NBC engagements.
- Predetermined fire techniques.

5-19. TRACKING AND LEADING EXERCISES

The gunner normally completes instruction in firing at stationary targets before he receives instruction in firing at moving targets. The technique of engaging a moving target differs from that of engaging a stationary target. The gun must be aimed ahead of the moving target a sufficient distance to cause the bullet and target to arrive at the aiming point at the same time. The distance is measured in target lengths. One target length as seen by the gunner is one lead. Leads are measured from the center of mass. The lead depends upon range, speed, and direction of movement of the target. To hit the target, the gunner aims at a point ahead of the target equal to the estimated number of leads, maintains the lead by tracking the target (manipulating the gun at the same angular speed as that of the target), and then fires. Fire is adjusted by observation of strike/tracer (Figure 5-36).

Figure 5-36. Lead technique.

a. **Tracking.** Tracking consists of maintaining correct alignment of the sights (with or without a lead) on a moving target by moving the gun at the same angular speed as that of the target.

(1) The gunner is required to aim at a prescribed point (center base to take advantage of the beaten zone) on the target and maintain that aim during uniform movement of the target. As instruction progresses, speeds used should differ for successive runs of the target. The speeds the 10-meter targets should move to represent speeds at various ranges are shown in Table 5-2.

TARGETS SPEEDS IN MILES PER HOUR	TARGET SPEEDS IN INCHES PER SECONDS CORRESPONDING TO–		
	300 M	500 M	900 M
7 1/2	6	4	2
15	12	8	4
30	24	15	9

Table 5-2. 10-meter moving target.

(2) The target handler must have practice in moving the target silhouette across the background at the varying speeds.

b. **Leading.** Mathematical computation or use of voluminous lead tables to obtain exact leads on a moving target are impractical in combat. The simple lead table shown in Table 5-3 gives the amount of lead necessary to hit a target moving at right angles (90 degrees) to direction to hit at speeds and ranges indicated.

SPEED IN MILES PER HOUR	RANGE OF TARGET		
	300 M	500 M	900 M
15	1/2 target length	1 target length	2 target lengths

Table 5-3. Lead table.

(1) The gunner must correct the lead as conditions change. If the target speed is 7 1/2 miles per hour, the amount of lead is half that shown in the table; at 30 miles per hour, double that shown, The angle at which the target is moving also alters the amount of lead taken. If the angle between line of fire and line of travel of the target is less than 45 degrees, use half the lead shown in the table.

(2) For targets moving directly toward the gun, the point of aim is placed on the center or the lower edge of the target, when possible. For targets moving directly away from the gun, the point of aim is placed on the center or upper edge of the target. Too much lead is better than too

little because the target runs into the fire; also, the observation of strike is easier. Intelligent use of the lead table includes immediate application of fire with estimated lead followed by necessary corrections based upon observation of strike/tracer.

c. **Tracking and Leading.** Combined tracking and leading exercises at 10 meters are used to gain proficiency in tracking the target. The gunner is required to repeat the tracking exercise while using a designated lead to simulate firing when his sights are properly aligned. As a further exercise in tracking and leading, the gunner may be required to track and lead moving targets at greater ranges. A vehicle can be run at right angles to the line of aim at ranges between 500 and 1,000 yards, and at varying speeds, averaging 15 miles per hour (Figure 5-37).

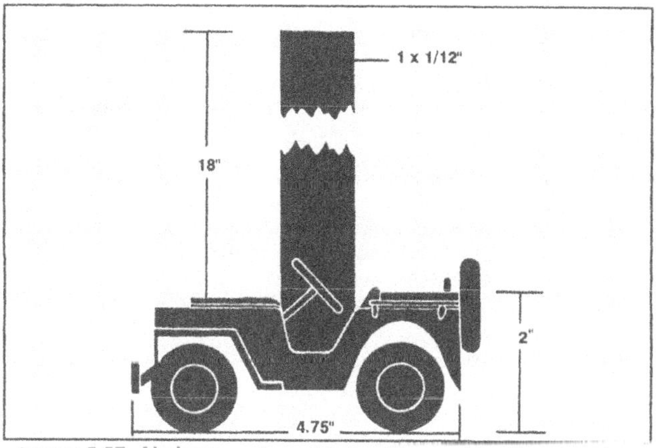

Figure 5-37. Aiming target used in tracking and leading exercise.

d. **Conducting the Lead Exercise (10-Meter).** The gunner is required to take a position at the gun, swing through the target's silhouette, and aim at a point ahead of the target equal to the prescribed lead from the center of mass. The gunner then directs the target handler to move the marking silhouette until the center of the target is at the point of aim. He repeats this procedure three times for each target lead announced. The target handler places his marking silhouette on the blank background, traces around it, and holds it in place for the gunner to aim, using the prescribed leads. Following the gunner's instructions, he moves the marking silhouette until the gunner commands HOLD. He then places a pencil dot at this point and returns the silhouette to the original position. This procedure is followed until the gunner has completed three tries for each target lead announced. The three pencil dots for each target should fit within a one-centimeter circle. The exercise should be conducted for varying left and right leads.

5-20. MOUNTED FIRING EXERCISE

The purpose of the mounted firing exercise is to teach the gunner techniques of firing the M2 MG from a mounted platform and to develop the gunner's ability to fire the M2 with it mounted on its primary carrier. (See Appendix C for details on the setup and conduct of fire.)

5-21. MOUNTED NBC FIRING EXERCISE

The probability of fighting mounted in a tactical environment that has been contaminated by NBC agents is very likely; therefore, gunners must be trained to engage targets while in a mounted NBC posture. (See Appendix C for details on the setup and conduct of fire.)

5-22. PREDETERMINED FIRING EXERCISE

The predetermined firing exercises are designed to instruct the gunners on preparing and using range cards during any visibility conditions. (See Appendix C for details on setup and conduct of fire.)

CHAPTER 6

COMBAT TECHNIQUES OF FIRE

Technique of fire is the method of delivering and controlling fire Each member of the machine gun crew must be trained in standard methods of applying fire, either as a crewmember or a gunner, and must perform his assigned task automatically and effectively. The simplest and most effective technique of delivering fire with the machine gun mounted on its ground or vehicular mount is to align the sights of the gun on the target and fire; this is called direct laying. At times, techniques of fire other than direct laying are more appropriate and effective. When delivering overhead fire or fire from position defilade, the gunner must use the appropriate technique.

Section I. FUNDAMENTALS

Before the machine gun can be employed to the best advantage using any firing technique, certain fundamentals must be understood and applied. These include:

- Characteristics of fire.
- Classes of fire.
- Fire control.
- Target types and methods of engagement.
- Overhead fire.
- Methods of laying the gun.
- Firing from position defilade.
- Final protective fires.

6-1. CHARACTERISTICS OF FIRE

The gunner's knowledge of his machine gun is not complete until he learns something of the action and effect of the projectiles when fired. This section discusses various characteristics of machine gun fire, including trajectory, cone of fire, and the beaten zone.

a. **Trajectory.** The trajectory is the curved path of the projectile in its flight from the muzzle of the weapon to its impact. The major factors that influence the trajectory are the velocity of the round, gravity, rotation of the round, and resistance of the air. The farther the round travels, the greater the curve of the trajectory. The highest point of the trajectory is called the maximum ordinate. This is a point approximately two-thirds of the distance from the gun to the target. The maximum ordinate increases as the range increases (Figure 6-1).

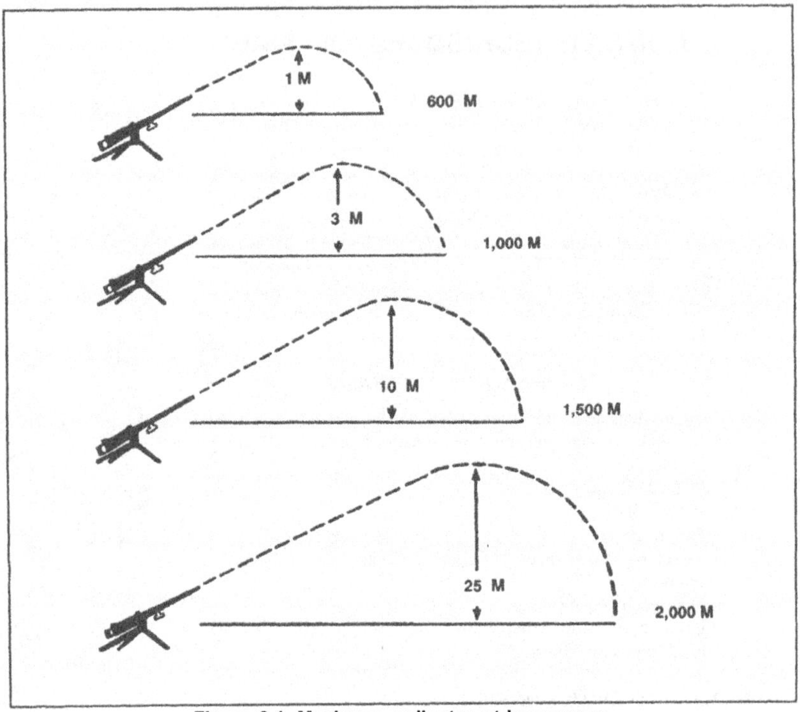

Figure 6-1. Maximum ordinates at key ranges.

b. **Cone of Fire.** When the weapon is fired automatically in bursts, all the rounds do not follow the same path. This is due to the vibrations of the gun and mount, variations in ammunition, and atmospheric conditions, which cause the rounds to follow a slightly different trajectory. This group of trajectories formed by a single burst is called the cone of fire (Figure 6-2).

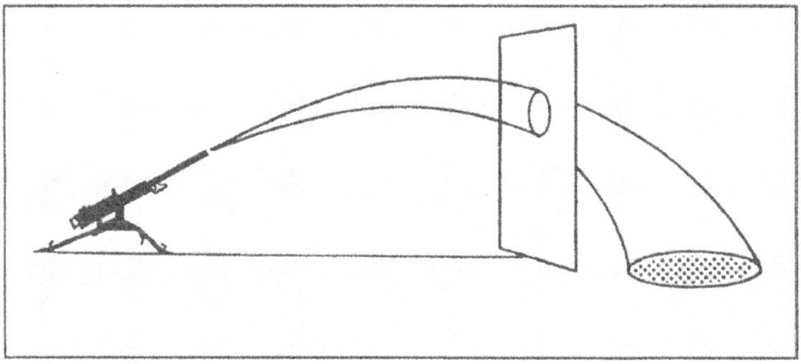

Figure 6-2. Cone of fire.

c. **Beaten Zone.** The beaten zone is an elliptical pattern formed by the cone of fire as it strikes the ground. The beaten zone is always about 2 meters in width.

(1) *Effect of range.* As the range to the target increases, the beaten zone becomes shorter and wider.

(2) *Effect of terrain.* The length of the beaten zone for any given range will vary according to the slope of the ground. On rising ground, the beaten zone becomes shorter but remains the same width. On ground that slopes away from the gun, the beaten zone becomes longer but remains the same width.

6-2. CLASSES OF FIRE

Machine gun fire is classified with respect to the ground (Figure 6-3, page 6-4), the target (Figure 6-4, page 6-5), and the gun (Figure 6-5, page 6-7).

a. **Classes of Fire with Respect to the Ground.**

(1) *Plunging fire.* Fire in which the angle of fall of the rounds (with reference to the slope of the ground) is such that the danger space is confined to the beaten zone, and the length of the beaten zone is materially shortened. Plunging fire is obtained when firing from high ground to low ground, when firing from low ground to high ground, and when firing at long ranges.

(2) *Grazing fire.* Grazing fire is fire in which the center of the cone of fire does not rise more than one meter above the ground. When firing over level or uniformly sloping terrain, the maximum extent of grazing fire obtainable is about 700 meters.

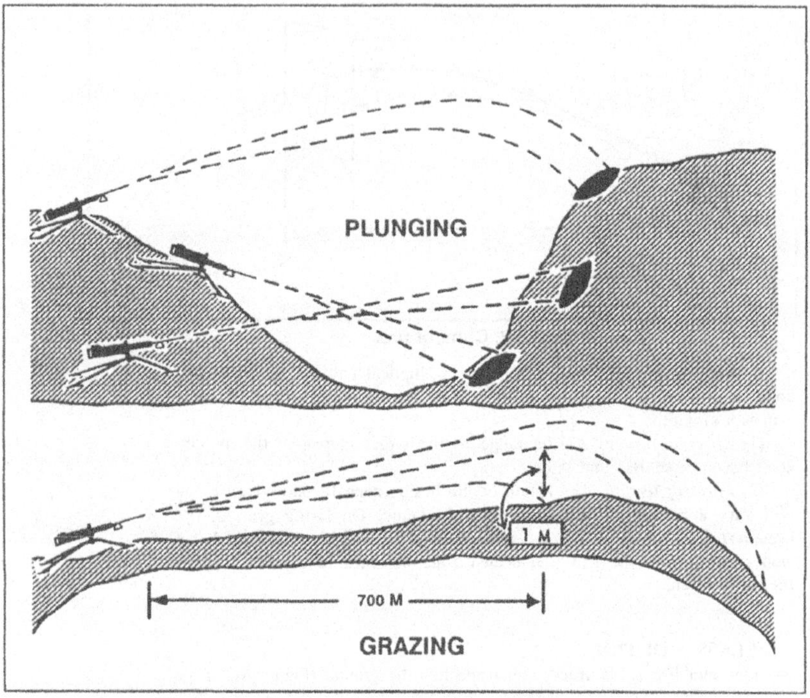

Figure 6-3. Classes of fire with respect to the ground.

b. **Classes of Fire with Respect to the Target.**

(1) *Frontal.* The long axis of the beaten zone is at a right angle to the long axis of the target.

(2) *Flanking.* Fire is delivered against the flank of a target.

(3) *Oblique.* The long axis of the beaten zone is at an angle (but not a right angle) to the long axis of the target.

(4) *Enfilade.* The long axis of the beaten zone coincides or nearly coincides with the long axis of the target. This class of fire is either frontal or flanking. It is the most desirable class of fire with respect to the target because it makes maximum use of the beaten zone.

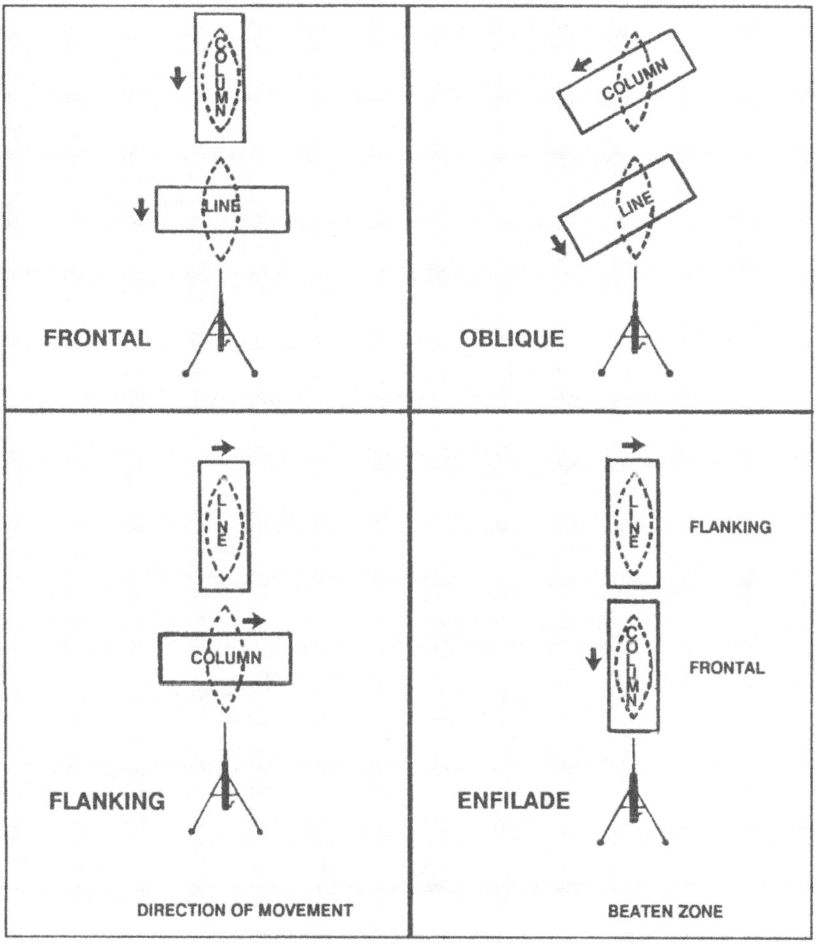

Figure 6-4. Classes of with respect to the target.

c. Classes of Fire with Respect to the Gun.

(1) *Fixed fire.* This is fire delivered on a point target with little or no manipulation needed. After the initial burst, the gunners will follow any change or movement of the target without command.

(2) *Traversing fire.* This is fire distributed against a wide target requiring successive changes in the direction of the gun. When engaging a wide target requiring traversing fire, the gunner should select successive aiming points throughout the target area. These aiming points should be close enough together to ensure adequate target coverage; however, they need not be so close as to be wasteful of ammunition by concentrating a heavy volume of fire in a small area. Two clicks on the traversing handwheel after each burst ensure coverage (2 clicks = 2 mils = constant width of beaten zone).

(3) *Searching fire.* This is fire delivered against a deep target or a target that has depth, requiring changes in elevation of the gun. The amount of elevation change depends upon the range and slope of the ground.

(4) *Traversing and searching fire.* This is fire delivered both in width and depth by changes in direction and elevation. It is employed against a target whose long axis is oblique to the direction of the fire.

(5) *Swinging traverse.* This is employed against targets that require major changes in direction but little or no change in elevation. Targets may be dense, wide, in close formations moving slowly toward or away from the gun, or vehicles or mounted troops moving across the front. The traversing slide lock lever is loosened enough to permit the gunner to swing the gun laterally.

(6) *Free gun.* This is when fire is delivered against moving targets that must be rapidly engaged with fast changes in both direction and elevation. Examples are aerial targets, vehicles, mounted troops, or infantry in relatively close formations moving rapidly toward or away from the gun position. To fire free gun on the M3 tripod mount, remove the T&E mechanism from the receiver and traversing bar and put it down. When firing swinging traverse and free gun, the weapon is normally fired at the cyclic rate of fire which is in excess of 450 rounds per minute. This consumes a lot of ammunition, and there is no beaten zone because each round seeks its own area of impact.

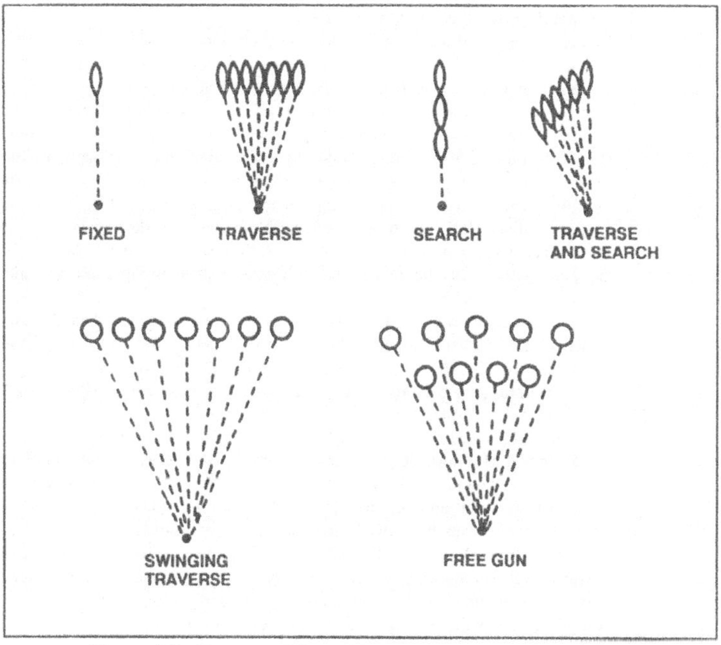

Figure 6-5. Classes of fire with respect to the gun.

Section II. FIRE CONTROL

Fire control of machine guns includes all operations connected with the preparation and actual application of effective fire on a target. It implies the-ability of the leader to-open fire at the instant he desires, to adjust the fire of the gun(s) on the target, to regulate the rate of fire, to shift from one target to another, and to cease firing. This ability to exercise proper fire control depends primarily on the discipline and the proper training of the crew. Failure to exercise fire control results in danger to friendly troops, loss of surprise, premature disclosure of positions, application of fire on unimportant targets, loss of time in adjusting fire, and waste of ammunition.

6-3. METHODS OF FIRE CONTROL

Of the several methods of controlling machine gun fire, the noise of battle will limit the use of some. Therefore, the leader must select the method or combination of methods that will best accomplish his purpose. The chain of fire control begins with the leader. He is responsible for both the technical and tactical employment of the gun and the training of the crew. He is responsible for passing on to the crewmembers all instructions and orders from his next higher leader regarding the situation and mission. He assigns sectors of fire and firing positions, designates targets to be engaged, adjusts fire, and ensures effective coverage of the targets.

a. **Oral.** Oral control is effective unless the leader is too far away from the gun crew(s), or the noise of battle makes it impossible for the gun crew(s) to hear him.

b. **Arm-and-Hand Signals.** This is an effective method when the gun crew(s) can see the leader. All crewmembers must understand the standard arm-and-hand signals used to control machine gun fire.

c. **Prearranged Signals.** These are either visual or sound signals such as pyrotechnics or blasts on a whistle. These signals should be included in the units' SOPs and must be clearly understood by all crewmembers.

d. **Personal Contact.** In many situations, the leader must move to individual crewmembers to issue orders. This method is used more than any other by small-unit leaders. The leader must use maximum cover and concealment to keep from disclosing the gun crew's position.

e. **Standing Operating Procedures.** SOPs cover actions the gun crews perform without command because they were developed and practiced during the training of the gun crews. Their application eliminates many commands and simplifies the leader's job of fire control.

6-4. TARGETS AND THEIR ENGAGEMENT

Targets presented to the machine gunners during combat will in most cases consist of enemy soldiers in various formations, which require distribution and concentration of fire. These targets have width and depth, and the application of machine gun fire is designed to completely cover the area in which the enemy is known or suspected to be. These targets may be easy to see or may be indistinct and difficult to locate.

a. When machine gun fire is under direct control of a leader, he designates the midpoint and flanks or ends of a target unless they are obvious to the gun crew(s). When a target other than a point target is engaged by two gunners, it is always divided. Each gunner applies his fire to that portion of the target corresponding to his position with relation to the other gun. Normally, each gunner engages one-half of the target;

however, gunners must be prepared to engage the entire target if necessary. Gunners continue to fire on the target until it is neutralized or until another signal is received from the leader.

b. The gunner's positions (including vehicular-mounted) should be numbered so each gunner will know which portion of a target he should engage. It should be emphasized that the positions are numbered – not the guns or gunners. To ensure that gunners react quickly and properly when they detect a target or when a target is designated by the leader, standard methods of applying fire to the various type targets are taught. These methods are the same for ground and vehicular-mounted guns. The following are the different types of targets and how they are engaged with the MG.

(1) Point targets are targets that require the use of a single aiming point. Enemy bunkers, weapon emplacements, vehicles, small groups of soldiers, and aerial targets such as helicopters or descending paratroopers are examples of point targets. A point target is engaged with fixed fire. If the target moves after the initial burst, the gun crew(s) keeps fire on the target by following its movement with the gun(s).

(2) Linear targets have sufficient width to require traversing fire and no more depth than can be effectively covered by the beaten zone. Linear targets are engaged with traversing fire.

(a) Two guns, normal division. The target is divided at the midpoint; the right gun engages the right half of the target, and the gun on the left engages the left half of the target. The point of initial lay and adjustment for both guns is at the midpoint of the target. After adjusting on the midpoint, the right gun traverses the right half of the target to include one aiming point beyond the last visible target flank and returns to the midpoint.

(b) Two guns, special division. If one portion of the target presents a greater threat than another, the target can be divided so fire is concentrated on that portion presenting the greatest threat. The special division of the target is accomplished by a subsequent fire command after firing begins. The gunners initially lay at the midpoint, regardless of the special division to be made, thus precluding confusion.

(c) One gun. A single gunner must engage the entire width of a linear target. The point of the initial lay and adjustment is on the midpoint, or that portion of the target presenting the greatest threat. The gunner traverses to either flank and then covers the remainder of the target (Figure 6-6, page 6-10).

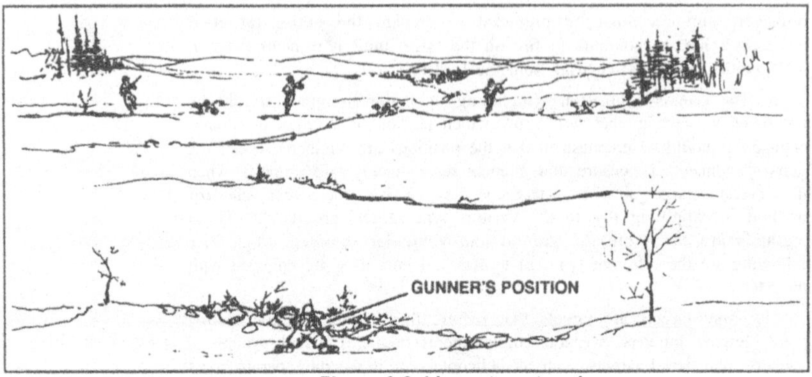

Figure 6-6. Linear target and one gun.

(3) Linear targets with depth are targets that have sufficient width to require traversing fire and depth which cannot be covered by the beaten zone. A combined change in direction and elevation (traversing and searching fire) is required to maintain effective fire on these targets (Figure 6-7). Linear targets with depth are engaged with traversing and searching fire. When range is announced, the range to the midpoint is given.

Figure 6-7. Linear target with depth.

(a) Two guns. The method of division, the point of initial lay and adjustment, and the extent of manipulation for both guns are the same as prescribed for linear targets. The gunners, however, apply enough search between each burst to ensure the center of the beaten zone is maintained at the center base of the target (Figure 6-8).

(b) One gun. A single gunner initially lays and adjusts on the midpoint of a linear target with depth unless some other portion of the target presents a greater threat. The gunner traverses and searches to the near flank, then he covers the entire target area (Figure 6-8).

(4) Deep targets have depth but very little width and can be effectively covered by searching fire (Figure 6-9, page 6-12). When the range is announced, it is given to the midpoint of the target.

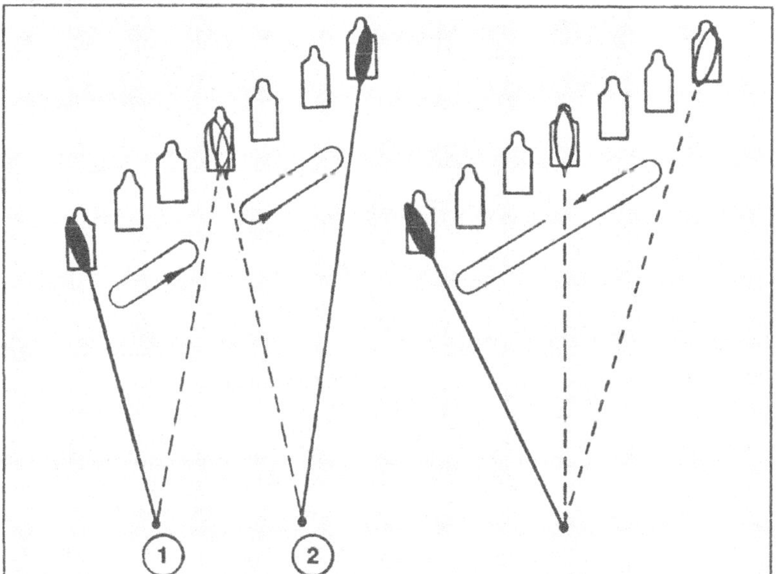

Figure 6-8. Engagement of linear targets with depth.

Figure 6-9. Deep target.

(a) Two guns. The point of initial lay of both guns is on the midpoint, which is also the point of division. Since enfilade fire is delivered, it is not necessary to adjust on the midpoint of the target because the long axis of the beaten zone will compensate for missing the midpoint. However, should the gunner's beaten zone be out of the lateral confines of the target, it will be necessary to adjust fires into the target area. After the initial bursts, the right gun searches to the near end of the target, and the left gun searches to the far end of the target. Both gunners then reverse their direction of search and return to the midpoint (Figure 6-10).

(b) One gun. A single gunner initially lays and fires at the midpoint of a deep target, unless another portion of the target presents a greater threat. The gunner immediately searches to the near end, then covers the entire target (Figure 6-10).

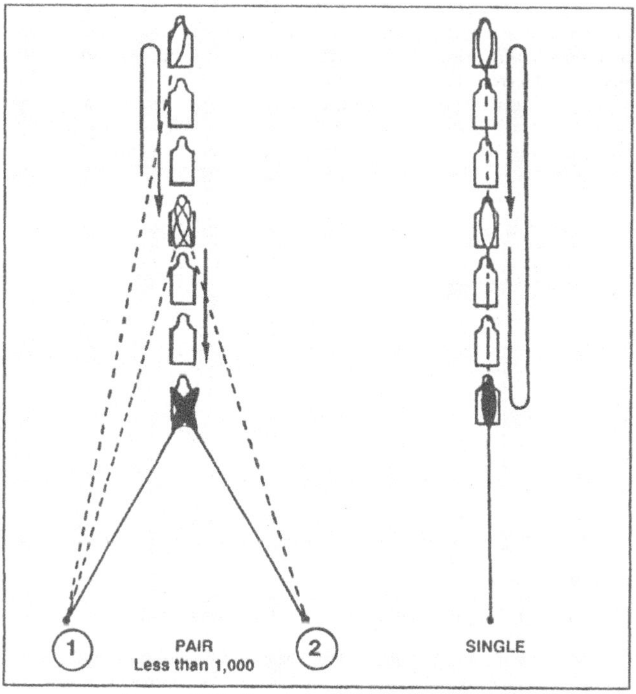

Figure 6-10. Engagement of deep targets.

(5) Area targets as discussed in this manual have considerable width and depth, and they require extensive traversing and searching fires. This type target exists when the enemy is known to be in a certain area, but his exact location is not known. A hilltop is a typical area target. The leader designates an area target by indicating to the gun crew(s) the width and depth of the target.

(a) Two guns. The target is divided at the center of mass; the right gun fires on the right half and the left gun fires on the left half. The point of initial lay and adjustment for both guns is on the center of mass. After adjusting on the center of mass, fire is distributed by determining the size of the beaten zones and applying direction and elevation changes that cause the most effective coverage of the target area. Both guns traverse and search their respective halves to the flanks, then return to the midpoint (Figure 6-11).

(b) One gun. A single gunner engages an area target by laying and adjusting on the center of mass, traversing and searching to either flank, then reversing the direction, traversing and searching to the other flank (Figure 6-11).

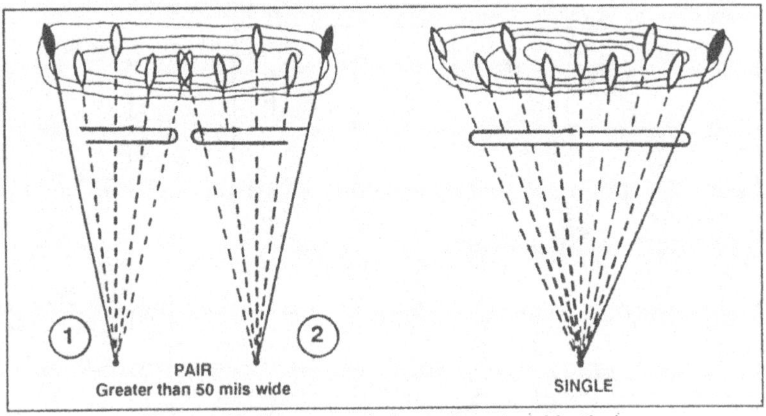

Figure 6-11. Engagement of area targets (objective).

NOTE: After the target is engaged in whatever formation it is in, the configuration of that target will change. The gunner must be trained to compensate for this change and still place effective fire on the target.

6-5. OVERHEAD FIRE

Overhead fire is fire delivered over the heads of friendly troops. A machine gun on a tripod is capable of delivering this type of fire because of the small and uniform dispersion of the cone of fire. In the attack, the use of overhead fire permits the machine gun to support the advance of rifle units. Sectors of fire allow the trainers to plan safe training while still incorporating the combat realities of overhead fires.

a. **Minimum Clearance.** The center of the cone of fire must clear the heads of the friendly troops by a prescribed distance (Figure 6-12). This distance, known as minimum clearance, is found by adding together the following elements:

- The height of a standing man, taken as 1.8 meters.
- Half the vertical dimension of the 100-percent cone of fire at the range to the troops.
- A margin of safety equal to the vertical distance which extends a 5-mil angle at the gun or 3 meters, whichever is greater.
- An additional allowance to compensate for a 15-percent error in range determination.

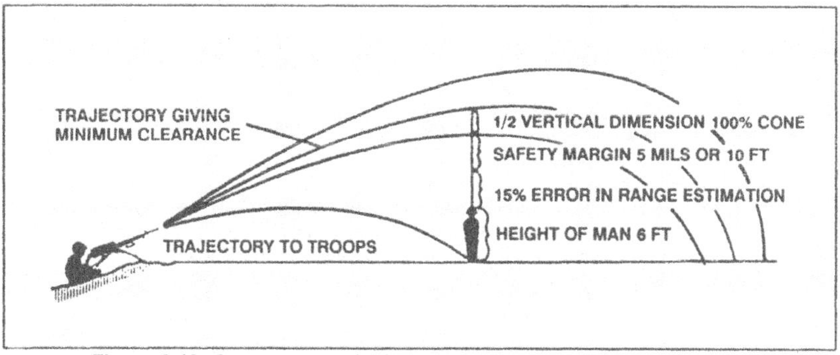

Figure 6-12. Components of minimum clearance.

b. **Safety Angles.** To obtain this minimum clearance, the gun is elevated so that the center of the cone of fire is raised from the feet-of the friendly troops to maintain clearance above their head. The amount of this elevation change is known as the safety angle. When the gun is fired from the tripod with the required safety angle, the center of impact determines the shortest range at which fire can be delivered over the heads of friendly troops. The range from the gun to the point of strike is called the corresponding range. When the ground is level or uniformly sloping between the gun and the target, the corresponding range for the safety angle used is obtained by converting the angle of elevation expressed in mils into range.

c. **Conditions.** Overhead fire is used only when the following conditions have been met:

(1) The safety limit has been determined and has been identified on the ground.

(2) The gun mount is firmly seated.

(3) Friendly troops have been notified, if at all possible, that fire is to be delivered over them.

(4) The rate of fire does not exceed 40 rounds per minute.

(5) The gun barrel is not badly worn. This condition is indicated by excessive muzzle blast.

d. **Uneven Terrain.** Level or uniformly sloping ground is seldom found in the field. This limits the use of firing tables and corresponding ranges in determining the limit of troop safety. In lieu of firing tables, a rule of thumb has been devised to give the gunner a simple method of checking for troop safety.

(1) The gunner's rule can be applied when the friendly troops are at least 350 meters in front of the gun position, and the range to the target is 850 meters or less (Figure 6-13).

(a) Lay the gun on the target with the correct sight setting to hit the target.

(b) Without disturbing the lay of the gun, set the rear sight at a range of 1,600 meters.

(c) Look through the sights and notice where the new line of aim strikes the ground. This is the limit of troop safety. When the feet of the friendly troops reach this point, fire must be lifted or shifted.

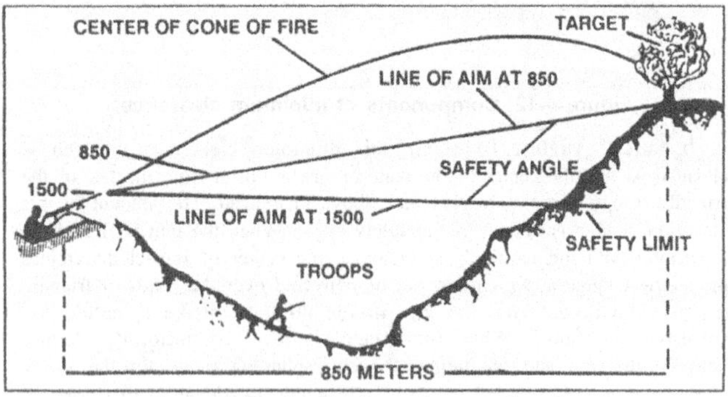

Figure 6-13. Application of gunner's rule.

(2) When the range to the target is greater than 850 meters, overhead fire should be delivered only in an emergency and then only out to a range in which either the tracers or the strike of the bullets can be seen by the gunner. In this situation the leader's rule applies (Figure 6-14).

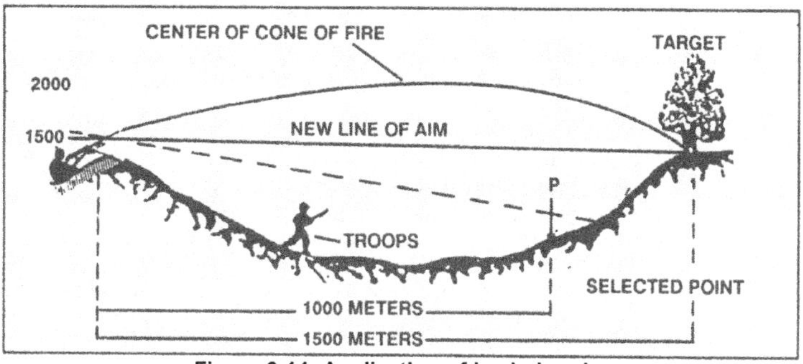

Figure 6-14. Application of leader's rule.

The platoon or section leader uses the leader's rule only when the target is greater than 850 meters. The rule is as follows:

(a) Select a point on the ground where it is believed friendly troops can advance with safety.

(b) Determine the range to this point by the most accurate means available.

(c) Lay the gun on the target with the correct sight setting to hit the target.

(d) Without disturbing the lay of the gun, set the rear sight to 1,600 meters, or the range to the target plus 500 meters, whichever is greater. Under no conditions should the sight setting be less than 1,500 meters.

(e) Note the point where the new line of aim strikes the ground.

- If it strikes at the selected point, that point marks the limit of safety.
- If it strikes short of the selected point, it is safe for troops to advance to the point where the line of aim strikes the ground and to an unknown point beyond. If it is desired to fire after friendly troops advance farther than the point where the line of aim strikes the ground, this farther point is determined by testing new selected points until the line of aim and the selected point coincide.

- If it clears the selected point, it is safe for the troops to advance to the selected point and to an unknown point beyond. If it is desired to have troops advance beyond the selected point, this farther point must be determined by testing new selected points until the line of aim and the selected point coincide. This point marks the line of safety.

e. **Precautions.** The following safety precautions *must* be observed in delivering overhead fire.

(1) Firmly emplace the tripod mount.

(2) Use depression stops to prevent the muzzle of the gun from accidentally being lowered below the safety limit.

(3) Do not deliver overhead fire through trees.

(4) Inform commanders of friendly troops when fire is to be delivered over their heads.

(5) Ensure that all members of the gun crew(s) are aware of the safety limit.

(6) Do not deliver overhead fire if the range from the gun to the target is less than 350 meters or more than 850 meters.

(7) Do not use a barrel that has excessive muzzle blast or is otherwise determined to be badly worn.

(8) Do not lay machine guns so their fire will cross at any point over the heads of friendly troops.

6-6. DEFILADE POSITIONS

To achieve maximum effectiveness, the machine gun must be employed using the technique of direct lay; however, at times it may be desirable to employ guns from defilade positions.

a. **Full Defilade.** A machine gun is in defilade when the gun and its crew are hidden from enemy ground observation by a land mass such as the crest of a hill. The position may be on the reverse side of the crest or the forward slope of the next higher ground (Figure 6-15). The gun must fire up and over the hill. Fire must be observed and adjusted by a crewmember who can observe the target from a position on a flank or to the rear of the gun (on higher ground). A defilade position allows little opportunity to engage new targets.

b. **Partial Defilade.** A machine gun is in partial defilade when a mask (usually the crest of a hill) provides the gun and gunner with some protection from enemy direct fire, but the gunner is able to engage the target using direct laying techniques. The gun is far enough up the slope so that the gunner can see the target through the sights but the lower portion of his body and lower portion of the gun are protected by the mask. Partial defilade positions are desirable when a fire mission cannot be accomplished from a defilade position.

c. **Advantages.** The gun and crew have cover and concealment from direct fire weapons. The crew has some freedom of movement in the vicinity of the gun position, and control and supply are facilitated. The characteristic smoke and flash of the gun are partially concealed from observation.

d. **Disadvantages.** Rapidly moving ground targets are not easily engaged because adjustment of fire must be made through an observer. Targets close to the mask usually cannot be engaged, and it is difficult to secure grazing fire for a final protective line.e.

e. **Position Selection.** The fire unit leader selects the location of the gun position. To select a position in partial defilade, he moves up the reverse side of the slope until he has the target in view above the mask when sighting at the height of the gunner's eye. To select a position in maximum defilade, he estimates the lowest point below the mask at which the gun can still engage the target without danger of hitting the mask.

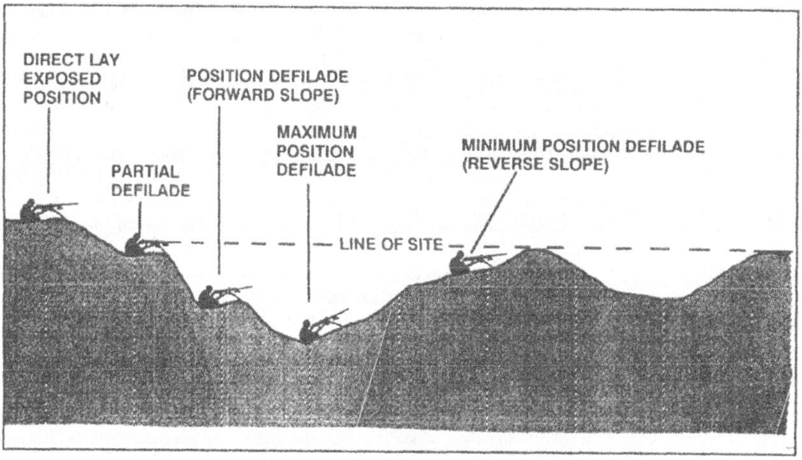

Figure 6-15. Minimum and maximum position defilade, partial defilade, and direct lay areas.

6-7. METHODS OF LAYING THE GUN FOR DEFILADE FIRING

The essential elements in engagement of a target from defilade position are direction, elevation, mask clearance, and adjustment of fire.

a. **Direction.** An observer places himself on the gun-target line in a position from which he can see the gun and the target. He aligns the gun approximately by having the gunner shift the mount. The gunner then loosens the traversing slide lock lever and, as directed by the observer, moves the gun right or left until it is aligned on the target; he then clamps it in that position. A prominent landmark, visible to the gunner through his sights, is selected as an aiming point. An aiming point on the gun-target line and at an equal or greater range than the target is desirable. However, an aiming point on the mask may be used. If the aiming point is on the gun-target line, the gun is laid on the aiming point and is thereby aligned for direction. If the aiming point is not on the gun-target line, the deflection is measured by binoculars or compass. This measured deflection is laid off with the gun.

b. **Elevation.** An aiming point visible from the gun position is selected (preferably a point at a greater range and at a higher elevation than the target) and the range to the target is determined. The leader, using binoculars, measures the vertical angle in roils from the aiming point to the base of the target. He then lays the gun on the aiming point with the sight set to hit the target. He directs the gunner to manipulate the gun through the number of mils measured. For example, in Figure 6-16, the range to the target is 1,300 meters. The angle read with the binoculars from the aiming point down to the base of the target is 12 mils. The sight is set at 1,300 meters, the gun laid on the aiming point, and the muzzle depressed 12 mils. If the aiming point is off the gun-target line, deflection in mils may be taken with the rear sight windage screw knob if it is not over 5 mils; otherwise, the deflection must be taken up on the traversing handwheel.

c. **Mask Clearance.** After the gun has been laid, determine if the entire cone of fire will clear the mask.

(1) *Visual method.* When the range to the mask is not more than 450 meters, mask clearance exists when the axis of the bore is elevated 7 mils or more above the gun-mask line. Mask clearance can be checked after the gun has been laid on the target by depressing the muzzle of the gun 2 mils and sighting along the bottom of the receiver and the barrel support. If this line of sight clears the mask, the clearance exists. Elevate 2 mils before firing.

(2) *Firing tables method.* Determine the range to the mask and obtain the corresponding angle of elevation for mask clearance from the firing

tables. The range corresponding to the angle of elevation is set on the gun sight. If the line of aim through the sight clears the mask, the clearance exists.

d. **Adjustment of Fire.** Under field conditions, even the most practical methods of laying the gun on the target quickly do not always result in the initial burst being on the target. For this reason, adjustment of fire on the target is essential. Creeping fire should be avoided. (See paragraph 6-10 for details on adjustment of fire.)

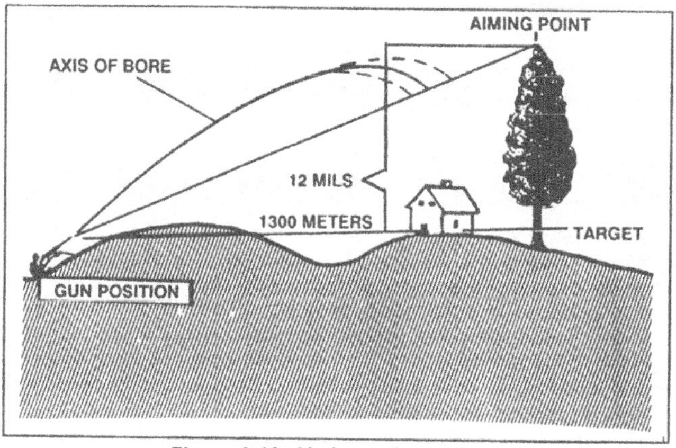

Figure 6-16. Aiming point method.

6-8. FINAL PROTECTIVE FIRES

These are types of fire that are placed on a predetermined line along which grazing fire is placed to stop an enemy assault. This fire is fixed in direction and elevation; however, a few roils of search are employed during firing to compensate for irregularities in the terrain. FPLs are always laid in using the extreme left or right of the tripod, causing the T&E to move to the extreme left or right on the traversing bar. The FPFs can be delivered in any visibility conditions. When terrain permits, final protective lines are assigned to machine guns along the forward line of troops as a part of the FPFs of the defending unit. The signal used to call for FPFs is normally prescribed in the company operation order. The authority to call for these fires may be delegated to the platoon leader of a forward rifle platoon. Final protective fires are ceased on order.

a. **Signals.** Arm-and-hand signals, voice commands, or pyrotechnic devices may be used in calling for these fires.

b. **Rates of Fire.** When firing FPFs, the rapid rate of fire is used unless it is obvious that a different rate is necessary to accomplish the mission. When engaging other preselected target areas, the rapid rate of fire is used until commanded to cease fire.

6-9. APPLICATION OF FIRE

To be effective, machine gun fire must be distributed over the entire target area. Improper distribution of fire results in gaps which allow the enemy to escape or use weapons against friendly positions without effective opposition.

a. The method of applying fire to a target is generally the same for either a single gun or a pair of guns. Direct laying is pointing the gun for direction and elevation so that the sights are aligned directly on the target. Fire is delivered in width, depth, or in a combination of the two. To distribute fire properly, the gunners must know where to aim, how to adjust their fire, and the direction to manipulate the gun. The gunner must aim, fire, and adjust on a certain point of the target. Binoculars may be used by the leader to facilitate fire adjustment.

b. The gunner ensures throughout his firing that the center of the beaten zone is maintained at the center base of the target for maximum effect from each burst of fire. When this is done, projectiles in the upper half of the cone of fire will pass through the target if it has height, and the projectiles in the lower half of the beaten zone may ricochet into the target (Figure 6-17).

c. The gunner must move his beaten zone in a certain direction over the target. The direction depends on the type of target and whether the target is engaged with a pair of guns or a single gun. When engaging targets other than point targets with a pair of guns, the targets are divided so that fire is evenly distributed throughout the target area. Fire delivered on point targets or a specific area of other target configurations is called concentrated fire.

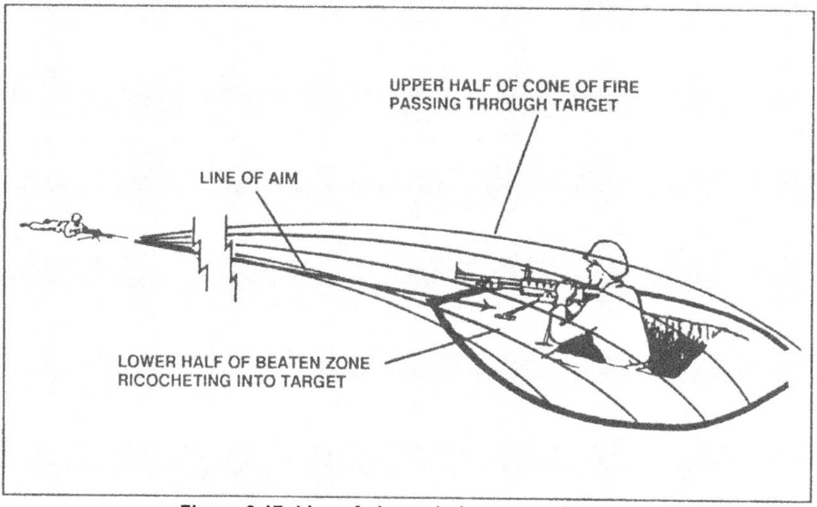

Figure 6-17. Line of aim and placement of center of beaten zone on target.

6-10. FIRE ADJUSTMENT

Machine gun fire is adjusted by observing the strike of the rounds, observing the flight of tracers, frequently re-laying the gun, or by a combination of these. Adjustment by observation of fire is the most important element of fire control if it is bold, aggressive, rapid, and continuous throughout the action.

a. The gunner is trained to observe and adjust his gun's fire without command. He is trained to anticipate the action of the enemy after the initial burst, and is prepared to shift his fire to cover any change in formation or movement of his target. If the gunner fails to accomplish this, the fire unit leader must promptly correct him by announcing or signaling subsequent fire commands. This responsibility to adjust fire continues through the chain of commands.

b. When subsequent fire commands are given, the gunner makes the required corrections and continues to engage the target without any further command to fire. If the gun is fired on the tripod mount, subsequent commands are given to make changes in direction, elevation, and the rate of fire. These changes are given orally as SHIFT RIGHT, SHIFT LEFT, ADD, or DROP. (For arm-and-hand signals see FM 21-60.) When making these announced changes,mils may be used to indicate the amount of desired shift; for example, SHIFT RIGHT 5 or SHIFT LEFT 7. When making changes in elevation, roils are not used, as it is normally difficult to determine just how high or low the center of the beaten zone is striking the ground in relation to the target.

(1) *Observation.* When firing on the 10-meter range, the strike of the bullets is visible on the target. When firing at greater distances, the strike of the bullets on the ground may cause dust to rise, which is visible to the gunner; however, during wet weather the strike cannot always be seen. In this event, the tracers will allow the gunner or crew leader to note the strike of the burst in relation to the target.

(2) *Adjustment.* Using the mil relation, one click of the traversing handwheel or elevating handwheel moves the strike of the bullet 1/2 inch on the target at a range of 10 meters.

(a) When firing on the 10-meter range, adjust by moving the shot group a required number of centimeters vertically or horizontally until the center of the group is on the aiming paster. Should the gunner's initial burst strike the target 2 centimeters to the left and 3 centimeters below the aiming paster, he adjusts his fire by traversing right 4 clicks and elevating 6 clicks before firing again.

(b) When firing on field targets, adjust by moving the burst into the target. One click (roil) on the traversing handwheel will move the strike 1/2 meter at 500 meters or 1 meter at 1,000 meters; however, the distance 1 click (roil) in the elevating handwheel will move the strike depends on the range to the target and the slope of the ground. The gunner determines the number of roils necessary to move the center of the strike into the target, and he manipulates the gun the required number of roils. This does not require the use of sights. For example, should the gunner fire on a target at 500 meters and observe the strike 10 meters to the right of the target and short about 50 meters, he would traverse the gun to the left 20 clicks (roils) and add one or more clicks (roils), depending on the slope of the ground.

(c) The gunner may use the adjusted aiming point method to adjust the fire. In this method the gunner must use his sights. He selects an aiming point that will place the next burst on the target. For example, should the gunner fire on a target at 500 meters and estimate that the strike is 20 meters short and 10 meters to the right of the target, he would rapidly select an aiming point approximately 20 meters beyond the target and 10 meters to the left of the target, lay on that aiming point, and fire.

6-11. ANTIAIRCRAFT GUNNERY

The MG can provide units with a self-defense capability against hostile low-flying, low-performance aircraft. These guns are employed in the air defense role as part of the unit's local defense. The MGs are not components of an integrated and coordinated air defense system. Unless otherwise directed, hostile aircraft within range of the gun (about 800 meters maximum effective range) should be engaged. The decision will be made by the commander. Typical targets are surveillance, reconnaissance, and liaison aircraft; troop carriers; helicopters; and drones.

a. **Engagement and Employment. The** mission is to impose maximum attrition upon the attacking enemy, such as low-flying, low-performance aircraft. Employment of MGs used for air defense is guided by the following defense design factors:

- Defense design should produce an equally balanced defense that is effective in all directions, unless a forced route of approach exists.
- Machine guns should be sited so that the maximum number of targets can be engaged, continuous fire can be delivered, and the most likely routes of approach are covered.
- Machine guns used to defend march columns should be interspersed in the convoy, with emphasis on the lead and rear elements (Figure 6-18).

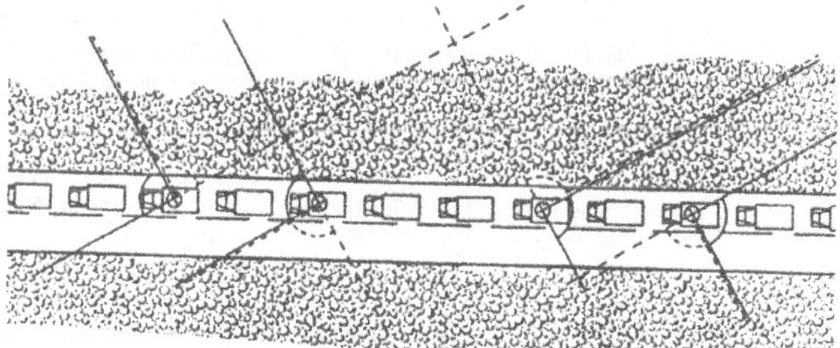

Figure 6-18. March column with four MGs (added).

b. **Target Selection and Engagement Control. These** actions depend upon visual means. The sites selected for the guns must provide maximum observation and unobstructed sectors of fire. Units furnished MGs in sufficient numbers should site them within mutual support distances of 90 to 360 meters. Each gun is assigned a primary and secondary sector of fire. Weapon crews maintain constant vigilance in their primary sectors of fire, regardless of the sector in which the guns are actually engaged.

Section III. LIMITED VISIBILITY CONDITIONS

The machine gun is provided with a stable tripod mount, M3, and a traversing and elevating mechanism. By manipulating the T&E mechanism, gun crews can record target data during good visibility and engage the same targets in poor visibility. This section provides guidance on machine gun firing techniques and terms used during limited visibility, which includes darkness, smoke, fog, rain, or snow.

6-12. DIFFICULTIES

Crewmembers encounter difficulties while defending during limited visibility, which preclude the use of many of the daylight techniques of engaging targets.

a. During limited visibility, the machine gunner's sector of responsibility cannot be observed in depth; therefore, targets are difficult or impossible to detect.

b. Visibility may be so limited that the leader cannot control the fires of his guns by selecting and directing fire on targets as he would during good visibility. Oral commands are not dependable, arm-and-hand signals may not be seen, and personal contact with the gunner is difficult.

c. At night, machine gunners have a tendency to fire indiscriminately at noises and suspected enemy locations.

To overcome these difficulties, special techniques must be developed for engaging targets and delivering preplanned fires by the use of range cards. (See Appendix E.)

6-13. TERMINOLOGY

The following terms must be familiar to MG crews for them to complete their missions in poor visibility.

a. **Sector of Fire.** An area (to be covered by fire) assigned to an individual or unit. Machine guns are normally assigned two sectors of fire, a primary and a secondary sector.

b. **Final Protective Line.** A predetermined line along which grazing fire is placed to stop an enemy assault. The FPL is fixed as to direction and elevation; however, a few roils of search are employed during firing to compensate for irregularities in the terrain. The FPL can be delivered regardless of visibility conditions. The FPL is always the inner limit of the primary sector, which is assigned close to the forward line of troops area. When terrain permits, FPLs are assigned to machine guns along the **FLOT** as a part of the final protective fires of the defending unit.

c. **Principal Direction of Fire.** A PDF is a priority direction of fire that marks a specific area assigned to a weapon. This area may extend from the gun position to the maximum effective range of the weapon and therefore is not fixed for elevation. Visible targets appearing in the PDF take priority over targets that may appear elsewhere in the sector. A PDF may be assigned to cover an area that provides good fields of fire, is a likely avenue of foot approach, or mutually supports an adjacent unit.

d. **Sector of Graze.** A wedge-shaped area formed by assigned sector limits that afford grazing fire (one meter high, maximum) from the muzzle of the weapon to the first major break in the terrain. The sector of graze is fired using swinging traverse in the primary sector of fire. It can be fired in the secondary sector in conjunction with field expedients by freeing the T&E mechanism and using the mount as a pivot. A sector ot graze can be delivered regardless of the condition of visibility.

e. **Area of Graze.** This is an area, other than the sector of graze, within a sector of fire that is covered by grazing fire. Grazing fire need not be continuous from the muzzle of the weapon to the area over which grazing fire is desired.

6-14. TARGET ENGAGEMENT

A gunner's ability to detect and identify targets during limited visibility will vary, depending upon the amount of natural and artificial light and the types and numbers of sensors used. All tracer ammunition allows a gunner to more effectively engage visible targets during limited visibility; it should be used when possible. Gunners must be trained to fire low initially and adjust up when engaging targets during limited visibility. This helps them overcome the tendency to fire high during these conditions. The types of point targets machine gunners will be concerned with during limited visibility, particularly at night, are enemy automatic weapons and assaulting enemy personnel.

a. Point targets such as automatic weapons may be identified during limited visibility by their muzzle flashes. To effectively engage these targets, fire should be delivered in a heavy volume and adjusted by observing the tracer stream.

b. During the final stage of an enemy assault, machine guns normally fire at personnel on an FPL; they may be assigned a PDF. Both are considered as final protective fires and should be planned for and coordinated as such. If individual enemy soldiers are observed in the proximity of the gun position, they must be neutralized by someone other than the machine gunner (by the other crewmembers or by security forces of the supported unit). The FPFs are fired according to the order or SOP, and the machine gunner is not allowed to stop firing them except in accordance with those orders or SOP.

6-15. FIRE CONTROL

During limited visibility, the leader cannot direct the fires of his guns as effectively as with good visibility. Consequently, initiative is required of the gunners. When targets within their sectors become visible to gunners, they must engage such targets without command and continue to fire until the targets have been neutralized. Gun crews engage targets only when they can identify them, unless ordered to do otherwise. For example, if one gunner detects a target and engages it, the other gunners will observe the area in which fire is being placed. They will add their fire only if they can identify the target or are ordered to place fire in the area.

6-16. PREPLANNED FIRES

In addition to engaging appropriate visible targets, the gunner must be able to deliver preplanned fires during limited visibility. These fires are used to cover target areas of tactical significance (such as routes, avenues of approach, anticipated enemy supporting weapons positions, and probable enemy assault positions) and to establish sectors of graze and final protective lines. For maximum effect in all preplanned target areas, grazing fire should be obtained when possible.

a. **Obtaining Maximum Extent of Grazing Fire Over Level or Uniformly Sloping Terrain.** The machine gunner sets the rear sights at 700 meters; selects a point on the ground, which he determines to be at a range of about 700 meters; and lays, fires, and adjusts on this point. If the gunner cannot obtain 700 meters of grazing fire because of a major break in the ground at a range of less than 700 meters, he places the range to the break on his sight and lays, fires, and adjusts at that point.

b. **Determining the Extent of Grazing Fire on the Final Protective Line.** The extent of grazing fire on the FPL is determined using the techniques described above. Any intermediate breaks in the terrain along this line that cannot be covered by grazing fire from a gun firing along the line is considered dead space.

c. **Determining the Extent of Grazing Fire in the Sector of Graze.** The ranges to the extent of grazing fire in a sector of graze are determined by observing the terrain and by observing the tracer stream from behind or from a flank of the gun position. Normally, the extent of grazing fire within this area will be much less than on an FPL and will form an irregular pattern.

d. **Determining the Amount of Grazing Fire in an Area of Graze.** The same procedures used in paragraph 6-8a are used in determining the extent of grazing fire in an area of graze. The ranges to areas of grazing fire are determined by observing the flight of tracer ammunition from behind or from the flank of the gun position. The gunner determines the lateral extent of areas of graze by selecting and engaging successive aiming points in the area believed to afford grazing fire, using the same range setting as when determining the range to the extent of grazing fire.

6-17. NBC CONSIDERATIONS

During this phase of training, the gunner is introduced to firing the machine gun while in MOPP, keeping in mind that engagement of some targets in MOPP is a qualification requirement. Firing weapons is only part of the overall NBC training. Soldiers must first be familiar with the NBC equipment, its use, and proper wear before they progress to learning the techniques of MOPP firing. Although there is no different technique required to fire the MG, there are certain fundamentals that may be slightly impaired.

a. **Immediate Action.** Under normal conditions, a gunner should be able to clear a stoppage in two to four seconds; however, under full MOPP, this may take a few seconds longer. Dry-fire practice under these conditions is necessary to reduce time and streamline actions. When practicing with the hood/mask and gloves, care must be taken not to snag or damage the gloves or dislodge the hood/mask during movement. Trainers should apply immediate action to a variety of stoppages during dry fire until the gunners are able to instinctively do it without compromising their NBC environment.

b. **Target Detection.** Techniques and principles of target detection and target acquisition still remain valid during NBC conditions, but considerations must be made for limiting factors imposed by MOPP equipment. For example, vision is limited to what can be seen through the mask's lens/faceplate. Peripheral vision is severely restricted. The lens/faceplate may be scratched or partly fogged, thus further restricting vision. Gunners requiring corrective lenses must be issued insert lenses before training. Scanning movements may be restricted by the hood/mask. Any of these factors could adversely affect the gunner's ability to quickly and accurately detect targets. Extra skill practice should be conducted.

c. **Efficient Performance.** The trainer must keep in mind that although movements are slowed, tasks take longer, and function checks, loading, unloading, and cleaning are affected by MOPP, it is a must that the gunner avoid damaging MOPP gear and risk possible exposure to lethal agents. Because of the great difference between no MOPP and MOPP4, gunners must be trained in all aspects of operation and maintenance of the weapon while practicing at the highest MOPP level. Only through repeated training and practice can the soldier be expected to perform all tasks efficiently.

CHAPTER 7

TRAIN THE TRAINER PROGRAM

The information in this chapter tells how to train the trainer, how to advise the trainer, and how to assist the trainer in preparing the crew for the three phases of MG marksmanship.

7-1. CONCEPT

The Train The Trainer Program must be planned properly and conclude with a certification program. In the planning phase, the instructor teaches the trainer to be flexible and thorough. He must plan alternate exercises in case weather or other constraints prohibit the originally scheduled training. Training must not follow rigid timetables that inhibit training and learning; instead, schedules should be established that provide sufficient time to correct mistakes and ensure learning. The following are factors that the trainers are taught to consider:

a. **Support.** Training requires support. The trainers must ensure that the training is conducted within the resource levels and that the training received justifies the material used. Public address systems should be used if the group is larger than a platoon-size element.

b. **Time.** Ample time must be allocated for each phase.

c. **Participants.** Trainers must consider whether or not the groups or individuals to be trained are capable of benefiting from the phases selected. (See paragraph 7-2a.)

d. **Safety.** Safety is the most important factor and therefore will be the main consideration. Trainers as well as the personnel to be trained will ensure all safety precautions are met. Anyone observing any unsafe act will immediately call CEASE FIRE or HALT to any training. A good way to emphasize safety is to give the students a test on the procedures. An example of a written examination on safety is shown in Figure 7-1, page 7-2.

SAFETY REQUIREMENT EXAM		
INSTRUCTIONS: Mark each answer as being true or false.		
T	F	
	✓	1. Firing can be done without a safety officer.
	✓	2. Firing can be done without correct headspace.
	✓	3. Firing can be done without correct timing.
✓		4. No gun should be loaded until the command to do so.
✓		5. Each gun is cleared and checked after it is fired.
✓		6. No one is allowed in front of the gun for any reason unless authorized by the OIC/NCOIC.
	✓	7. Guns can be fired outside safety areas.
	✓	8. Guns can be pointed at other personnel while clearing them.
✓		9. Firing stops immediately upon the command CEASE FIRE.
✓		10. Personnel will stay in designated safety areas until the command is given to leave by OIC/NCOIC.
Answer key: 1. F; 2. F; 3. F; 4. T; 5. T; 6. T; 7. F; 8. F; 9. T; 10. T.		

Figure 7-1. Safety examination.

7-2. TRAINER CERTIFICATION PROGRAM

The Trainer Certification Program is designed to build pride, confidence, and overall working/teaching knowledge of the functions, employment, and overall training of the MG. The trainer must know how to do all the tasks he is going to be teaching; the certification program determines his capability to do this. The program is conducted in three phases.

a. **Phase I: Basic Fundamentals of Marksmanship.** In this phase the instructor must ensure the trainers being trained perform dry fire techniques; prepare firing positions, fighting positions, and range cards; and manipulate the T&E mechanism. At the end of this phase, they will be given a hands-on examination to test their basic knowledge. This test may be developed by the OIC/NCOIC of the MG committee.

b. **Phase II: Basic Marksmanship.** In this phase, the trainer moves one step closer to becoming a skilled gunner. He must learn the correct procedures to zero the weapon; he must master the 10-meter target paster; and once this is accomplished, he must be able to engage targets on a transition range. He must show his mastery in this phase by engaging targets at different ranges. Upon completion, the trainer is ready to move to the advanced stage of marksmanship.

c. **Phase III. Advanced Gunnery.** This is the final phase that the trainer has to achieve to standard.

(1) First, he must understand the procedures using the techniques of advanced gunnery.

(2) He must then learn tracking, which consists of maintaining correct alignment of the sights on a moving target by moving the gun at the same angular speed as that of the target.

(3) He will learn fire distribution next, which means he must be able to distribute fire over the entire target area.

(4) The last stage the trainer must learn is the correct assault fire techniques. During this stage, the gunner will normally fire on an FPL after the enemy has assaulted. He may also be given a PDF to fire at. Another assault method the trainer learns is to fire the weapon while it is mounted on a moving M113.

d. **Recertification Program.** All MG instructors are required to be recertified on a semi-annual basis. This will consist of the trainer being required to present a selected period of instruction within the current POI. The selection will be made by the CDR/OIC/NCOIC.

NOTE: Documentation of the results of the Trainer Recertification Program and the requirements must be maintained on file.

7-3. RESPONSIBILITIES AND DUTIES OF THE TRAINER

The trainer must be present during the planning and during any instruction given. The success of the preparation and instruction of all training depends upon the thoroughness with which the trainer performs the following duties:

- He must assist the gunner in targeting the MG.
- He should require the gunner to inspect his equipment and MG.
- He should explain and require the gunner to explain the exercise that he is about to perform.
- He should ensure the gunner's sight picture is accurate.
- He must observe the gunner's position, grip, and manipulation during any firing.
- He must show the gunner how to adjust his fire by observation.
- He must point out errors and explain their effect on the exercise.

NOTE: Safety is a must during the entire certification program. The OICs/NCOICs must ensure that no unsafe acts are tolerated.

* APPENDIX A

SAFETY

Safety precautions must be observed during all marksmanship training. This appendix recommends safety precautions for the ranges described in this manual. Range safety requirements will vary with the different courses of fire.

A-1. TRAINING RANGES

The following safety precautions will be observed when using any training range.

a. A red flag will be displayed at the entrance or in some other prominent location on the range during firing.

b. Firing limits will be marked with red and white striped poles visible to all firers.

c. Firing lanes must be clearly marked for easy identification.

d. All communication equipment, such as microphones, PA systems, loudspeakers, and radios, must be in good working condition.

e. When not in use, all MGs will be kept in a prescribed area with proper safeguards.

f. No smoking will be permitted near ammunition, explosives, or flammables. Firefighting equipment must be installed in ammunition shed or area.

g. Hearing protection devices will be worn by all personnel during firing.

h. Obstructions will never be placed in the muzzles of guns about to be fired.

i. Always assume that guns are loaded until they have been thoroughly examined and found to contain no ammunition.

A-2. RANGE PROCEDURES

Safety requirements must be followed by all personnel on the range before, during, and after firing. It is the responsibility of the safety officer, OIC, and safety NCO to ensure all procedures are met. They are as follows:

a. **Before Firing**.

(1) All prescribed roadblocks and barriers will be closed and necessary guards posted.

(2) All guns will be checked to ensure that they are clear of ammunition and obstructions, and that the covers are up to show they are cleared.

(3) All individuals will be briefed on the firing limits of the range and firing lanes.

(4) Range clearance will be obtained from the installation range-control office.

(5) The downrange area will be checked before firing to ensure that all personnel and equipment are clear of the area.

(6) The NCOIC or range safety officer places a complete first aid kit on the range.

(7) The NCOIC or range safety officer places medical personnel on the range or nearby to enable quick contact.

* (8) Due to the potential injury hazards associated with improper headspace and timing, either the NCOIC or RSO will assume personal responsibility for ensuring that all weapons are set for proper headspace and timing. He will do this before emplacing weapons on the firing line. This applies to all manually fired weapons. It does not apply to mechanically fired weapons.

*** WARNING**

Before placing any manually fired weapon on the firing line, personally check each one for proper headspace and timing.

(9) Soldiers handle guns only on the command of the tower operator or the OIC.

(10) Soldiers draw ammunition only on command of the OIC. When using two or more lots of ammunition, the OIC keeps the lots separate. He identifies ammunition by lot numbers in case of accident or malfunction. Soldiers should retain original packaging for possible repacking of ammunition.

(11) Protect all ammunition from the rays of the sun. Keep oil away from ammunition.

(12) Avoid moving forward of the firing line except with permission of the tower operator, range safety officer, or OIC.

b. **During Firing**.

(1) Should anyone note an unsafe condition, that person will immediately call CEASE FIRE. Firing resumes only when the OIC so directs.

(2) All personnel know the danger of moving forward of the firing line to score their targets. Before the firing line is cleared, and before anyone can go forward, the OIC or range safety officer clears all machine guns.

(3) In the event of a runaway gun, the firer should keep the barrel pointed downrange. He should leave the cover latched.

c. **During Darkness**.

(1) The NCOIC or range safety officer checks the downrange area before firing. He ensures that the area is clear of personnel and equipment. To make sure, he asks three times over a public address system, "Is anyone downrange?" pausing long enough each time to permit a response.

(2) A blinking red light must be used in addition to the red flag. It should be displayed at the entrance to the range or at some other prominent location.

(3) Two red lights will be mounted on the striped poles marking the limits of fire. They must be visible to all firers.

(4) No one will move from his position until told to do so by the OIC.

d. **After Firing**.

(1) Safety personnel will inspect all guns to ensure that they are clear. All guns on the firing line will be rodded whether they were fired or not. A check will be conducted to determine if any brass, links, or live ammunition is in possession of the troops.

(2) When guns have been cleared, they will be kept in a prescribed area with the bolt lock to the rear, and the feed cover raised.

A-3. WEAPONS MAINTENANCE

Before disassembly can be conducted, the gun must be cleared as discussed in Chapter 1.

a. Make sure the bolt is forward before removing the backplate assembly.

b. Never attempt to cock the gun while the backplate is off and the driving spring assembly is in place. If the backplate is off and the driving spring assembly is compressed, the retaining pin on the driving spring rod can slip from its seat in the sideplate and could cause serious injury to anyone behind the gun.

c. When lifting the receiver, barrel, or the tripod, hold it with a firm grip to avoid dropping it and possibly injuring someone.

APPENDIX B

TRAINING AIDS AND DEVICES

The use of devices in marksmanship training programs are important factors because they may allow the gunner to get an idea of what actual combat is like. The training devices designed for the MG are the M19 blank firing attachment, the multiple integrated liner engagement system, the sighting bar, and the M3 recoil amplifier barrel used with plastic ammunition.

B-1. BLANK FIRING ATTACHMENT

The M19 BFA was developed to permit the MG to fire the M1A1 blank cartridge in the automatic fire mode. The BFA is an easy to install, reliable device that allows the MG to be used more realistically during an FTX. The M19 BFA weighs 15.5 pounds and can be installed using either a crescent wrench or the blank/live round discriminator (a component of the BFA that has been designed to serve as a wrench). The design of the M19 will not allow a live round to be loaded while the blank/live round discriminator cover is in position. Normal headspace and timing must be made when firing blank rounds using the M19 (Figure B-1).

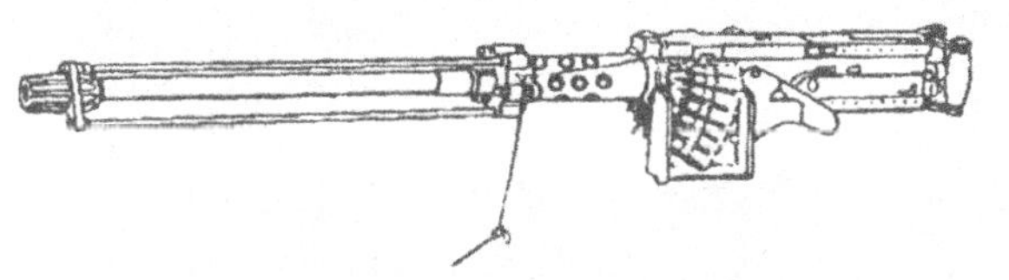

Figure B-1. M19 BFA mounted on MG.

> **WARNING**
> **The flash of the M1A1 round fired with the M19 BFA at night is extremely bright. It could cause temporary night blindness, and night vision devices may be temporarily disabled.**

B-2. MULTIPLE INTEGRATED LASER ENGAGEMENT SYSTEM

The MILES transmitter has been developed for the MG as well as for many other weapons. It consists of a lightweight laser transmitter that sends coded hit/kill messages whenever blank rounds are fired. The MILES transmitter is normally used as part of an M113 APC MILES kit; however, it can also be used on truck- or ground-mounted machine guns with modifications of the detector straps and the combat vehicle kill indicator (CVKI) light (Figure B-2).

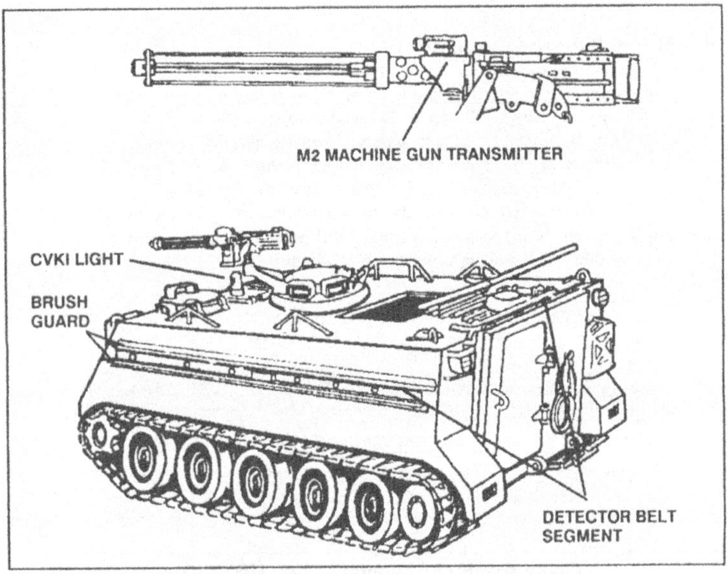

Figure B-2. Components of the M113 APC MILES kit.

a. Inspecting and Servicing the MILES Transmitter. Remove any dirt or oil from the lens using lens paper or a soft, dry cloth. Take care not to scratch the lens. Inspect the foam microphone cover; if it is wet or caked with dirt or blank firing residue, clean or replace it. Inspect the transmitter for indications of damage that would prevent normal operations (Figure B-3). Clean all surfaces. Ensure that the orange key is present. Use DA Form 2404 to report any damage to the transmitter.

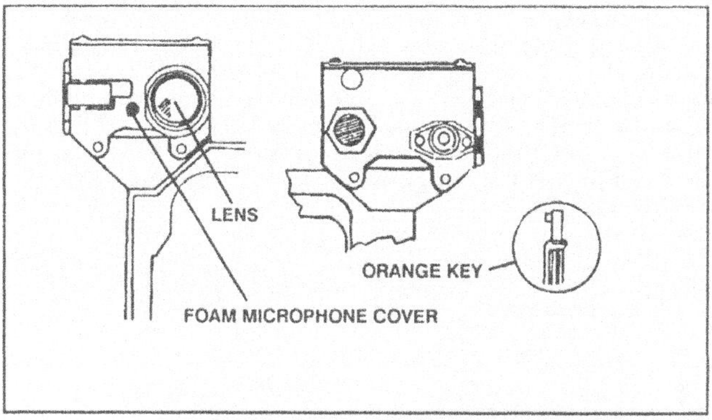

Figure B-3. Inspection of the MILES transmitter.

b. **Placing the Battery into the Transmitter.** Use the correct battery (NSN 6135-01-063-1978, BA 3090/U, 9V, alkaline); it will last about 100 hours. Flip open the latch; open the battery door and insert the battery; press the door closed; then press the latch closed (Figure B-4).

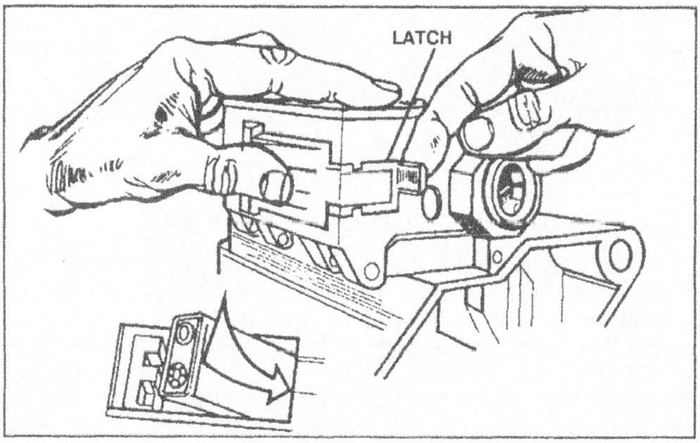

Figure B-4. Battery insertion.

c. **Attaching the Transmitter to the Machine Gun.** Attach the machine gun to its mount. Attach the M19 BFA to the machine gun. Unscrew the knob from the side of the transmitter bracket; swing the bottom plate down; and place the transmitter on the barrel support, flush against the receiver. The lip on the back of the mounting bracket must sit behind the barrel cooling bracket. Swing the bottom plate back up against the transmitter mounting bracket and tighten the knob securely with your hand (Figure B-5).

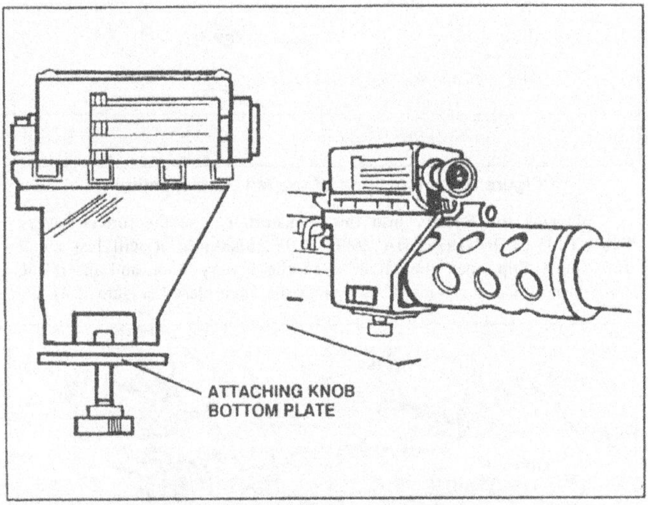

Figure B-5. MILES transmitter attachment.

d. **Operating the Transmitter.** Ensure that the M19 BFA is correctly mounted. Turn the orange key to WEAPON ON. Load the machine gun with M1A1 blank ammunition only (Figure B-6). Fire normally. The sound of the blanks firing will trigger the transmitter. The transmitter will work only as long as the supply of blank ammunition lasts. Once a day, use a clean cloth to remove blank fire residue from the transmitter lens. To test the transmitter, fire a shot burst and watch the firing lamp (Figure B-7). If no light, replace the battery and test again. If still no light, replace the transmitter. Use DA Form 2404 to report the information.

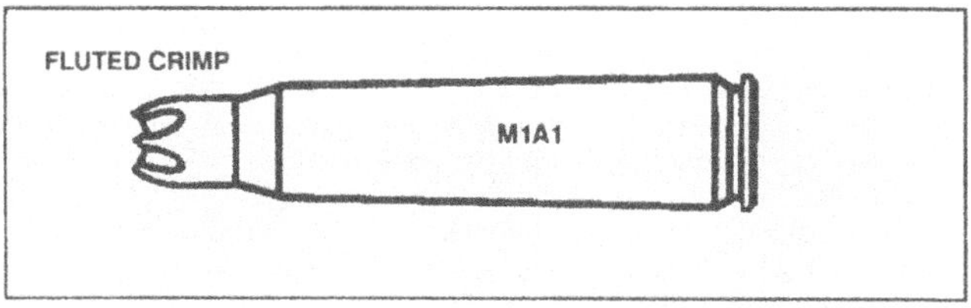

Figure B-6. Proper blank ammunition to use with M19 BFA.

WARNING

Do not use ball ammunition or M1 blank ammunition with the BFA. Use only M1A1 blanks with the fluted, crimped case mouth.

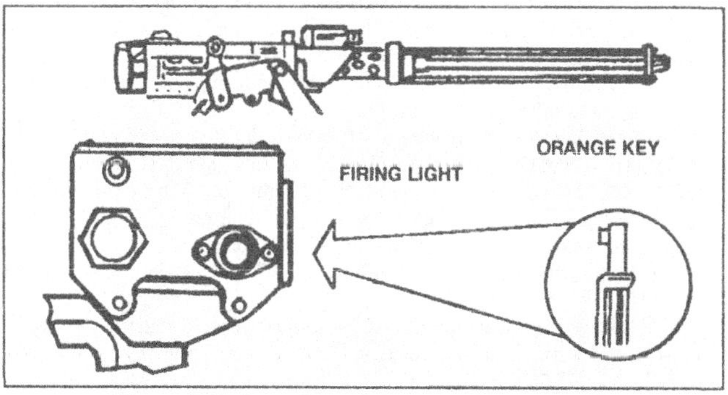

Figure B-7. Operation of the MILES transmitter.

e. **Aligning the Transmitter to the Machine Gun.** The transmitter must be aligned in the dry fire mode, using the dry fire trigger cable. To align the transmitter, you need a soldier wearing a MILES helmet and torso harness. The gun should already be mounted on a tripod and have the T&E attached.

(1) Position the soldier 50 to 75 meters away from the machine gun. Have him place a green controller key in his MILES receptacle and turn the key to the CONTROLLER position.

(2) Connect the trigger cable assembly to the receptacle on the rear of the MILES transmitter. Use the controller key to reset the system.

(3) Insert an orange weapon key into the transmitter receptacle and turn to WEAPON ON.

(4) Fire at the soldier by pressing the trigger cable button. Traverse the gun left and right, up and down, until the soldier's buzzer signals a "near miss." Hold the gun in that position.

(5) While he is being fired on, have the soldier move to his right until his buzzer stops. Have him mark that location.

(6) Have the soldier move to his left until the buzzer stops again. Have him mark that location.

(7) Have the soldier estimate the center between the two marks, and stand there.

(8) Cease fire. Without moving the barrel of the machine gun, adjust the sights until the soldier is seen through the rear sight aperture.

(9) Begin firing again and move the barrel up and down, noticing where the buzzer stops. Estimate the center of this up and down area and adjust the sight elevation to that point.

(10) Remove the trigger cable assembly and screw the protective cover onto the transmitter receptacle.

f. **Resetting the Transmitter After a Kill.** Remove the orange key from the transmitter. Use it to silence the "kill" indicator and the CVKI light. The transmitter will not fire with the key removed. When the controller resets the "kill" indicator, turn the transmitter to WEAPON ON and reinsert the orange key. Continue the mission.

g. **Using Precautions.** The following safety considerations apply to the M1A1 blank cartridge:

(1) Never fire blanks directly at personnel within 20 meters of the gun.

(2) Never fire blanks without wearing hearing protection.

(3) Never fire blanks when the temperature is below 0 degrees or above 120 degrees Fahrenheit.

(4) Never stand directly to the side of the weapon muzzle when blanks are fired.

(6) Never fire blanks in bursts in excess of 20 rounds.

(7) Always allow a cooling period between bursts to avoid overheating the gun.

B-3. SIGHTING BAR

The sighting bar is a locally fabricated device used for practicing the sighting and aiming exercise (Figure B-8). Paragraph 5-9 explains its use. This exercise requires the soldiers to look through the peephole and center on the front sight blade. The gunner adjusts the peephole until he is able to align the front sight blade in the center of it. The instructor will then review and critique the gunner on sight alignment. The device is also equipped with a scaled target for practicing sight picture. Once sight alignment is mastered, the gunner will be required to show a correct sight picture. This is done by adjusting the scaled target until the front sight plate is bottom center of it. Again the instructor will review and critique the gunner. This exercise will continue until the gunner has mastered the technique of sighting and aiming; this requires a lot of practice.

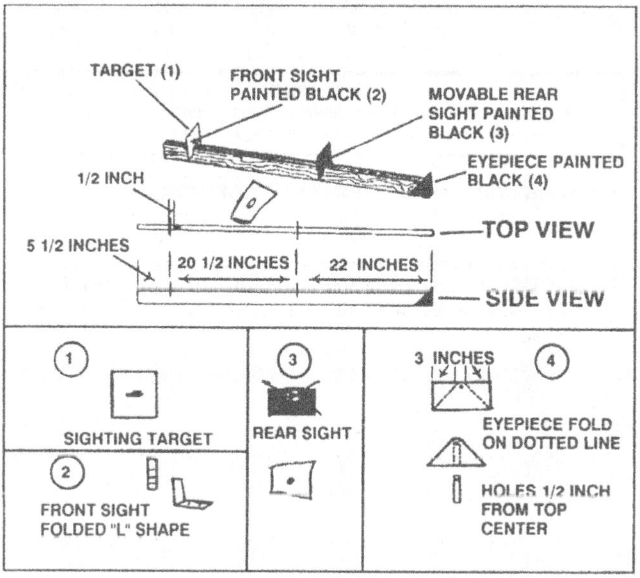

Figure B-8. Sighting bar.

B-4. SHORT RANGE TRAINING AMMUNITION

To augment the use of the caliber .50 ammunition, a new generation of SRTA (the M858 ball and the M860 tracer) has been developed.

a. The caliber .50 ball and tracer cartridges have configurations and dimensions similar to those of the corresponding service rounds; however, the base of the cartridge, which encloses the primer, is made of aluminum. The rest of the case is-made of a blue-colored plastic material. The tip of the tracer projectile is colored red. The complete round weighs about 1.05 ounces and is 5.19 inches long, which is slightly shorter than the service round. The cartridges are linked with M9-type metallic links for use with the MG. Muzzle velocity of the projectile is about 4,000 feet per second. The light weight of the projectile, about 50 grains, and its low sectional density cause its velocity to drop rapidly.

b. The SRTA allows training in small local training areas without fixed training facilities, in MOUT facilities, and in combat training theaters. With its 150-meter usable range, the SRTA can be used on grouping, zeroing, and 10-meter scaled silhouette firing.

> **WARNING**
> **The SRTA projectile causes considerable damage out to 150 meters and is considered dangerous out to its maximum range of 700 meters.**

B-5. M3 RECOIL AMPLIFIER BARREL ASSEMBLY

An M3 RABA (Figure B-9) is used with the machine gun for firing the training ammunition. It is assembled to the gun in the same manner as the standard barrel. The RABA provides the means to boost the power that is required to recoil the barrel and effectively cycle the weapon when relatively underpowered M858 SRTA or M860 SRTA-T are fired in the machine gun.

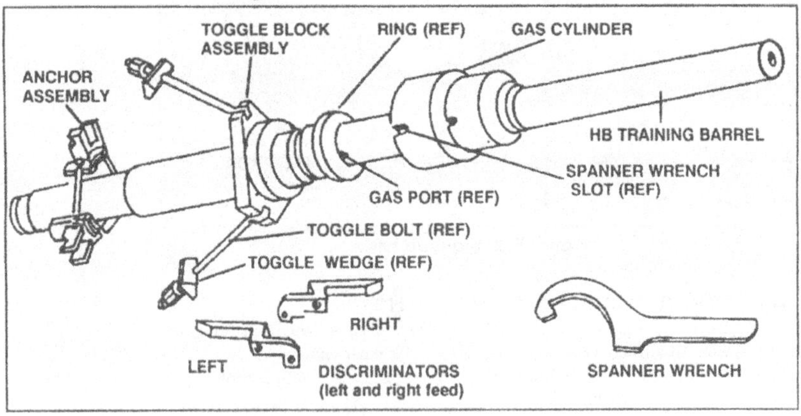

Figure B-9. M3 RABA.

APPENDIX C

RANGES AND TRAINING EXERCISES

This appendix provides guidance for US Army machine gun exercises. It is designed to be a working guide for trainers and commanders by explaining range layouts, conduct of firing exercises, and qualification standards. The two basic firing ranges will be the multipurpose MG transition range and the multipurpose range complex.

* Section I
MULTIPURPOSE MACHINE GUN TRANSITION RANGE

All machine gun firing can be conducted on the multipurpose MG transition range. It allows for complete firing of the 10-meter course as well as transition, day, night, and NBC firing. The firing area has 10 lanes; its layout is shown in Figure C-1. Details on the setup and target configuration are described in FM 25-7. The minimum personnel required to operate the range are as follows: OIC, NCOIC, safety officer, ammunition NCO, tower operator, lane NCOs, primary instructor, and concurrent training instructors. Local policy may require more personnel.

* The targets listed in this section are only recommendations. Units may alter the types of targets and ranges used depending on availability.

- *For night qualification*, units with the AN/PEQ-2A, ANPAS 13 or the AN/TVS-5 upgraded with the third-generation tube should use Appendix G tables and qualification standards.
- *For day qualification*, leaders will use this section until the STRAC (AR 350-38) changes, at which time they will use only Appendix G.

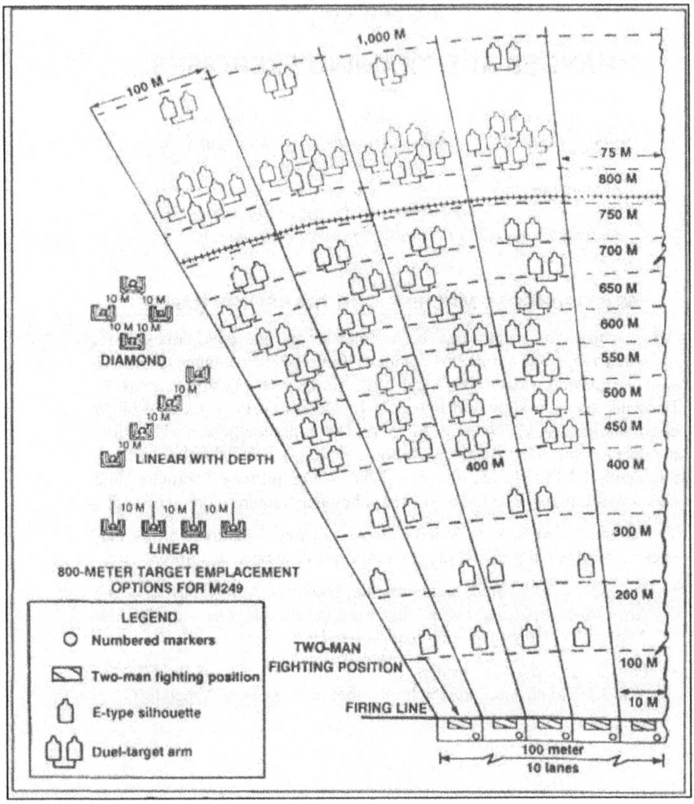

Figure C-1. Multipurpose machine gun transition range layout.

C-1. TEN-METER RANGE LAYOUT

The standard, basic 10-meter range can accommodate a unit of 200 to 250 soldiers at a time; however, concurrent training may be required. This range (Figure C-2) can be used to zero the M249, M60, and M2 machine guns, and to fire the 10-meter portion of qualification. It is also used to familiarize soldiers with the characteristics, noise, and recoil of the weapon. This is the range used to practice target observation and adjustment of fire, to practice MG traversing and searching, to develop speed during operation, and to obtain an accurate burst.

a. **Characteristics.** The following data applies to the standard 10-meter range.

Number of firing positions	**20 lanes.**
Firing lane width	**60 meters (3 meters per lane).**
Target area width	**60 meters at the farthest targets (10 meters).**
Firing point configuration	**Numbered markers on slightly elevated and sodded ground; brass deflectors between lanes.**
Target configuration	**Standard 10-meter machine gun target stretched over a wooden frame, one on each lane.**
Associated facilities	**Standard facilities, public address system, and bleachers.**

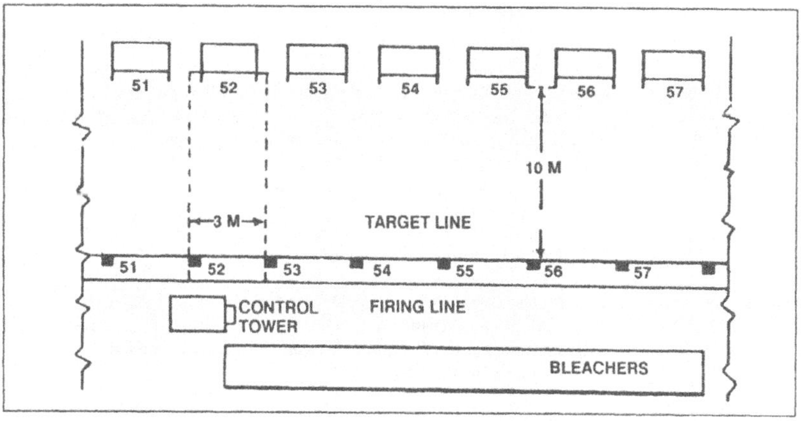

Figure C-2. Ten-meter range layout.

b. **Setup of Range.** The 10-meter range should meet the following requirements:

(1) The firing line should be long enough to emplace 20 MGs, allowing 3 meters between positions. For control, each position should be numbered.

(2) The target line should be 10 meters in front of the firing line. The MG marksmanship target is a paper target pasted onto target cloth that is stretched over a wooden frame. One target is set up for each position and is numbered to correspond with the numbered position.

(3) The instruction site for this range is the bleachers, which should be to the rear of the firing line.

(4) The control tower is located to the immediate rear and center of the firing line.

c. **Personnel Required.** Each range must be staffed with the following personnel:

- One officer in charge.
- One safety officer.
- One principal instructor.
- One assistant instructor for every 10 students.
- One ammunition NCO.
- One tower operator.
- Medical personnel.

d. **Equipment Required.** Although more equipment may be required by local range regulations, safety regulations, or unit SOPs, the minimum equipment required to operate the range is as follows:

- One public address system.
- One MG for each firing lane and one extra gun for every five lanes as a backup.
- One cleaning rod per assistant instructor.
- One scorecard per soldier.
- One asbestos glove for every two weapons.
- One M3 tripod for each lane.
- One caliber .50 pintle for each lane.
- One caliber .50 T&E mechanism for each lane.
- Medical evacuation capability.
- Communication equipment (wire or radio as required to operate the range).

* C-2. TEN-METER FIRING EXERCISES

The purpose of 10-meter firing is to develop skills in the delivery of initial bursts on target. When conducted properly, it will train gunners in the basic skills of zeroing, controlling the burst, traversing, and traversing and searching techniques. This paragraph discusses firing the 10-meter exercise on the multipurpose transition range; however, it is conducted the same on a basic 10-meter range except that the groups will be broken down to fit the firing lanes. The course is designed to fire the tripod-mounted M2. The basic MG target is used for all 10-meter firing and it allows four gunners to use it for practice or two gunners for practice and qualification. Those units wanting to fire qualification right after practice should use section A for practice and B for qualification for the first gunner, and sections C and D for the second gunner. Ten-meter firing will be conducted IAW Firing Table I (Figure C-3).

TASK	CONDITIONS TARGET/SITUATION	AMMO	STANDARD_____
1. Zero the M2 machine gun at 10 meters.	Instruct the gunner to engage scoring paster 1 in his assigned section with 6 single rounds in two 3-round shot groups and then paster 2 with the remaining rounds.	12 single rounds of 12.7-mm	Gunner must impact 4 of 6 rounds of the 3-round shot groups within paster 2 of his assigned section.______
2. Engage a 10-meter target using controlled-burst techniques.	Instruct the gunner to engage paster 3 in his assigned section using controlled bursts, then engage paster 4 in the same manner. Gunner will use 5- to 7-round bursts for this engagement.	two 7-round belts of 12.7-mm	Gunner must impact one controlled burst on pasters 3 and 4 in his assigned section. (No score required.)____
3. Engage scoring pasters 5 through 6 using the traverse and search technique of fire.	Instruct the gunner to engage pasters 5 through 6 using traverse and search technique of fire. Gunner will use 5- to 7-round bursts for this engagement.	35-round belt of 12.7-mm	Gunner must impact one round on each scoring paster of 5 through 6 in his assigned section. (No score required.)____
4. Engage scoring pasters 7 through 8 using the traverse and search technique of fire.	Instruct the gunner to engage pasters 7 through 8 using traverse and search technique of fire. Gunner will use 5- to 7-round bursts for this engagement.	56-round belt of 12.7-mm	Gunner must impact one round on each scoring paster of 7 through 8 in his assigned section ______. (No score required.)____
5. Engage linear target at 10 meters using traverse fire technique.	Instruct the gunner to engage pasters 1 through 4 in his assigned section using traverse fire. Gunner will use 5- to 7-round bursts for this engagement.	28-round belt of 12.7-mm	Gunner must impact at least four rounds on each scoring paster 1 through 4 in his assigned section ______.
6. Engage scoring pasters 7 through 8 using the traverse and search technique of fire.	Instruct the gunner to engage pasters 7 through 8 using traverse and search technique of fire. Gunner will use 5- to 7-round bursts for this engagement.	56-round belt of 12.7-mm	Gunner must impact at least four rounds on each scoring paster 7 through 8 in his assigned section ______.
7. Engage scoring pasters 5 through 6 using the traverse and search technique of fire.	Instruct the gunner to engage pasters 5 through 6 using traverse and search technique of fire. Gunner will use 5- to 7-round bursts for this engagement.	35-round belt of 12.7-mm	Gunner must impact at least four rounds on each scoring paster 5 through 6 in his assigned section ______.

*** Figure C-3. Ten-meter firing table (Firing Table I).**

a. **Objective**. In order for the M2 gunner to become proficient with the machine gun, the following objectives must be considered:

- Obtain an accurate initial burst.
- Master the technique of controlled burst firing.
- Traverse and search the machine gun effectively.
- Observe and adjust fire.
- Operate with speed.

b. **Organization**. The unit is organized into groups of 10. Each group is assigned a firing order number. One order becomes gunners and the other assistant gunners. The gunners and assistant gunners are assigned lanes and required to set up their guns and perform pre-fire checks.

(1) Ten-meter practice fire (Tasks 1 through 4).

(a) The gunner is required to set his rear sight at 500 yards. On command, the first order zeros using scoring pasters 1 and 2 of their assigned section.

(b) At the completion of zeroing, the gunners are required to engage the next two aiming pasters (scoring pasters 3 and 4 of the same section), using controlled bursts, and then area targets 5 through 6 and 7 through 8, using traverse and search techniques. Time is allowed between bursts to permit gunners to observe and adjust their fire.

(c) After the first order has fired, the second order fires the course in the same manner using the next assigned section. After both orders have fired, the guns are cleared.

(2) Ten-meter qualification (Tasks 5 through 7).

(a) The gunner is then required to set his rear sight at 550 yards. On command, the first order fires at scoring pasters 1 through 4 of their assigned section.

(b) At the completion of their linear engagement, commands are given to require the gunners to engage area targets 5 through 6 and then 7 through 8, using traverse and search techniques. Time is allowed between bursts to permit gunners to observe and adjust their fire.

(c) After the first order has fired qualification, the second order fires the course in the same manner using the next assigned section. After both orders have fired, the guns are cleared.

(2) *Ten-meter qualification (Tasks 5 through 7).*

(a) The gunner is then required to set his rear sight at 550 yards. On command, the first order fires at scoring pasters 1 through 4 of their assigned section.

(b) At the completion of their linear engagement, commands are given to require the gunners to engage area targets 5 through 6 and then 7 through 8, using traverse and search techniques. Time is allowed between bursts to permit gunners to observe and adjust their fire.

(c) After the first order has fired qualification, the second order fires the course in the same manner using the next assigned section. After both orders have fired, the guns are cleared.

c. **Target Analysis.** The exercises use the basic machine gun target (FSN 6920-078-5128) (Figure C-4). The following explanation of the target, including the size of the aiming pasters and scoring spaces, will aid in zeroing the M2 and will facilitate control during firing exercises.

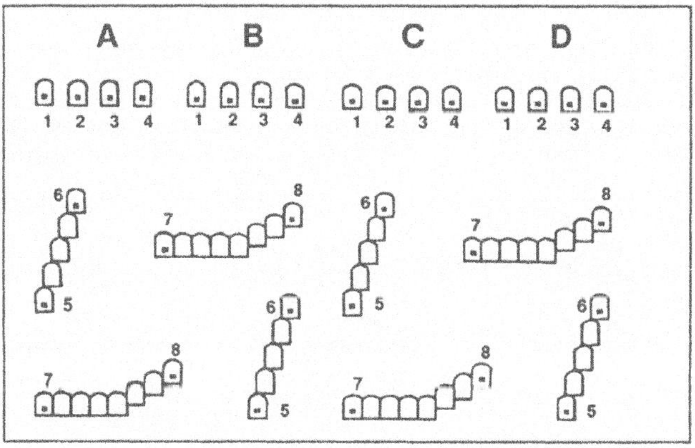

Figure C-4. Basic machine gun target.

(1) The target consists of four sections lettered A, B, C, and D. Each section has four point targets numbered 1, 2, 3, and 4; and two sets of area targets numbered 5 through 6 and 7 through 8. Each space is 4 centimeters wide and 5 centimeters high. The black aiming pasters within some numbered scoring spaces are 1-centimeter squares. Firing at targets 1 through 4 allows the gunner to use the traverse-fire technique

targets 5 through 6 and 7 through 8 give them practice in traverse and search techniques.

(a) Point targets. Point targets on the basic machine gun target are considered to be pasters 1 through 4 of section A, B, C, and D. Firing at point targets exposes the gunner to zeroing techniques and controlled-burst fire techniques. Targets 1 through 4 will also be used as a linear target for qualification.

(b) Area targets. Area targets on the basic machine gun target are considered to be pasters 5 through 6 and 7 through 8 of section A, B, C, and D. Target group 5 through 6 exposes the gunner to T&E manipulation when using the tripod mount and body position changes to engage targets in depth (elbow position changes). This causes him to use a series of aiming points to disburse fire across the target when using the tripod. Target group 7 through 8 exposes the gunner to position changes to engage linear targets with depth. It causes him to control the burst length from the weapon, use a series of aiming points, and disburse fire across the target.

(2) Targets are analyzed and scored to determine the gunner's proficiency and to see if more training is needed in any of the fundamentals of M2 gunnery. During firing with a properly zeroed weapon, a target is best analyzed by considering the common errors of M2 gunnery (Figure C-5).

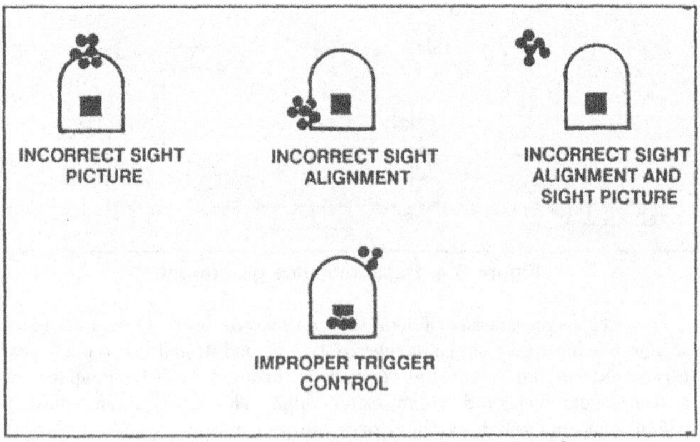

Figure C-5. Common errors found on the basic MG target.

NOTE: Large shot groups are usually caused by incorrect position and grip; small shot groups outside of the scoring space are usually caused by incorrect sight alignment, sight picture, or zero.

d. **The 10-Meter Zero.** Ten-meter zeroing is accomplished by adjusting the rear sight until the strike of the projectile coincides with the point of aim at a given range. On the 10-meter range, the elevation scale must be set at 500 yards so that it can impact on the desired aiming point. Before zeroing the weapon, the gunner must first center the rear sight for windage by aligning the index with the center of the windage scale.

e. **Controlled Burst Firing.** This firing exercise reinforces the dry-fire experience and allows the gunner practice in firing of controlled bursts and in providing the most accurate and tightest shot groups obtainable with the M2.

f. **Ammunition Issue.** The ammunition for the firing of this exercise will be broken down at the ammunition point and issued to each assistant gunner as he is assigned a firing point. The total ammunition for this exercise is 236 rounds (117 for practice and 119 for qualification). Issue will consist of the following rounds or belts:

- Task 1, 12 single rounds
- Task 2, two 7-round belts.
- Task 3, one 35-round belt.
- Task 4, one 56-round belt.
- Task 5, one 28-round belt.
- Task 6, one 56-round belt.
- Task 7, one 35-round belt.

g. **Firing Sequence.** Firing of Table 1 will be conducted in the following manner. These procedures pertain to firing with a single gunner using sections A and B; if there is a second gunner, he will use C and D.

(1) *Task 1, Zero.*

(a) The gunner will fire 3 rounds, single shot, at paster A1.

(b) The gunner will then move down range to observe the shot group and triangulate it. No adjustments to the weapon will be made at this time.

(c) The gunner will then fire another 3 single rounds at A1.

(d) The gunner then goes down range to observe the shot group, triangulate it, and make necessary adjustments to his weapon.

(e) The gunner repeats steps (c) and (d), but shoots at paster A2.

NOTE: If the gunner is able to zero his weapon using 9 rounds, have him use the remaining 3 to confirm his zero. If the gunner is unable to zero within 12 rounds, remove him from the firing line and give him remedial training.

(2) *Task 2, Controlled Burst Firing.*

(a) The gunner will fire the first burst of 5 to 7 rounds at paster 3.

(b) The gunner then goes down range to observe, marks the projectile holes, and analyzes his burst.

(c) The gunner repeats steps (a) and (b), but fires at paster 4 of same section.

NOTE: The primary purpose of this task is to get the gunner to fire 5- to 7-round bursts. Impact on the target is desired, but not necessary to complete this task. The gunner will fire a 5- to 7-round burst at each paster.

(3) *Task* 3, *Traverse and Search Fire.*

(a) The gunner will engage pasters 5 through 6, section A, firing a 5- to 7-round burst for each paster, using the traverse and search technique.

(b) The gunner will then move down range to observe and analyze his targets.

NOTE: These tasks are fired from the tripod firing position. If the gunner shows difficulty in manipulating the weapon, remove him from the firing line and give him remedial training.

(4) *Task 4, Traverse and Search Fire.*

(a) The gunner will engage pasters 7 through 8, section A, firing a 5- to 7-round burst at each paster, using the traverse and search technique.

(b) The gunner will then move down range to observe and analyze his targets.

(5) *Task 5, Traverse Fire.*

(a) The gunner will engage pasters 1 through 4, section B, firing a 5- to 7-round burst at each paster, using the traverse technique.

(b) The gunner will then move down range to observe and analyze his targets.

(6) *Task 6, Traverse and Search Fire.*

(a) The gunner will engage pasters 7 through 8, section B, firing a 5- to 7-round burst at each paster, using the traverse and search technique.

(b) The gunner will then move down range to observe and analyze his targets.

(7) *Task 7, Traverse and Search Fire.*

(a) The gunner will engage pasters 5 through 6, section B, firing a 5- to 7-round burst at each paster, using the traverse and search technique.

(b) The gunner will then move down range to observe and analyze his targets.

DANGER

Before any gunner goes down range, the lane NCO must ensure that the weapon has been properly cleared. Anyone observing any unsafe acts should immediately call a cease fire and notify range personnel.

h. **Scoring Procedures.** When scoring the 10-meter target, all scoring spaces are scored (1 through 4, 5 through 6, and 7 through 8). One point is given for each round impacting within each space. Rounds touching the boundary of a scoring space are considered hits, but they can be counted in only one scoring space. When firing at 1 through 4 with 28 rounds, the maximum score is 28 points. In group 5 through 6, five scoring spaces are engaged with 35 rounds with a possible score of 35 points. When firing groups 7 through 8, eight scoring spaces are engaged with 56 rounds with a possible score of 56 points.

(1) The total possible score of both groups is 119 points. A minimum of 84 points is required to pass the 10-meter course of fire.

(2) Soldiers failing to achieve minimum standards must be retrained and retested in a dry-fire mode until proficiency is demonstrated. They may not move on to transition firing. Soldiers should then re-fire the 10-meter portion with close supervision and coaching to ensure that the fundamentals are applied properly during live fire.

C-3. TRANSITION FIRING EXERCISES

Transition firing of the M2 machine gun will teach the gunner some techniques of fire that he may encounter in combat situations. The gunner will field zero his weapon and also incorporate the techniques of fire during limited visibility and NBC environments. The gunner needs to be aware that during certain situations his capabilities will be degraded. He is expected to compensate for these situations by applying all available techniques. Within this training, the gunner will be required to apply all the fundamentals of gunnery learned in preparatory gunnery training and 10-meter firing. Instructors should encourage gunners to perform immediate action if a stoppage occurs during fire. This procedure may be modified if local policies require the gunners to notify the range personnel first.

a. **Objectives.** The objectives of transition fire are —

- To engage targets at long ranges with the tripod-mounted machine gun.
- To understand the characteristics of fire.
- Ž To field zero the M2 machine gun.
- Ž To use range estimation to determine the distance to targets.
- Ž To apply the method of adjusted point of aim.

b. **Organization.** Organization of a unit for transition firing is conducted the same as for 10-meter firing. Field zeroing is the first firing task of transition day-fire phase. The transition day-fire portion of the course must be fired to become a qualified M2 gunner.

c. **Ammunition Issue.** The ammunition for the firing of this exercise will be broken down at the ammunition point and issued to each assistant gunner as he is assigned a firing point. The total ammunition for this exercise is 182 rounds, and issue will consist of two belts; one 28-round belt and one 154-round belt.

d. **Firing Sequence.** Firing of Table II (Figure C-6) will be conducted in the following manner. These procedures pertain to firing with a single gunner using a tripod-mounted M2. If there is a second gunner, he will use the same procedures.

(1) *Task 1, Field Zero.*

(a) The gunner must first center the rear sight in the same manner as sight setting for 10-meter firing. He selects the appropriate range mark with his elevation knob.

(b) The gunner will load one 28-round belt of ammunition and fire a burst of 5 to 7 rounds at the 550-meter, double E-type silhouette.

(c) If the impact of the rounds (beaten zone) is over the target, the gunner has probably fired with an incorrect sight picture (too high on the target). He needs to relay the gun on the original point of aim and then he relays back on the target and fires again. If the gunner again fails to zero with a proper sight picture, he should have the unit armorer inspect the weapon before continuing the zeroing procedure.

(d) The gunner will observe the beaten zone and make adjustments to the sights so the round will impact on the target.

(e) After adjustments are made, the gunner will then relay on the center base of the target and fire another 5- to 7-round burst.

(f) The gunner will repeat steps c, d, and e with the remaining rounds.

NOTE: If the gunner is unable to zero within 28 rounds, he is removed from the firing line and given remedial training.

(2) *Task 2 through Task 8, Single and Multiple Engagements.*

(a) The gunner will load one 154-round belt of ammunition and fire bursts of 5 to 7 rounds at double E-type silhouettes at ranges of 800, 400, 700, and 1,000 meters at vehicular targets. He will also engage multiple double E-type silhouettes targets at 400 to 700; 550 and 800; and 400, 550, and 1,000 meters respectively.

(b) The gunner will observe the beaten zone and make adjustments as necessary to hit the targets.

(c) After firing is completed, weapons will be cleared and the gunner critiqued. If there is a second gunner to fire, the entire procedure will be repeated.

NOTE: After both gunners have fired for practice, they will then be assigned new lanes and issued more ammunition for record firing. If possible, the gunner should keep the same weapon he used for practice fire. The only change in the above procedures will be in step (c); here, the gunner will be scored along with the critique. If the gunner is unable to meet standard, he is removed from the firing line, given remedial training, and re-fired.

DANGER

Before any gunner changes lanes, the lane NCO must ensure that the weapon has been properly cleared. Anyone observing any unsafe acts should immediately call a cease fire and notify range personnel.

TASK	CONDITIONS TARGET/SITUATION	AMMO	STANDARD
1. Zero the M2 machine gun.	Gunner is in stationary tripod firing position and engages a 550-meter, double E-type silhouette. Gunner will use 5-to 7-round bursts for this engagement.	28-round belt of 12.7-mm.	Gunner must impact one burst on the 550-meter target.
2. Engage a single, double E-type silhouette.	Gunner is in stationary tripod firing position and engages an 800-meter, double E-type silhouette. Gunner will use 5-to 7-round bursts for this engagement.	14 rounds of 12.7-mm.	Gunner must impact one burst on the 800-meter target within 20 seconds.

Figure C-6. Practice/Qualification Table II.

TASK	CONDITIONS TARGET/SITUATION	AMMO	STANDARD
3. Engage a single. double E-type silhouette.	Gunner is in stationary tripod firing position and engages a 400-meter, double E-type silhouette. Gunner will use 5-to 7-round bursts for this engagement.	14 rounds of 12.7-mm	Gunner must impact one burst on the 400-meter target within 20 seconds.
4. Engage a single. double E-type silhouette.	Gunner is in stationary tripod firing position and engages a 700-meter, double E-type silhouette. Gunner will use 5-to 7-round bursts for this engagement.	14 rounds of 12.7-mm	Gunner must impact one burst on the 700-meter target within 25 seconds.
5. Engage a single. double E-type silhouette.	Gunner is in stationary tripod firing position and engages a 1,000-meter, vehicle type silhouette. Gunner will use 5-to 7-round bursts for this engagement.	14 rounds of 12.7-mm	Gunner must impact one burst on the 1,000-meter target within 25 seconds.
6. Engage multiple double E-type silhouettes.	Gunner is in stationary tripod firing position and engages 400-and 700-meter, double E-type silhouettes. Gunner will use 5-to 7-round bursts for this engagement.	28 rounds of 12.7-mm.	Gunner must impact one burst on each target within 35 seconds.
7. Engage multiple double E-type silhouettes.	Gunner is in stationary tripod firing position and engages 550-and 800-meter, double E-type silhouettes. Gunner will use 5-to 7-round bursts for this engagement.	28 rounds of 12.7-mm.	Gunner must impact one burst on each target within 35 seconds.
8. Engage multiple double E-type silhouettes.	Gunner is in stationary tripod firing position and engages 400-, 550-and 1,000 meter, double E-type and vehicle silhouettes. Gunner will use 5-to 7-round bursts for this engagement.	42 rounds of 12.7-mm.	Gunner must impact one burst on each target within 45 seconds.

Figure C-6. Practice/Qualification Table II, continued.

e. **Scoring Procedures.** One point is given for each target hit with an additional 2-point bonus if the target is hit on the first burst. The total possible points for day fire is 33. A minimum 7 points (7 out of 11 exposures) is required to meet standard on the day-fire course.

C-4. NBC FIRING

Since NBC plays an important part in our preparation for war on the modern battlefield, it is important that each soldier is prepared to accomplish the mission even if the area is contaminated and he must wear protective gear.

a. **Objectives.** The objectives of this training are:

- To teach the gunners to take appropriate action when notified of a contaminated area.
- Ž To identify targets while in a contaminated area.
- Ž To engage targets while in appropriate NBC gear.

b. **Organization.** The unit is organized the same as in paragraph C-2b, except both the gunner and assistant gunner must wear protective masks. Protective clothing is optional.

c. **Ammunition Issue.** The ammunition for the firing of this exercise will be broken down at the ammunition point and issued to each assistant gunner as he is assigned a firing point. The total ammunition for this exercise is 182 rounds, and issue will consist of two belts; one 28-round belt and one 154-round belt.

d. **Firing Sequence.** Firing of Table III (Figure C-7, pages C-16 and C-17) will be conducted in the following manner. These procedures pertain to firing with a single gunner using a tripod-mounted M2. If there is a second gunner, he will use the same procedures.

(1) *Task 1, Field Zero.*

(a) The gunner must first center the rear sight in the same manner as setting the sights for 10-meter firing. He selects the appropriate range mark with his elevation knob.

(b) The gunner will load one 28-round belt of ammunition and fire a burst of 5 to 7 rounds at the 550-meter, double E-type silhouette.

(c) If the impact of the rounds (beaten zone) is over the target, the gunner has probably fired with an incorrect sight picture (too high on the target). He should re-fire concentrating on the proper point of aim and sight picture. If the impact is still off, have the unit armorer correct the calibration and the gunner repeat the zeroing procedure.

(d) The gunner will observe the beaten zone and make adjustments to the sights so the round will impact on the target.

(e) After adjustments are made, the gunner will then relay on the center base of the target and fire another 5- to 7-round burst at the same target.

(f) The gunner will repeat steps c, d, and e with the remaining rounds.

NOTE: If the gunner is unable to zero within 28 rounds, he is removed from the firing line and given remedial training.

(2) *Task 2 through Task 8, Single and Multiple Engagements.*

(a) The gunner will load one 154-round belt of ammunition and fire bursts of 5 to 7 rounds at double E-type silhouettes at ranges of 800,400,

700, and a 1,000-meter vehicular target. He will also engage multiple, double E-type silhouettes targets at 400 and 700; 550 and 800; and 400, 500, and 1,000 meters, respectively.

(b) The gunner will observe the beaten zone and make adjustments as necessary to hit the targets.

(c) After firing is completed, weapons will be cleared and the gunner critiqued. If there is a second gunner to fire, the entire procedure will be repeated.

NOTE: After both gunners have fired for practice, they will then be assigned new lanes and issued more ammunition for record firing. If possible, the gunner should keep the same weapon he used for practice fire. The only change in the above procedures will be in the last step; here, the gunner will be scored along with the critique. If the gunner is unable to meet standard, he is removed from the firing line, given remedial training, and re-fired.

DANGER

Before any gunner changes lanes, the lane NCO must ensure that the weapon has been properly cleared. Anyone observing any unsafe acts should immediately call a cease fire and notify range personnel.

e. **Scoring Procedures.** One point is given for each target hit with an additional 2-point bonus if the target is hit on the first burst. The total possible points for day fire is 33. A minimum of 7 points (7 out of 11 exposures) is required to meet standard on the NBC fire course.

TASK	CONDITIONS TARGET/SITUATION	AMMO	STANDARD
1. Zero the M2 machine gun.	Gunner is in stationary tripod firing position and engages a 550-meter, double E-type silhouette. Gunner will use 5-to 7-round bursts for this engagement.	28-round belt of 12.7-mm.	Gunner must impact one burst on the scaled, 550-meter target.
2. Engage a single, double E-type silhouette at 800 meters while wearing a protective mask.	Gunner is in stationary tripod firing position and engages a 800-meter, double E-type silhouette. Gunner will use 5-to 7-round bursts for this engagement.	14 rounds of 12.7-mm.	Gunner must impact one burst on the 800-meter target within 20 seconds.

Figure C-7. NBC Qualification Table III.

TASK	CONDITIONS TARGET/SITUATION	AMMO	STANDARD
3. Engage a single, double E-type silhouette at 400 meters while wearing a protective mask.	Gunner is in stationary tripod firing position and engages a 400-meter, double E-type silhouette. Gunner will use 5-to 7-round bursts for this engagement.	14 rounds of 12.7-mm.	Gunner must impact one burst on the 400-meter target within 20 seconds.
4. Engage a single, double E-type silhouette at 700 meters while wearing a protective mask.	Gunner is in stationary tripod firing position and engages a 700-meter, double E-type silhouette. Gunner will use 5-to 7-round bursts for this engagement.	14 rounds of 12.7-mm.	Gunner must impact one burst on the 700-meter target within 20 seconds.
5. Engage a single, double E-type silhouette at 1,000 meters while wearing a protective mask.	Gunner is in stationary tripod firing position and engages a 1,000-meter, vehicle type silhouette. Gunner will use 5-to 7-round bursts for this engagement.	14 rounds of 12.7-mm.	Gunner must impact one burst on the 1,000 meter target within 25 seconds.
6. Engage multiple, double E-type silhouettes at 400 and 700 meters while wearing a protective mask.	Gunner is in stationary tripod firing position and engages 400-and 700-meter, double E-type silhouettes. Gunner will use 5-to 7-round bursts for this engagement.	28 rounds of 12.7-mm.	Gunner must impact one burst on each target within 35 seconds.
7. Engage multiple, double E-type silhouettes at 550 and 800 meters while wearing the protective mask.	Gunner is in stationary tripod firing position and engages 550-and 800-meter, double E-type silhouettes. Gunner will use 5- to 7-round bursts for this engagement.	28 rounds of 12.7-mm.	Gunner must impact one burst on each target within 35 seconds.
8. Engage multiple, double E-type silhouettes at 400, 500, and 1,000 meters while wearing the protective mask.	Gunner is in stationary tripod firing position and engages scaled 400-,500-, and 1,000-meter, double E-type and vehicle type silhouettes. Gunner will use 5-to 7-round bursts for this engagement.	42 rounds of 12.7-mm.	Gunner must impact one burst on each target within 45 seconds.

Figure C-7. NBC Qualification Table III, continued.

C-5. NIGHT-FIRE EXERCISE

The night-fire exercise gives the soldiers the practical application of engaging targets using the AN/TVS-5 at night or during limited visibility.

a. **Objectives.** The objectives of this training are:

- To teach the gunners to zero the AN/TVS-5 to the M2.
- To engage targets at different ranges using the AN/TVS-5.
- To detect targets using the AN/TVS-5.

b. **Organization.** Organization of a unit for night firing is conducted the same as day fire. AN/TVS-5 zeroing is the first firing task of the night-fire phase. The night-fire portion of the course is also a requirement to become a qualified M2 gunner.

c. **Ammunition Issue.** The ammunition for the firing of this exercise will be broken down at the ammunition point and issued to each assistant gunner as he is assigned a firing point. The total ammunition for this exercise is 180 rounds, and issue will consist of the following rounds or belts:

- Task 1, 12 single rounds and two 7-round belts
- Task 2 through Task 8, one 154-round belt.

d. **Firing Sequence.** Firing of Table IV (Figure C-8) will be conducted in the following manner; these procedures pertain to a single gunner using a tripod-mounted M2 with an AN/TVS-5 If there is a second gunner, he will use the same procedures.

(1) *Task 1, Zero the AN/TVS-5.*

(a) The gunner must first mount the AN/TVS-5 to the weapon and place it into operation. Once the device is mounted, the gunner will fire a 7-round burst to seat the device, then fire another 7-round burst and tighten the device to make sure it is settled.

(b) The gunner must then center the reticle pattern in the field of view of the device. He then places the reticle aiming point on the 50-meter zero target aim point and fires three single rounds.

NOTE: Each click of the azimuth or elevation adjustment actuator moves the strike of the round 1/2 inch at 50 meters. One click of adjustment moves the reticle one square of the target at 50 meters.

(c) If the impact of the round (beaten zone) is over the target, the gunner has probably fired with an incorrect sight picture (too high on the target). He should re-fire concentrating on the proper point of aim and AN/TVS-5 sight picture. If the gunner again fails to zero with a proper device sight picture, he should inspect the sight calibration of the device.

(d) The gunner will observe the beaten zone and make adjustments to the sights so the strike of the round impacts on the target.

(e) After adjustments are made, the gunner will then fire another round at the same target until zero is obtained.

NOTE: If the gunner is able to zero his weapon using 12 rounds, he is then ready to continue the course. If not, he is removed from the firing line and given remedial training.

(2) *Task 2 through Task 8, Single and Multiple Engagements.*

(a) The gunner will load one 154-round belt of ammunition and fire bursts of 5 to 7 rounds at double E-type silhouettes at ranges of 800, 400, 700, and a 1,000-meter vehicular target. He will also engage multiple, double E-type silhouettes targets at 400 to 700; 550 and 800; and 400, 500, and 1,000 meters, respectively.

(b) The gunner will observe the beaten zone and make adjustments as necessary to hit the targets.

(c) After firing is completed, weapons will be cleared and the gunner critiqued. If there is a second gunner to fire, the entire procedure will be repeated.

NOTE: After both gunners have fired for practice, they will then be assigned new lanes and issued more ammunition for record firing. If possible, the gunner should keep the same weapon he used for practice fire. The only change in the above procedures will be in the last step; here, the gunner will be scored along with the critique. If the gunner is unable to meet standard, he is removed from the firing line, given remedial training, and re-fired at the commander's discretion.

DANGER

Before any gunner changes lanes, the lane NCO must ensure that the weapon has been properly cleared. Anyone observing any unsafe acts should immediately call a cease fire and notify range personnel.

e. **Scoring Procedures.** One point is given for each target hit with an additional 2-point bonus if the target is hit on the first burst. The total possible points for night fire is 33. A minimum 7 points (7 out of 11 exposures) is required to meet standard on the night-fire course.

NOTE: Firing Table IV is set up for gunners to engage targets out to 1,000 meters under moonlight. However, if visibility is limited by other conditions, then the commander may use his discretion to alter the ranges for better visibility.

TASK	CONDITIONS TARGET/SITUATION	AMMO	STANDARD
1. Zero the AN/TVS-5 to the M2 machine gun.	Gunner is in stationary tripod firing position at 50 meters and engages an M16A1/A2 zero target. Gunner will use single shots for this engagement. Two bursts will be used to spot the device.	12 single 12.7-mm rounds and two 7-round belts.	Gunner must impact 4 of 6 rounds of two 3-round shot groups within a 4-cm circle IAW FM 23-26.
2. Engage a single, double E-type silhouette at 800 meters while using the AN/TVS-5.	Gunner is in stationary tripod firing position and engages an 800-meter, double E-type silhouette. Gunner will use 5-to 7-round bursts for this engagement.	14 rounds of 12.7-mm.	Gunner must impact one burst on the 800-meter target within 20 seconds.
3. Engage a single, double E-type silhouette at 400 meters while using the AN/TVS-5.	Gunner is in stationary tripod firing position and engages a 400-meter, double E-type silhouette. Gunner will use 5-to 7-round bursts for this engagement.	14 rounds of 12.7-mm.	Gunner must impact one burst on the 400-meter target within 20 seconds.
4. Engage a single, double E-type silhouette at 700 meters while using the AN/TVS-5.	Gunner is in stationary tripod firing position and engages a 700-meter, double E-type silhouette. Gunner will use 5-to 7-round bursts for this engagement.	14 rounds of 12.7-mm.	Gunner must impact one burst on the 700-meter target within 25 seconds.
5. Engage a single, double E-type silhouette at 1,000 meters while using the AN/TVS-5.	Gunner is in stationary tripod firing position and engages a 1,000-meter, double E-type silhouette. Gunner will use 5- to 7-round bursts for this engagement.	14 rounds of 12.7-mm.	Gunner must impact one burst on the 1,000-meter target within 25 seconds.
6. Engage multiple, double E-type silhouettes at 400 and 700 meters while using the AN/TVS-5.	Gunner is in stationary tripod firing position and engages 400-and 700-meter, double E-type silhouettes. Gunner will use 5- to 7-round bursts for this engagement.	28 rounds of 12.7-mm.	Gunner must impact one burst on each target within 35 seconds.
7. Engage multiple, double E-type silhouettes at 550 and 800 meters while using the AN/TVS-5.	Gunner is in stationary tripod firing position and engages 550- and 800-meter, double E-type silhouettes. Gunner will use 5- to 7-round bursts for this engagement.	28 rounds of 12.7-mm.	Gunner must impact one burst on each target within 35 seconds.
8. Engage multiple, double E-type silhouettes at 400, 550, and1,000 meters while using the AN/TVS-5.	Gunner is in stationary tripod firing position and engages 400-, 550-, and 1,000-meter, double E-type silhouettes. Gunner will use 5- to 7-round bursts for this engagement.	42 rounds of 12.7-mm.	Gunner must impact one burst on each target within 45 seconds.

Figure C-8. Night Fire Qualification Table IV.

* C-6. QUALIFICATION STANDARDS

To qualify on the M2, the gunner must achieve the minimum standards on the firing tables.

a. **Firing Table I**. On this table, a gunner can earn one point for each round that hits in each space. He can earn up to 119 points on this table, but must earn at least 12 points (12 out of 17 pasters).

b. **Firing Tables II, III, and IV**. On each of these tables, the gunner can earn one point for each target hit plus a two-point bonus if he hits each target with the first burst. He can score up to 33 points on each of these tables, but to qualify, he must score at least 23 points on each.

c. **All Tables**. The gunner firing tables I through IV can earn up to 218 points. However, he must earn at least 12 points on Table I plus at least 23 points on each of the other three firing tables. Thus, he needs at least 12 + (3 x 23) = 12 + 69 = 81 points to qualify:

FIRING TABLE	POINTS NEEDED TO QUALIFY
Firing Table I	12
Firing Table II	23
Firing Table III	23
Firing Table IV	+ 23
	81 TOTAL POINTS NEEDED TO QUALIFY ON M2

d. **Ratings**. The grader rates each gunner as follows:

MAXIMUM	218
EXPERT	196 TO 217
FIRST CLASS	174 TO 195
SECOND CLASS	153 TO 173
UNQUALIFIED	152 OR LESS

e. **Scorecard**. This change prescribes a new edition of DA Form 7007-R. This blank, reproducible form may be copied from the back of this manual onto 8 1/2 by 11-inch paper. It is also available on the Army Electronic Library (AEL) CD-ROM (EM0001) and at the USAPA website located at (http://www.usapa.army.mil).

MACHINE GUN SCORECARD FOR M2

For use of this form, see FM 23-65, Change 1, or the back of this form. The proponent agency is TRADOC.

DATA REQUIRED BY THE PRIVACY ACT OF 1974

AUTHORITY: 10 USC 3012(g)/Executive order 9397.
PRINCIPAL PURPOSE: To aid individual training on targets at various ranges.
ROUTINE USE: To evaluate individual proficiency. SSN is used for positive identification purposes only.
DISCLOSURE: Voluntary. However, mass rating and scoring require some tracking method. Therefore, only those soldiers who provide their SSNs can receive scores or qualify on weapons or equipment.

1a. LAST NAME	1b. FIRST NAME	1c. MI	2. DATE *(YYYYMMDD)*
LEMAY	CURTIS	X	20021211

3. SSN	4. UNIT	5. LANE
222-22-2222	2/29TH INF	4

EXAMPLE

	6a. FIRING TABLE I			7a. FIRING TABLE II					8a. FIRING TABLE III					9a. FIRING TABLE IV				
TSK	RANGE (M)	HIT	PTS	RANGE (M)	TIME	HIT	PTS	BON	RANGE (M)	TIME	HIT	PTS	BON	RANGE (M)	TIME	HIT	PTS	BON
1	10	NA	NA	550	None	NA	NA	NA	550	None	NA	NA	NA	550	None	NA	NA	NA
2	10	NA	NA	800	20 Sec	X	1	2	800	20 Sec	X	1	2	800	20 Sec	X	1	2
3	10	NA	NA	400	20 Sec	X	1	2	400	20 Sec	X	1	0	400	20 Sec	X	1	2
4	10	NA	NA	700	25 Sec	X	1	2	700	25 Sec	X	1	2	700	25 Sec	X	1	2
5	10	X	20	1,000	25 Sec	X	1	2	1,000	25 Sec	X	1	2	1,000	25 Sec	X	1	2
6	10	X	30	400	35 Sec	X	1	2	400	35 Sec	X	1	2	400	35 Sec	X	1	2
				700		X	1	2	700		X	1	2	700		X	1	2
7	10	X	45	550	35 Sec	X	1	2	550	35 Sec	X	1	0	550	35 Sec	X	1	2
				800		X	1	2	800		X	1	2	800		X	1	2
8	NA	NA	NA	400	45 Sec	X	1	2	400	45 Sec	X	1	2	400	45 Sec	X	1	2
				550		X	1	0	550		X	1	2	550		X	1	2
				1,000		X	1	2	1,000		X	1	2	1,000		X	1	2

6b. SUBTOTAL FIRING TABLE I	7b. SUBTOTAL FIRING TABLE II	8b. SUBTOTAL FIRING TABLE III	9b. SUBTOTAL FIRING TABLE IV
95	31	29	33

10. TOTAL SCORE	11. RATING
188	GUNNER FIRST CLASS
12a. OIC's PRINTED OR TYPED NAME	**13a. GRADER'S PRINTED OR TYPED NAME**
SSG DWIGHT X. EISENHOWER	SSG EDWIN X. ROMMEL
12b. OIC'S SIGNATURE	**13b. GRADER'S SIGNATURE**
Dwight X. Eisenhower	Edwin X. Rommel

DA FORM 7007-R, JAN 2002 DA FORM 7007-R, JUN 1991, IS OBSOLETE. USAPA V1.00

Figure C-9. Example completed DA Form 7007-R (front).

The following procedure will be used to fill out the scorecard.

1. NAME: Enter last name , first name, and middle initial.
2. SSN: Enter gunner's social security number.
3. UNIT: Enter unit designation.
4. DATE: Enter date of firing.
5. LANE: Enter the lane number for the firing point of your gunner.
6. HIT: When firing Table I, enter the number of rounds impacting within the scoring space in task numbers 5, 6, and 7. For Tables II, III, and IV, enter an X when target is hit.
7. POINTS: Enter 1 point for each round impacting within the scoring space in task numbers 5, 6, and 7 in Table I. Gunners will receive 1 point for each target hit in Tables II, III, and IV.
8. BONUS: Entry is used only for Tables II, III, and IV. Two bonus points are awarded when gunner hits the target on initial burst of each target exposure.
9. TOTAL SCORE: Enter the total of combined hits, points, and bonus scores.

NOTE: It is possible that a gunner may score the required points outlined for qualification. However the gunner must also score 70 percent on each firing table before he is considered a qualified gunner.

MAXIMUM	218
EXPERT	196-218
GUNNER FIRST CLASS	174-195
GUNNER SECOND CLASS	153-173
UNQUALIFIED	152 and below

Back, DA Form 7007-R, Jun 91

Figure C-10. Example DA Form 7007-R, back.

Section II. MULTIPURPOSE RANGE COMPLEX

The multipurpose range complex is designed to conduct armor and mobilized machine gun firing. Detailed setup and target configuration is described in FM 25-7; its layout is shown in Figure C-11. This section also discusses the mounted firing exercise, requirements for crews, rating procedures, and standards.

C-7. DESIGN SPECIFICS

The tower and administrative facilities must be positioned so they do not impede tactical maneuver onto or off of the range.

a. Hardened emplacements are for aerial gunnery exercises in addition to tank and BFV gunnery.

b. System calibration targets must be provided at ranges of 950, 1,200, and 1,500 meters, as shown in Figure C-11.

c. Gunnery tasks that require the use of dud-producing ammunition cannot be fired on the range proper. Provisions for these tasks must be made in impact areas adjacent to the range.

d. Double target mechanisms are recommended for use in the first six target groupings for BFV training, or for positioning throughout the range as needed to support local training requirements.

e. The administrative area is not shown in exact location or scale.

f. Additional defilade positions maybe required for BFV gunnery.

g. The addition of battle positions and positioning of personnel targets closer to the baseline may be necessary to permit BFV dismounted infantry training.

Carrier 2 will upload and wait at the ammunition point. When the course is clear, carrier 2 will move to the initial fire point followed by carrier 4 and proceed as stated. Carrier 3 should now be at the ammunition point for uploading.

(1) *Task 1, Conduct Prefire Inspection.*

(a) The crew will move forward to the initial firing position. Once the carrier is in position, the crew will perform a prefire inspection by setting the headspace and timing, and load a 7-round belt of ammunition.

(b) The gunner will then fire the 7-round belt at a 450-meter target to ensure the machine gun is operational.

NOTE: If the machine gun fails to fire or fires sluggishly, recheck the headspace and timing and attempt to fire again. If the situation continues, clear the weapon and have the carrier move off the firing course and report to the armorer to have the problem corrected.

(c) Upon completion of Task 1, the gunner will clear the weapon and await further instructions.

NOTE: To accomplish Task 2 and Task 3, the gunner will load one 28-round belt of ammunition.

(2) *Task 2, Engage Stationary Target From a Mounted Stationary Firing Position.*

(a) The gunner will engage a 450-meter stationary personnel target from the initial firing location.

(b) The gunner will use a 5- to 7-round burst for this engagement.

(c) The gunner will observe the beaten zone and make adjustments as necessary to hit the target.

(d) Upon completion of Task 2, the gunner will await further instructions.

(3) *Task 3, Engage Stationary Target From a Stationary Firing Position.*

(a) From the initial firing location, the gunner will engage an 850-meter stationary vehicle target.

(b) The gunner will use a 5- to 7-round burst for this engagement.

(c) The gunner will observe the beaten zone and make adjustments as necessary to hit the target.

(d) Upon completion of Task 3, the gunner will await further instructions.

NOTE: On command from the grader, the crew will move the carrier to the next predetermined firing location, load one 28-round belt, and conduct Task 4.

(4) *Task 4, Engage Moving and Stationary Targets From a Mounted Stationary Firing Position.*

(a) From the second predetermined firing location, the gunner will engage one moving vehicle target at a range of 800 meters and one stationary vehicle target at a range of 1,000 meters.

(b) The gunner will use a 5-to 7-round burst for this engagement.

(c) The gunner will observe the beaten zone and make adjustments as necessary to hit the targets.

(d) Upon completion of firing Task 4, the gunner will await further instructions.

NOTE: When conducting Task 5, the gunner will load one 28-round belt and engage targets from a moving carrier along a predetermined route.

(5) *Task 5, Engage Stationary Targets While Firing From a Moving Carrier.*

(a) The carrier will be moving along the predetermined firing route, at which time the gunner will engage stationary vehicle targets at ranges of 300 and 500 meters.

(b) The gunner will use a 5- to 7-round burst for this engagement.

(c) The gunner will observe the beaten zone and make adjustments as necessary to hit the targets.

(d) Upon completion of Task 5, the gunner will await further instructions.

NOTE: On command from the grader, the crew will move the carrier to the third predetermined firing location, load one 28-round belt, and conduct Task 6.

(6) *Task 6, Engage a Stationary and a Moving Target From a Mounted Stationary Firing Position.*

(a) From the third predetermined firing location, the gunner will engage one 600-meter personnel target and one 800-meter moving vehicle target from a stationary firing position.

(b) The gunner will use a 5- to 7-round burst for this engagement.

(c) The gunner will observe the beaten zone and make adjustments as necessary to hit the targets.

(d) Upon completion of Task 6, the gunner will await further instructions.

NOTE: From the same firing location as Task 6, the crew will be placed in a chemical environment and be required to go to MOPP4.

(7) *Task 7, React in Chemical Environment.*

(a) Upon command, the crew will be required to react to a chemical agent.

(b) Upon completion of Task 7, the gunner will await further instructions.

NOTE: To accomplish Task 8 through Task 10, the gunner will load one 84-round belt of ammunition.

(8) *Task 8, Engage Stationary Target From a Mounted Stationary Firing Position While in MOPP4.*

(a) From the same firing location as Task 6, the gunner will engage a stationary target at a range of 500 meters from a mounted stationary firing position while in MOPP4.

(b) The gunner will use a 5- to 7-round burst for this engagement.

(c) The gunner will observe the beaten zone and make adjustments as necessary to hit the target.

(d) Upon completion of Task 8, the gunner will await further instructions.

(9) *Task 9, Engage Moving and Stationary Targets From a Mounted Stationary Firing Position While in MOPP4.*

(a) From the same firing location, the gunner will engage one 800-meter moving vehicle target and personnel targets at 1,000 meters.

(b) The gunner will use a 5- to 7-round burst for this engagement.

(c) The gunner will observe the beaten zone and make adjustments as necessary to hit the targets.

(d) Upon completion of Task 9, the gunner will await further instructions.

(10) *Task 10, Engage Multiple Stationary Targets in a Mounted Stationary Firing Position While in MOPP4.*

(a) From the same firing location, the gunner will engage one 300-meter personnel target and two stationary vehicle targets at ranges of 500 and 700 meters.

(b) The gunner will use a 5- to 7-round burst for this engagement.

(c) The gunner will observe the beaten zone and make necessary adjustments to hit the target.

(d) Upon completion of firing Task 10, the gunner will clear his machine gun, have it inspected by the safety officer, move from the firing line, and turn in any excess ammunition. Then the crew will be critiqued.

TASK	CONDITIONS TARGET/SITUATION	AMMO	STANDARD
1. Conduct prefire inspection.	At a stationary position given a 450-meter stationary target, headspace and timing gauge, unit's organic vehicle, and M2 machine gun.	7-round belt of 12.7-mm.	Conduct prefire inspection and fire 7 rounds to confirm operation of machine gun.
2. Engage stationary target from a stationary firing position.	Gunner is in a stationary mounted firing position and engages a 450-meter stationary personnel target. Gunner must use 5- to 7-round bursts for this engagement.	14 rounds of 12.7-mm.	Gunner must impact on target.
3. Engage stationary target from a stationary firing position.	Gunner is in a stationary mounted firing position and engages a 850-meter stationary vehicle. Gunner must use 5- to 7-round bursts for this engagement.	14 rounds of 12.7-mm.	Gunner must impact on target.
4. Engage moving and stationary targets from another stationary firing position.	Gunner is in a stationary mounted firing position and engages a moving vehicle at 800 meters and stationary vehicle at 1,000 meters. Gunner must use 5- to 7-round bursts for this engagement.	28-round belt of 12.7-mm.	Gunner must impact on each target.
5.Engage stationary targets while firing from a moving vehicle.	While the gunner's vehicle is moving, the gunner engages stationary vehicles at 300 and 500 meters. gunner must use 5- to 7-round bursts for this engagement.	28-round belt of 12.7-mm.	Gunner must impact on each target.
6. Engages a stationary and moving target from a stationary firing position.	Gunner is in a stationary firing position and engages a 600-meter personnel and 800-meter moving vehicle targets. gunner will use 5- to 7-round bursts for this engagement.	28-round belt of 12.7-mm.	Gunner must impact on each target.
7. React to chemical agent.	While in a stationary position, given protective mask and protective clothing, crew must react to chemical agent.	N/A	In accordance with FM 3-3.
8. Engage stationary target from a stationary firing position while in an upgraded NBC posture.	Gunner is in a stationary firing position and engages a 500-meter vehicle target. gunner will use 5- to 7-round bursts for this engagement.	14 rounds of 12.7-mm.	Gunner must impact on target.

Figure C-12. Mounted Firing Exercise Table VI.

TASK	CONDITIONS TARGET/SITUATION	AMMO	STANDARD
9. Engage moving and stationary targets while in a stationary firing position in an upgraded NBC posture.	Gunner will engage 800-meter moving vehicle target and personnel targets at 1,000 meters. Gunner will use 5- to 7-round bursts for this engagement.	28 rounds of 12.7-mm.	Gunner must impact on targets.
10. Engage multiple stationary targets while in a stationary firing position in an upgraded NBC posture.	Gunner will engage 300-meter personnel targets, 500-and 700-meter vehicle targets. Gunner will use 5- to 7-round bursts for this engagement.	42 rounds of 12.7-mm.	Gunner must impact on each target.

Figure C-12. Mounted Firing Exercise Table VI, continued.

e. **Scoring Procedures.** The mounted firing exercise is graded on a Go/No Go basis. Commanders may increase the difficulty of any or all tasks to align the exercise with the unit's mission.

C-9. PREDETERMINED FIRING EXERCISE

The predetermined firing exercise is for the gun crew that has demonstrated proficiency during the basic phase of gunnery. Emphasis will be on developing range cards and confirming range-card data during day and night firing.

a. **Objectives.** The objective of this training is to reinforce what was developed in the fundamental gunnery phases. It is designed to increase the effectiveness of the M2 MG crew by building their confidence to quickly and accurately deliver a large volume of fire on a prescribed target.

b. **Organization.** The unit is assembled in the bleachers, given instructions, and briefed on training that will be conducted while they are on the range. After briefing, they will be organized into gun crews and moved to firing lanes. Lanes will be used IAW local range policies.

c. **Ammunition.** This exercise requires 168 rounds of 12.7-mm linked ammunition. The gunner is allotted two bursts per target during the day phase and one burst per target during the night phase. Each gunner will be issued two belts of ammunition; one 112-round belt for the day phase and one 56-round belt for the night phase.

d **Firing Sequence.** The sequence of firing will be conducted IAW Firing Table VII (Figure C-13, page 33). The suggested sequence of firing is as follows:

(1) *Day phase.*

(a) Task 1, Prepare a Range Card. Once the gunner is assigned his firing point, he must prepare a range card for that position. Range cards must be prepared IAW paragraph E-2.

(b) Task 2, Obtain Direction and Elevation Readings for the Final Protective Line. During this task, the gunner is allotted 28 rounds to obtain and record the direction and elevation reading of his FPL.

(c) Task 3, Obtain Direction and Elevation Readings for Point Targets. The gunner will be required to obtain and record direction and elevation readings for point targets located at 400, 600, 800, and 1,000 meters. He is allotted 56 rounds for this task.

(d) Task 4, Obtain Direction and Elevation Readings for Linear Targets. The gunner will be required to obtain and record direction and elevation readings for linear targets located at 600 and 800 meters. He is allotted 28 rounds for this task.

(2) *Night phase.*

NOTE: Each crew is scored as a group in the night phase. (The day phase is not scored.) The crew applies the data obtained during the day and engages their targets. Each crew receives 10 points for each target engagement. A minimum of 40 out of a possible 70 is required. This exercise is not a requirement for qualification; however, commanders can use this training to test their gun crews' proficiency.

(a) Task 5, Engage Point Targets Using Range-Card Data. The gunner must engage point targets located at 400, 600, 800, and 1,000 meters using range-card data. He is allotted 28 rounds for this task. The grader will announce the sequence of engagements.

(b) Task 6, Engage Area Targets Using Range-Card Data. The gunner must engage linear targets located at 600 and 800 meters using range-card data. He is allotted 14 rounds for this task. The grader will announce the sequence of engagements.

(c) Task 7, Fire Final Protective Line. The gunner will fire his FPL to obtain grazing fire. He is allotted 14 rounds for this task.

e. **Conduct of Firing.** The gunner, assistant, and leader will prepare a range card for that position. Once complete, each crew will be given 168 rounds as prescribed in a predetermined firing table.

(1) Each gunner initially lays on his target by using the dry-fire technique. Each crew is issued 168 rounds of ammunition, and the gunner is ordered to load by the group NCO. When all the gunners of a group are ready, the NCO announces, "Up," to the officer in charge.

(2) When all groups are ready to fire, the OIC announces, "Give me an 'Up' when you have engaged all targets and have obtained the correct data to all targets. You are clear to fire."

(3) When the gunner has correctly engaged his target (FPL), the leader records the information from the T&E mechanism and traversing bar onto a range card. Then, the assistant gunner becomes the gunner, the gunner becomes the leader, and the leader becomes the assistant gunner.

(4) As each member of the crew becomes the gunner, he fires at a different preselected target (order number 2, a linear target; order number 3, a point target).

(5) After each crew member has been a gunner, assistant gunner, and leader and all data on the three targets have been obtained, the group NCO clears and checks the machine guns of his group and announces to the OIC, "Group cleared and checked."

TASK	CONDITIONS TARGET/SITUATION	AMMO	STANDARD
1. Prepare a range card for the M2 machine gun.	The gunner is given a range card, a selected firing position, M2 machine gun, M3 tripod, and silhouette targets located from 400 to 1,000 meters.	N/A	Gunner must prepare a range card IAW Appendix E, FM 23-65.
2. Determine and record direction and elevation for final protective line.	The gunner is allotted 12.7-mm linked ammo, and a sector of fire with limits.	28 rounds of 12.7-mm.	Gunner must record the direction and elevation for his range card.
3. Determine and record direction and elevation for point targets.	The gunner is allotted 12.7-mm linked ammo, four E-type silhouettes located at 400, 600, 800, and 1,000 meters.	56 rounds of 12.7-mm.	Gunner must impact one burst on each target.
4. Determine and record direction and elevation for area targets.	The gunner is allotted 12.7-mm linked ammo, two E-type silhouettes located at 600 and 800 meters.	28 rounds of 12.7-mm.	Gunner must impact one burst on each target.
5. Engage point targets using range card data.	The gunner is allotted 12.7-mm linked ammo, the same firing position, and his prepared range card.	28 rounds of 12.7-mm.	Gunner must impact one burst on each target.

Figure C-13. Predetermined Firing Table VII.

TASK	CONDITIONS TARGET/SITUATION	AMMO	STANDARD
6. Engage area targets using range card data.	The gunner is allotted 12.7-mm linked ammo, and his prepared range card.	14 rounds of 12.7-mm.	Gunner must impact one burst on each target.
7. Fire final protective line.	The gunner is allotted 12.7-mm linked ammo, and his prepared range card.	14 rounds of 12.7-mm.	Gunner must obtain grazing fire when firing his FPL.

Figure C-13. Predetermined Firing Table VII, continued.

APPENDIX D

FIGHTING POSITIONS

Fighting positions are stable platforms from which a gunner can engage the enemy. Fighting positions are dug as often as possible, but for the most part only when a unit is in the defense and has the time. This appendix discusses the construction of only two fighting positions: the one-, two-, or three-man, tripod-mounted position; and the M63 antiaircraft emplacement.

D-1. CONSTRUCTION CONSIDERATIONS

When constructing a fighting position, the leader and gunner should consider factors that affect the position, such as cover, concealment, fields of fire, size, and improvements.

a. The cover in a machine gun fighting position protects the gun crew from frontal small-arms fire and from fragments of high explosive shells impacting within 3 to 5 meters of the position. The frontal parapet must be at least one M16 rifle length thick and high enough to hide the helmets of the soldiers in the position. Overhead cover built out of sandbags stacked on logs will protect from shell fragments. The logs should be at least 6 inches thick, with at least two layers of sandbags over them (Figure D-1).

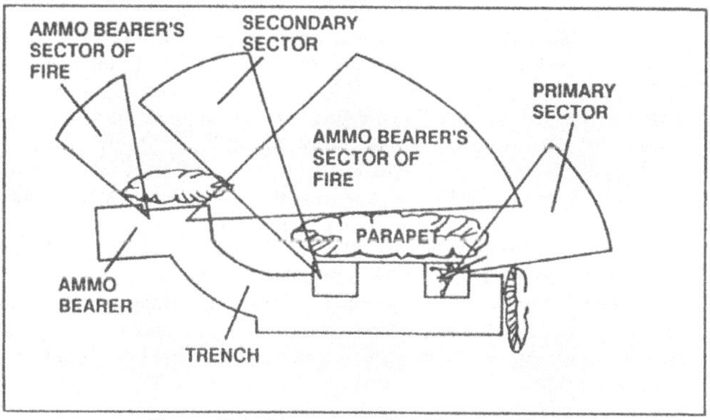

Figure D-1. Three-man fighting position.

b. Concealment hides the machine gun fighting position from direct observation by the enemy, who will make every attempt to locate and destroy heavy machine gun positions early in the battle. The position must be made to blend with its surroundings with both natural and man-made camouflage. Excess dirt from the initial digging should be moved away from the position, and the sod used to re-cover the overhead protection. From the enemy side, the position should not be visible at 35 meters or less (hand grenade range). It should never be obvious even at close range. Use sod or dampen the ground in front of the muzzle to reduce the dust cloud caused by firing. Consider what the position looks like from the air and camouflage it to blend with the surrounding terrain.

c. Selective clearing that does not destroy natural camouflage may be necessary to provide good observation and fields of fire in both primary and secondary sectors of fire.

d. The position should be armpit deep, be wide enough to allow two soldiers with load-bearing equipment to move freely, and have two distinct firing platforms.

e. Improvements to the position may include a grenade sump, a sloping floor with shallow trench for drainage, and a rear parapet for protection against shell fragments and small-arms fire from the rear (other friendly positions or supporting fire from armored personnel carriers).

D-2. CONSTRUCTION OF A TRIPOD-MOUNTED POSITION

After being assigned a sector of fire with an FPL or a PDF, the gun crew begins constructing the fighting position. The tripod is placed in position first and marked so that the weapon will be pointed in the general direction of the target area. A preliminary sketched range card is drawn to show the limits of the sector. The gun crew then outlines the shape of the platform and hole to include the area for the frontal cover in the ground (Figure D-2). The crew then starts digging out the platforms. When they get about 4 to 6 inches deep, the MG is put in place to cover the primary sector of fire until construction is complete.

a. When assigned an FPL, emplace the gun by locking the traversing slide to the extreme left or right of the traversing bar, depending on which side of the primary sector the FPL is on. Align the barrel on the FPL by shifting the tripod.

NOTE: No direction entry is needed in the data section of the range card for the FPL.

b. When assigned a PDF, emplace the gun by locking the traversing slide at the center of the traversing bar. Shift the tripod and gun until the barrel is aimed at the center of the sector. Check coverage of the sector limits by traversing the gun fully left and right.

Figure D-2. Planning the fighting position.

NOTE: In the data section of the range card, record the direction and elevation data of the PDF and the sector limits from the T&E mechanism.

c. The crew digging the hole uses the dug-up dirt to build up the cover – first for frontal cover and then for flank and rear cover. They dig the hole deep enough to protect the crew and still let the gunner shoot with comfort (usually about armpit deep) (Figure D-3). They fix the tripod legs in place by digging, sandbagging, or staking them down. This will ensure that the gun does not shift during firing, which would render the range card data useless.

Figure D-3. Digging the fighting position.

(1) The crew digs three trench-shaped grenade sumps at various points where the crew can kick grenades into them (Figure D-4).

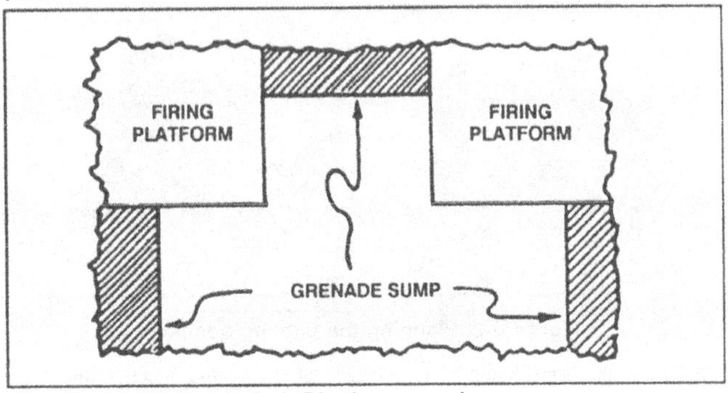

Figure D-4. Digging grenade sumps.

(2) When a position does not have a secondary sector of fire, the crew digs only half of the position (Figure D-5).

Figure D-5. Half of a position.

(3) When a position has both a primary and secondary sector, the crew prepares two firing platforms. The crew prepares overhead cover for a machine gun position like that of a two-man small arms fighting position. Time and material permitting, overhead cover should extend to cover the firing platforms (Figure D-6). Failure to properly construct overhead cover can result in reduced fields of fire, inability to mount NVDs, or problems in reloading. Proper construction of overhead cover is critical to survival.

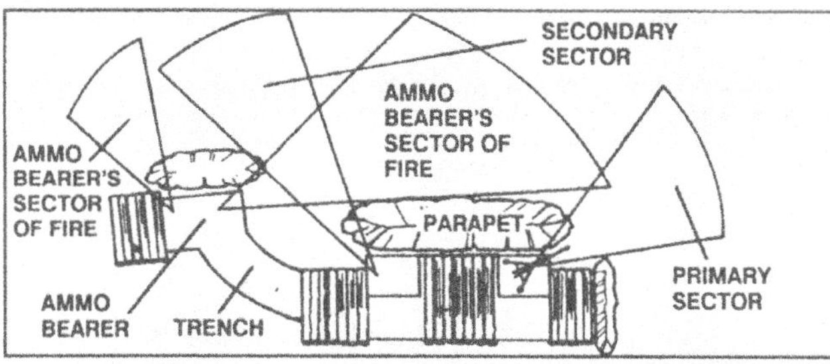

Figure D-6. Two firing platforms with overhead cover.

(4) When there is a three-man crew for the machine gun, the ammunition bearer digs a one-man fighting position to the flank of the gun position so that he can see and shoot to the oblique. This will allow him to cover the front of the machine gun's position (Figure D-7).

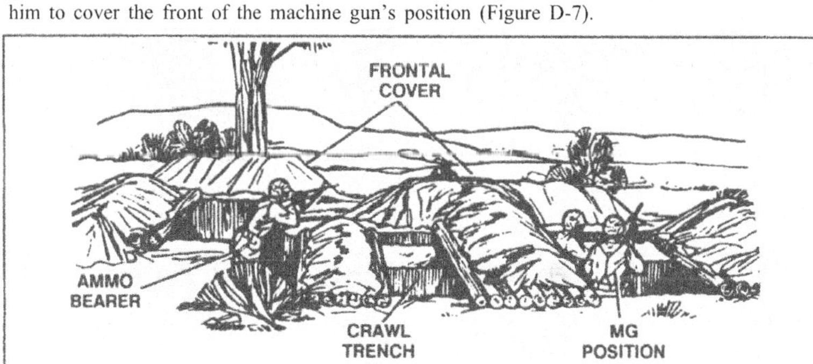

Figure D-7. Ammo bearer covering the front.

D-3. CONSTRUCTION OF AN ANTIAIRCRAFT EMPLACEMENT

Place the machine gun on the M63 mount alongside the designated location of the gun position. If needed, it can fire at any time during the construction process. Construct a circular position, with sufficient room to traverse the gun 360 degrees (normally, three and a half M16 lengths in diameter and one M16 length deep). Use the dirt dug out of the circular hole to construct a wall of sandbags around the position. Using sandbags, build a stable platform for the mount. Stake down the legs of the mount and place sandbags over them. This will prevent the mount from tipping backward when low-level targets are engaged. Include the M63-mounted machine gun's fire in the plan for defending the unit against ground attack (Figure D-8).

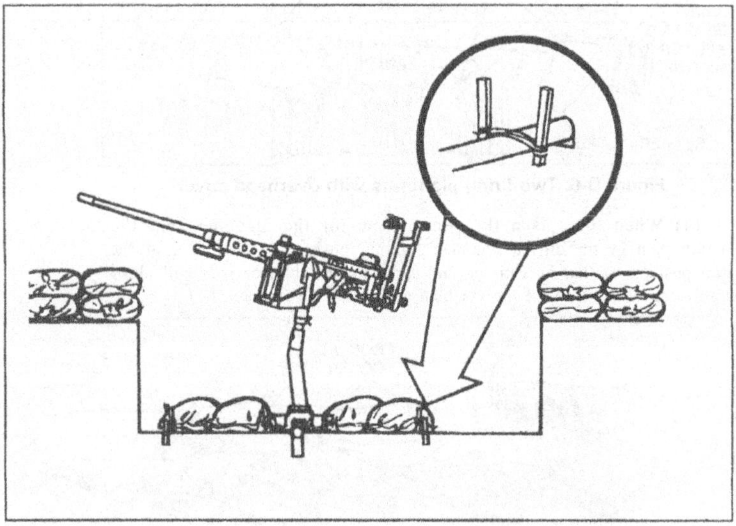

Figure D-8. Open pit-type antiaircraft position.

APPENDIX E

RANGE CARDS

A range card is a sketch or diagram of the terrain that a weapon is assigned to cover by fire. It shows possible target areas and terrain features plotted in relation to a firing position. The information on a range card is used for planning and controlling fire, for rapidly detecting and engaging targets, and for orienting replacement personnel or units. Therefore DA Form 5517-R, should be used to record the information.

E-1. RANGE CARD SECTIONS

Each machine gunner normally prepares range cards for his fighting position. He prepares one for each primary, alternate, and supplementary position designated in the defense and for any static position when enemy contact is possible; for example, a position in an assembly area. Each range card contains, as a minimum, the following information:

a. The symbol for the weapon covering the sector.

b. The azimuth (degrees) and distances (meters) of the firing position from an easily recognizable terrain feature. (This serves as an easy reference to locate the firing position.) If there is no easily recognizable terrain feature, an eight-digit grid maybe used.

c. The boundaries of the area assigned to be covered by observation and fire.

d. Areas where targets are likely to appear (engagement areas) and the range, azimuth, and elevation to them from the firing position.

e. Dead space (areas that cannot be observed or covered by fire).

f. The direction of magnetic north when the range card is properly oriented.

g. Identification data.

- Unit designation (no higher than company).
- Time and date of preparation.
- Firing position (primary, alternate, or supplementary).

E-2. PREPARATION

The following steps are taken to prepare a range card. (See Figure E-1, page E-3, for an example of a completed range card).

a. Draw the symbol for the MG in the lower center of the range card.

b. Show the sector of fire by drawing solid lines from the weapon symbol to the left and right limits. Sketch any easily recognizable terrain features that can be used to identify the sector. In the data section, indicate the magnetic azimuth and the range to the far limits of the sectors of fire.

NOTE: The left and right limits are labeled one and two, respectively, in the sector sketch and a circle drawn around each number. The left and right limits should be the first items drawn and labeled in the sector sketch and the data section of the range card. (The azimuth and range are not placed on the solid lines for left or right limit.)

c. Place target reference points at the location designated by the platoon leader or unit SOP, and at any other locations where a target is likely to appear. Number each TRP and likely target locations in the sketch section of the range card.

d. Draw a maximum engagement line across the sector of fire for the MG. These lines depict the maximum ranges at which a target can be effectively engaged.

e. Show dead space or areas where targets cannot be engaged with direct fire by drawing diagonal lines across the areas and writing the words DEAD SPACE.

f. Draw a magnetic north arrow on the range card to orient it with the terrain. Then add identification data – unit designation (no higher than company level), time and date of preparation, and type of position (primary, alternate, or supplementary).

g. Enter the information for the weapon reference point in the remarks block on the range card.

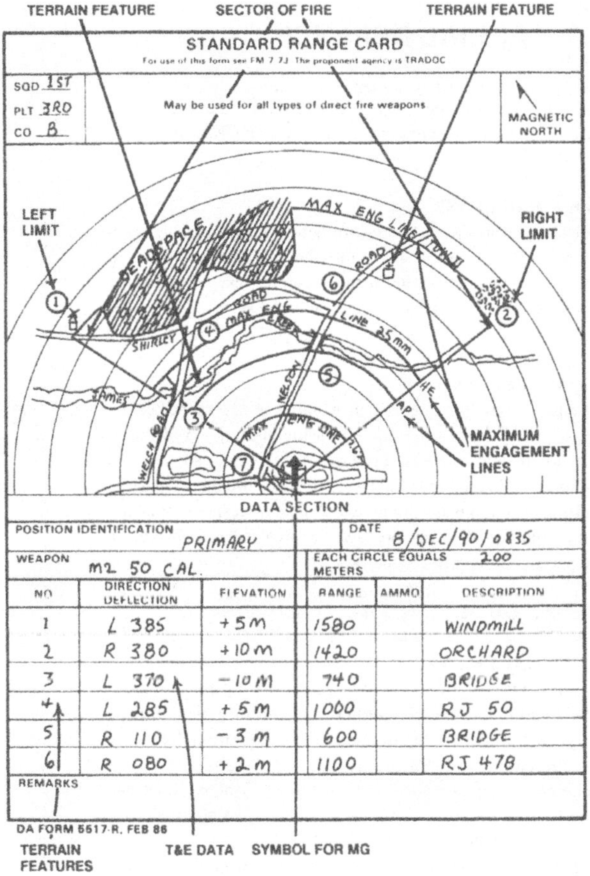

STANDARD RANGE CARD

For use of this form see FM 7-7J. The proponent agency is TRADOC

SQD 1ST
PLT 3RD
CO B

May be used for all types of direct fire weapons

DATA SECTION

POSITION IDENTIFICATION PRIMARY

DATE 8/DEC/90/0835

WEAPON M2 50 CAL.

EACH CIRCLE EQUALS 200 METERS

NO	DIRECTION DEFLECTION	ELEVATION	RANGE	AMMO	DESCRIPTION
1	L 385	+5m	1580		WINDMILL
2	R 380	+10m	1420		ORCHARD
3	L 370	-10m	740		BRIDGE
4	L 285	+5m	1000		RJ 50
5	R 110	-3m	600		BRIDGE
6	R 080	+2m	1100		RJ 478

REMARKS

DA FORM 5517-R, FEB 86

Figure E-1. Range card.

* APPENDIX F

AERIAL DEFENSE

The two methods for defending against aerial attacks are passive air defense and active air defense. Passive measures are taken to avoid air detection; active measures combat air attacks.

F-1. PASSIVE AIR DEFENSE

Passive air defense measures are a first line of defense. They include troop or vehicle dispersion, concealment and camouflage, and observation and reporting. Attack avoidance is based on the reasoning that what can be seen from the air can be destroyed and what cannot be seen will probably not be attacked.

F-2. ACTIVE AIR DEFENSE

Active air defense comes into play when a unit has been detected by hostile aircraft or when ordered to interdict hostile aircraft. Volume of fire is the key to its effectiveness. Because the action is so fast, the response must be coordinated and tightly controlled, usually at platoon level. At company level, too many are involved, making response time too slow. At section and squad level, too few are involved to achieve the volume of fire needed for an effective defense.

NOTE: All aircraft engagement firing is done at the cyclic rate.

F-3. AERIAL ENGAGEMENT

In engaging aerial targets, there are two requirements that must be considered: lead and line. Superelevation (compensation made for the pull of gravity on the projectile) is another consideration for some weapons, but the caliber .50 MG projectile is basically flat out to 800 meters. Therefore the gunner does not have to worry about it. He can instead observe the tracers crossing the target course line and make the required adjustment from them.

a. **Lead.** Leading the aircraft is the compensation made for its speed when aiming the weapon so the projectile and the aircraft will arrive at the same spot at the same time. The required lead for a given aircraft is always calculated as a certain number of lengths of that aircraft. A high performance aircraft will need a greater lead factor than a slow one. The gunner should be taught to estimate leads by projecting images of the fuselage in the front of the aircraft.

(1) If the gunner is able to track an aircraft and lead the air craft with a constant number of leads, there will be two points along the target course line where he will attain hits. As long as the gunner understands this principle, he can bring effective fire on an aircraft.

* (2) There are many different types of hostile aircraft (FM 44-80). To simplify things for the gunner, all aircraft are classified into two types: high performance and low performance. Any aircraft that has an estimated speed of more than 150 knots is classified as high performance and anything slower is low performance.

b. **Line.** Line is the requirement that the projectile must intersect the target course line of the aircraft. If the projectiles do not pass through this target course line, the aircraft cannot be engaged.

F-4. FLY-THROUGH TECHNIQUE

As an aerial target moves along its target course line, the lead required to engage the target changes because the range and angle between the aircraft and the gun position are changing. The lead required increases from the initial sighting (A in Figure F-1) to midpoint (B in Figure F-1). The required lead is at its maximum at midpoint.

a. As the aircraft moves beyond the midpoint on the target course line, the lead needed begins to decrease. Therefore, by leading an aerial target by less than the maximum required lead, there will be two points along the target course line where the constant lead will be the correct lead and the aircraft will fly through the projectiles.

b. In instructing the fly-through method of engagement, do not teach the gunners to bring continuous fire from points A through E, teach them to bring fire at points A through B and again before point D through the second fly through. As the gunner becomes proficient in this method, he will learn to adjust his lead and bring longer and more accurate fire on the target.

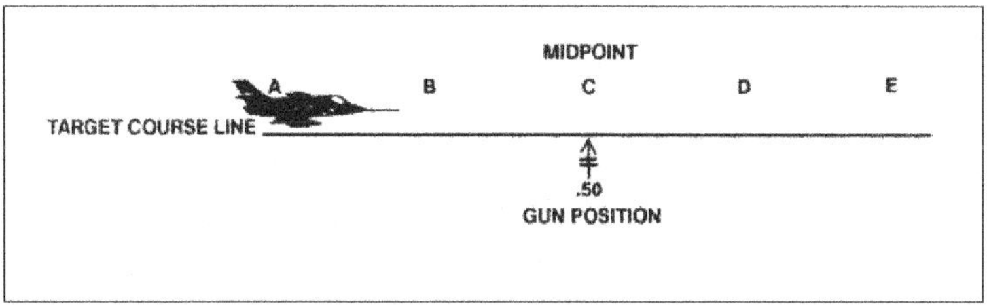

Figure F-1. Aerial target engagement.

* APPENDIX G

NIGHT OPTICS

Weapons-mounted night optics are less versatile than goggles with regard to situational awareness. However, they offer greater magnification and resolution at crew-served weapons ranges.

The graduated reticles on night optics offer the flexibility of engaging at various ranges. Laser pointers fix on one range, so the firer must "aim off" before engaging targets at a different range.

Night optics weigh more than goggles--they are awkward to move with and employ. Using night optics requires that the firer make specific adjustments to firing positions. Also, night optics might work poorly in some ambient light or thermal conditions.

Despite their drawbacks, properly boresighted night optics used in the right conditions offer units an extraordinary option: to engage targets beyond the range of opposing force weapons at night.

Section I. AN/PAS-13 (V3) HEAVY WEAPON THERMAL SIGHT

The AN/PAS-13 (V3) heavy weapon thermal sight (HWTS) (Figure G-1, page G-2) is a silent, lightweight, compact, durable, battery-powered, infrared imaging sensor that consumes little battery power. The self-contained infrared (IR) imaging sensor in this sight helps the firer acquire the target in low visibility conditions.

G-1. EQUIPMENT DESCRIPTION AND OPERATION

The HWTS works well at night and in the daytime. The telescope's IR sensor receives infrared light, converts it to digital data, processes it, and then displays it digitally as an infrared image for the user.

a. **Components**. The HWTS has two functional groups: the basic sensor and the telescope. See Figure G-1, page G-2, for equipment data.

(1) ***Basic Sensor***. The scanner reflects the IR light it receives from the telescope onto the detective (sensor) assembly. The assembly senses IR light, converts it into a video image, then conditions the video for display on the LED array. The LED array illuminates both the IR image and the reticle. This light image reflects off the scanner, which forms the actual image the firer sees in the eyepiece.

(2) ***Telescope***. The telescope receives IR light from an intended target and its surroundings, and then magnifies and projects this light onto the scanner on the basic sensor.

b. **Compatibility**. The heavy weapon thermal sight fits the weapons shown in Figure G-2:

- M2 .50-caliber machine gun.
- M24 sniper rifle.
- MK 19 grenade machine gun.
- Squad leader's M16.

c. **Operational Modes**. To place the HWTS in operational mode--

(1) If you have an AN/PAS-13 (V3), see TM 11-5855-302-12&P.

(2) If you have an AN/PAS-13 (OMNI), see TM 11-5855-312-10.

Weight:	4.5 pounds
Field of View (FOV):	
Narrow	3 degrees
Wide	9 degrees
Telescopic Magnification:	
Narrow FOV	10X
Wide FOV	3.3X
Dimensions:	
Length:	18 inches
Width:	6.25 inches
Height:	6.25 inches
Power Consumption:	
ON mode	12 watts
STANDBY mode	5.5 watts
Battery Life:	
71% STANDBY mode	10 hours
EMERGENCY mode	3.5 hours

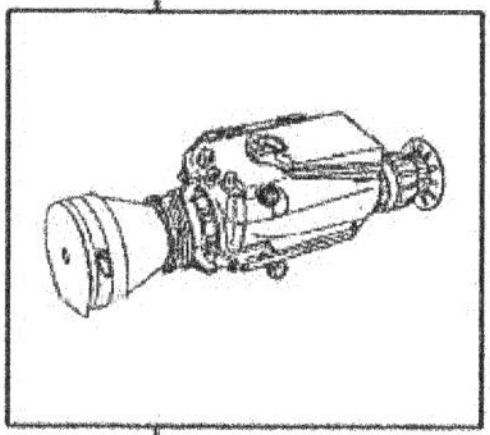

Figure G-1. Heavy weapon thermal sight.

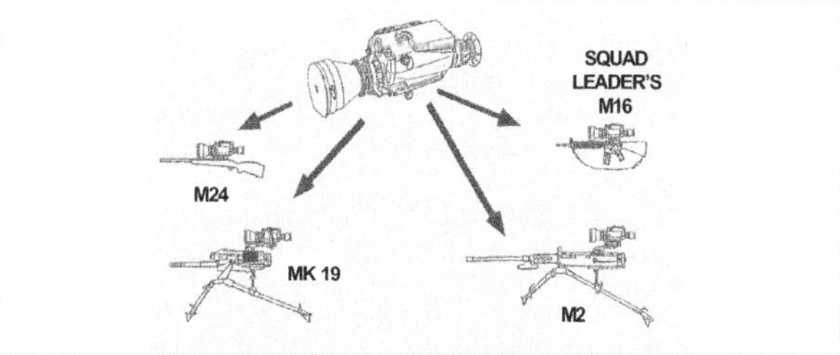

Figure G-2. Weapons that use the AN/PAS-13 (V3) HWTS.

G-2. FUNDAMENTALS OF MARKSMANSHIP

The fundamentals of marksmanship remain about the same for the AN/PAS-13 (V3) heavy weapon thermal sight (HWTS) as they do for the

AN/PVS-4. (TM 11-5855-302-12&P discusses HWTS reticle selection and point of aim.)

a. **Firing Position**.

(1) ***Sitting***. When using the tripod in the low or high position, sit directly behind the gun between the trail legs of the tripod. Extend your legs under the tripod, cross them, or brace your feet on the tripod. Regardless of which you choose, place your elbows on the inside of your thighs for support. Place your right eye in the HWTS eyepiece.

(2) ***Standing***. With the M2 mounted on a vehicle, stand with both hands on the control grips and your thumbs resting on the trigger. Keep your elbows against your body, your body forward, and your chest against your hands to brace the gun. Place your right eye in the HWTS eyepiece.

(3) ***Kneeling***. When using the M2 in a fighting or hasty tripod-mounted position, kneel and grasp the control grips with your thumbs on the trigger. Place your eye in the HWTS eyepiece.

(4) ***Prone***. Use the prone position when firing from a tripod that is sitting in a low position. Lie on the ground directly behind the gun and spread your legs a comfortable distance apart, with your toes pointed outward. Rest your left elbow on the ground and grasp the elevating handwheel of the T&E mechanism with your left hand. Grasp the right spade grip with your right hand, with the right thumb in position to press the trigger.

b. **Aiming**.

(1) Determine the range to the target. The aiming point of the M2 reticle depends on the range to the target. To determine range, you can use a range card, estimate range, use the reticle's aiming box, or use TRPs. The HWTS has two M2 reticles--a wide and a narrow field of view.

(2) Use the wide or narrow field of view to scan, to help determine range, and to engage targets.

(3) Use the indicators on the reticle to determine range, or use other methods of range determination.

(4) Set the mounting bracket range selector to NEAR for any target less than 1,400 meters and to FAR for any target beyond 1,400 meters.

(5) Using the T&E mechanism, sight the reticle aiming point on the target (bottom center).

(6) Take the slack out of the T&E by holding the M2 down and to one side. A loose T&E mechanism can move the HWTS reticle off the target's aiming point.

NOTE: Squad leader: observe the impact of the round to help the gunner adjust the sight to bring the strike of the round onto the target.

c. **Breath Control**. This fundamental of marksmanship does not change.

d. **Trigger Squeeze**. This fundamental of marksmanship does not change.

WARNING

***Do not* press your eye hard against the HWTS eyepiece while engaging a target. The M2 kicks back slightly and could harm you. Secure the M2 as well as you can to increase accuracy and to help prevent injury.**

e. **Boresighting**. This paragraph provides guidance on boresighting. For zeroing procedures, see TM 11-5855-302-12&P and TM 11-5855-312-10. (This stays in the pocket of the sight's soft storage case.) Boresight as follows:

(1) Place the M2 in the ready-to-fire position 10 meters from the boresight target offset. To avoid a failed boresight zero, level the target and weapon before you adjust the laser.

NOTE: The HWTS will not focus in on a silhouette copied to plain paper. To create a boresight target offset, use a 25-meter M16 zero target and attach to it a copy of the boresight target offset.

(2) Mount the HWTS and boresight the weapon. Set the M2 mount to the NEAR position.

(3) Use the reticle select switch to change to the M2 reticle. The reticle type displays in the upper right corner on the raster.

(4) Make adjustments with the T&E until the bore light moves onto the bore light aiming point on the boresight target offset.

(5) Adjust the reticle with the reticle adjuster on the right side of the HWTS. Remember to hold the adjuster down for a count of five. Press and hold the adjuster up to move the reticle up, down to move the reticle down, forward to move the reticle left, and back to move the reticle right.

(6) Use the vertical and horizontal reticle adjuster to move the 1,000-meter aiming point of the reticle to the HWTS aiming point on the boresight target offset.

(7) Recheck the bore laser point of aim and make sure the reticle remains in the center mass of the HWTS aim point on the boresight target offset.

TRAINING STRATEGIES:	When using night optics, M2 gunners use night initial training strategies. Additional nonfiring exercises also apply.

f. **Qualifications**. Changes to M2 qualifications accommodate new night vision sights and laser pointers. Using the new tables reduces the time and ammunition required to train and qualify gun crews. Lack of ranges and other resources previously limited qualification options. The new tables increase the opportunities for qualification and the flexibility in the process of qualification. Simpler scoring and grading procedures allow gunners to maintain a high level of proficiency.

(1) With the availability of night vision sights and laser pointers, the current night standards are not challenging. Taking into consideration the "own the night" concept, gunners need to qualify at night to the same or a greater standard than we do during the day. Using these proposed qualification tables and training strategies, the gunner can do so. Each night practice and qualification table has three variations, to reflect the different range, weapon, and sight combinations. Each day table has a variation for each type target.

(2) Use the training strategies and preliminary marksmanship instruction previously described to prepare for day practice and qualification tables. Graders need to prepare, and to ensure ranges are prepared, in accordance to the latest published manuals on the subject. After the gunner trains, he should use Firing Table I then Firing Table II. If he fails to meet the standards in Firing Table I, he must repeat the training and fire the table again. Gunners *must* pass the practice table before moving on to the qualification table.

(3) The gunner must pass day qualification before proceeding to night qualification training.

g. **Nonfiring Exercise**. Use this exercise to train the gunner to use the T&E in order to sight the laser on target.

(1) Place the M2 in the ready-to-fire configuration, with the sight mounted on the M2, and with the M2 mounted on the M3 tripod.

(2) Have the gunner close his left eye and look into the sight with his right eye.

(3) Tell the gunner to focus on a target in either the narrow or wide field of view.

(4) Tell the gunner to close both eyes. After he closes his eyes, you, the trainer, must turn and change the setting on the sight.

(5) Tell the gunner to achieve the same sight picture he had before you reset it.

(6) Look through the sight to ensure the gunner has acquired a clear sight picture. If not, show him how to get one.

(7) Repeat the exercise until the gunner achieves proficiency with the sight.

h. **T&E Manipulation and Aiming Exercise**. Use this exercise to train the gunner to manipulate the T&E, acquire targets, and properly sight the target based on the target range. This exercise also trains the gunner to use the mounting bracket to adjust for range.

(1) Set up two to three targets that the gunner can acquire with the HWTS using the T&E.

(2) Regardless of the actual range to the target, tell the gunner to sight in on the first target and give him the range you want it to be.

(3) Ensure the gunner sets the mounting bracket to the correct setting and uses the T&E correctly.

(4) Time the gunner as he conducts this exercise, and create a competition between the gunners in training.

i. **Practice Qualification.** Use practice qualification, modeled after the day qualification scenario, to transition the gunner into engaging multiple timed targets at various ranges.

j. **Qualification.** Use qualification to assess the gunner's ability to engage targets on a timed scenario. If he does not meet the standard during the night practice table, train him again. He must meet the standard on the practice table before firing the qualification table. Conduct the nonfiring exercise as follows:

(1) Place the M2, with HWTS mounted, in the ready-to-fire configuration on the M3 tripod.

(2) Have the gunner close his left eye and look into the HWTS with his right eye.

(3) Tell the gunner to focus on a target in either the narrow or wide field of view.

(4) Tell the gunner to close both eyes. After he closes them, change the setting on the sight.

(5) Tell the gunner to open his eyes and to achieve the same sight picture he had before you reset it.

(6) Look through the sight to ensure the gunner has acquired a clear sight picture. If not, show him how to get one.

(7) Repeat the exercise until the gunner reaches proficiency with the sight.

Section II. AN/TVS-5

Gunners use the AN/TVS-5 (Figure G-3, page G-7) a portable, battery-operated electro-optical instrument, to observe and aim weapons fire at night. The AN/TVS 5 amplifies reflected light such as moonlight, starlight, and sky glow, making the scene clearly visible to the operator. The sight *does not* emit visible or infrared light (except from the eyepiece) that the enemy can detect. By using this device, the gunner can observe the area, and then detect and engage any suitable target. The quality of the zero determines the usefulness of the weapon. Accurate zeroing requires practice. Figure G-3 also provides equipment data for the AN/TVS-5.

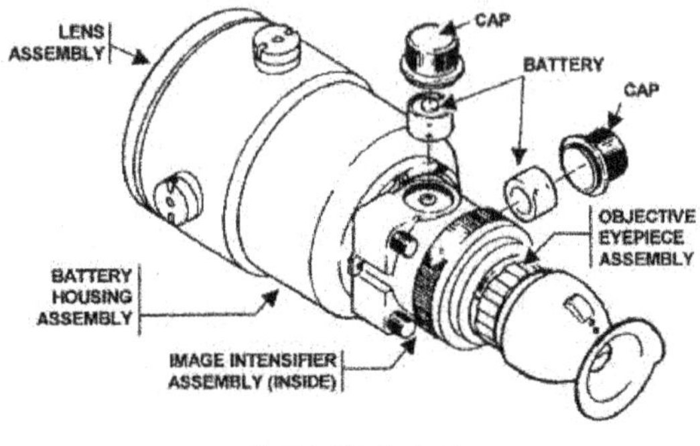

EQUIPMENT DATA

Magnification	5.6X
Field of view	9 degrees
Eyepiece focus	+3 to -6 diopters
Objective lens focus	25 meters to infinity
Range	1,000 to 1,200 meters for vehicle targets (under ideal conditions)
Reticle adjustment	+/-2.5 degrees (in ¼-mil increments)
Operating conditions	-65 F (-54C) to +125 F (+52C)
Battery type	BA-5567/U (2 each)
Voltage	3.0 volts DC
Shelf life	One year at 70 F (21C)
Operating life	About 50 hours at normal temperature

Figure G-3. AN/TVS-5 components and data.

CAUTION

Handle the AN/TVS-5, a precision electro-optical instrument, carefully at all times.

G-3. OPERATION

The AN/TVS-5 has the following controls and indicators:

a. Use the ON-OFF/TUBE BRIGHTNESS control to apply power to the sight and to control the brightness of the image-intensifier tube. This control also enables the ON-OFF RETICLE BRIGHTNESS control to function.

b. Use the ON-OFF RETICLE BRIGHTNESS to apply power to the reticle and to control the brightness of the reticle.

c. Use the OBJECTIVE FOCUS RING to adjust the range focus from 25 meters to infinity.

d. Use the DIOPTER FOCUS RING to focus the eyepiece.

e. Use the DIOPTER INDICATOR to learn the current direction of rotation of the DIOPTER FOCUS RING for plus and minus diopters.

f. Use the RETICLE ELEVATION ADJUSTMENT ACTUATOR to adjust the reticle up and down. Each click moves the strike of the round 1 inch at 100 meters.

g. Use the RETICLE AZIMUTH ADJUSTMENT ACTUATOR to adjust the reticle right or left. Each click moves the strike of the round 1 inch at 100 meters.

WARNING

- ***Do not*** **press the eye guard except with your eye area and then only to operate the sight. Used improperly, the sight emits illumination that the enemy can detect.**
- ***Do not*** **use the sight without the eye guard attached, or you may receive an injury when the weapon recoils.**
- **Use care when discarding the batteries. Their contents cause extreme irritation to the eyes and to oral and nasal passages. To prevent explosion, avoid burning the batteries.**
- ***Do not*** **short-circuit the batteries.**
- ***Do not*** **recharge the batteries.**
- **Remove batteries before storing the night vision sight.**
- **Always replace both mercury batteries at the same time.**

G-4. MOUNTING AND DISMOUNTING PROCEDURES

The AN/TVS-5 mounts and dismounts as follows:

a. **Mounting Procedures**. If the mounting bracket (Figure G-4) has not been installed already, install it now as described in TM 11-5855-214-10.

(1) Align the scribe line on the sight with the scribe line on the bracket.

(2) Place the sight in the groove of the bracket and tighten the lever screw clockwise.

(3) Secure the lever screw with lacing wire or tape to ensure the sight does not vibrate loose.

(4) Seat the device. After firing the initial burst, retighten the lever screw to ensure a secure mount for the sight. If unable to fire at this time, lightly shake the sight to ensure it is mounted correctly.

b. **Dismounting Procedures**.

(1) Remove the lacing wire or tape from the lever screw.

(2) Loosen the lever screw until the sight comes free, and then lift it off the bracket.

(3) Remove the batteries and place the sight in its carrying case.

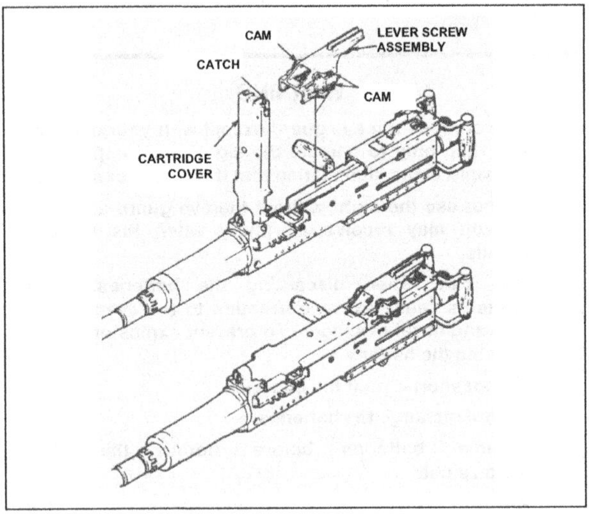

Figure G-4. Installation of M2 mounting bracket assembly.

G-5. FUNDAMENTALS OF MARKSMANSHIP

Except for aiming, the fundamentals of marksmanship for the AN/TVS-5 are the same as those for the AN/PAS-13 heavy weapon thermal sight.

a. Determine the range to the target (Figure G-5, page G-10). The range setting on the HWTS depends on the range to the target. To determine range to target, use any of several methods: range cards, range estimation techniques, the upper portion of the AN/TVS-5 reticle, or TRPs.

b. After determining range, use the aiming point designated for that range on the M2 reticle (Figure G-6, page G-12 shows the old reticle; Figure G-7, page G-13 shows the new reticle). If using the HWTS bracket and the M2 reticle, adjust the bracket to the nearest settings. Zero the sight on this setting and use the appropriate dot for the distance to the target as indicated in the reticle. For ranges beyond 1,200 meters, change the bracket to the far setting.

- Change "4" to 1,400 meters.
- Change "6" to 1,500 meters.
- Change "8" to 1,600 meters.
- Change "10" to 1,700 meters.
- Change "12" to 1,800 meters.

c. Bring the aiming point of the AN/TVS-5 onto the target using the T&E as explained in Section V.

d. Hold the weapon tight against the T&E, press the trigger, and adjust fire as needed.

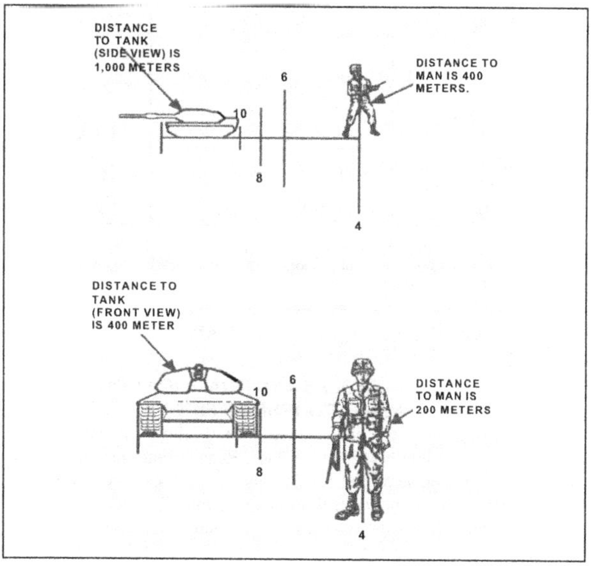

Figure G-5. Range estimation for M2 (old reticle).

G-6. BORESIGHTING PROCEDURES

Boresighting the AN/TVS-5 to the M2 requires the following steps:

a. Place the M2 in the ready-to-fire configuration, with the AN/TVS-5 mounted, 25 meters from the bore-light-offset zero targets. Ensure the weapon and target sit level before making any adjustments.

b. Adjust the T&E until the bore light moves onto the bore light aiming point on the boresight target offset.

c. Use the vertical and horizontal reticle adjusters to move the 1,000-meter aiming point on the reticle to the AN/TVS-5 aiming point.

d. Recheck the bore light aiming point and the AN/TVS-5 aiming points to ensure they remain center mass of the target.

G-7. AMMUNITION, NIGHT TRAINING STRATEGY

Table G-1 shows STRAC changes effective with this change publication.

EVENT	CURRENT STRAC BALL/MIX	REC'D CHGS	TOTAL BALL/MIX
10M ZERO	12 BALL	0	12 BALL
10M RECORD	119 BALL	0	119 BALL
TRANSITION ZERO, PRACTICE	50 MIX	ELIMINATE	0
TRANSITION RECORD	154 MIX	ELIMINATE	0
NIGHT ZERO, PRACTICE, RECORD	182 MIX	ELIMINATE	0
AG INSTR FIRE	54 MIX	ELIMINATE	0
TABLE I	0	140 MIX	140 MIX
TABLE II	0	140 MIX	140 MIX
TABLE III	0	140 MIX	140 MIX
TABLE IV	0	140 MIX	140 MIX
TOTAL	131 / 440 = 571	560 MIX	131/560 = 691

NOTE: The above STRAC supports one qualification. Allocations increase with the frequency of qualifications required. Check the firing unit's force activity designator (FAD).

Table G-1. Changes to STRAC effective with this publication.

Section III. TABLES AND QUALIFICATION, NIGHT COURSE OF FIRE

Improved firing tables and qualification methods allow the gunner to track his performance while he fires the tables. Requalifying gunners need not repeat the entire course of fire. This section standardizes the four tables to lead the gunner on a gradual path to qualification. The gunner must pass each firing table before he moves on to the next one. The tables in this section require 691 rounds to qualify a gunner--93 more than the old tables required, and 120 more than the STRAC previously authorized. The guidance in this appendix supersedes the STRAC. Though the new tables allow more rounds, they cut the time allowed to qualify. However, they recommend a band of ranges rather than a specific range for each target. This adds considerable flexibility. The ten-meter firing exercise is only a means of training--it is not a qualification table.

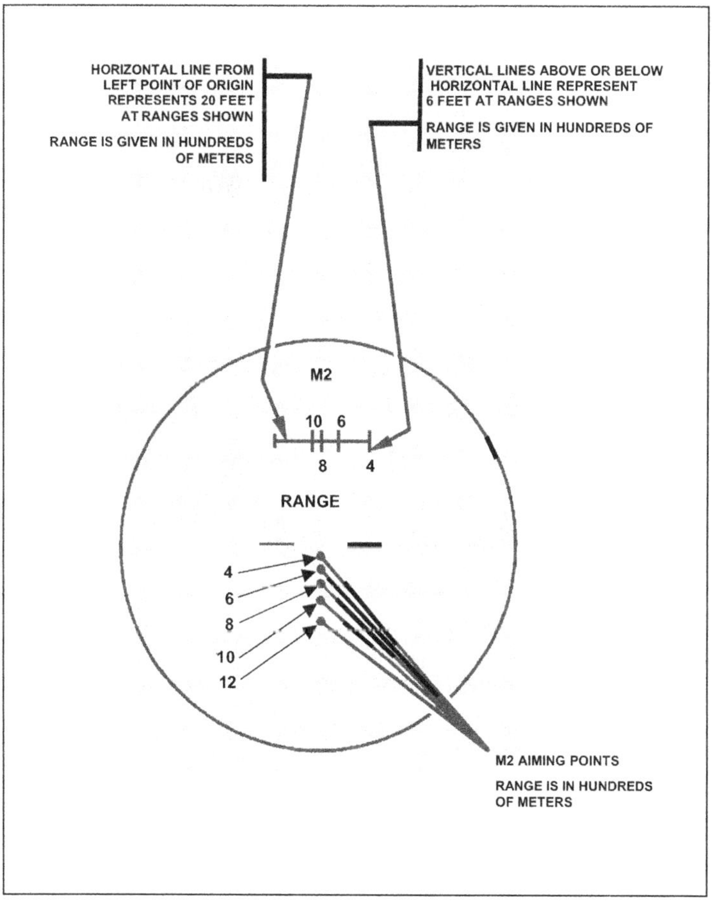

Figure G-6. Aiming points for the M2 (old reticle).

G-8. DIFFERENCES BETWEEN PRACTICE AND QUALIFICATION TABLES

Practice tables allow thirty extra seconds for each engagement. Also, infantry gun crews can practice in the daytime, but must qualify in both the daytime and at night.

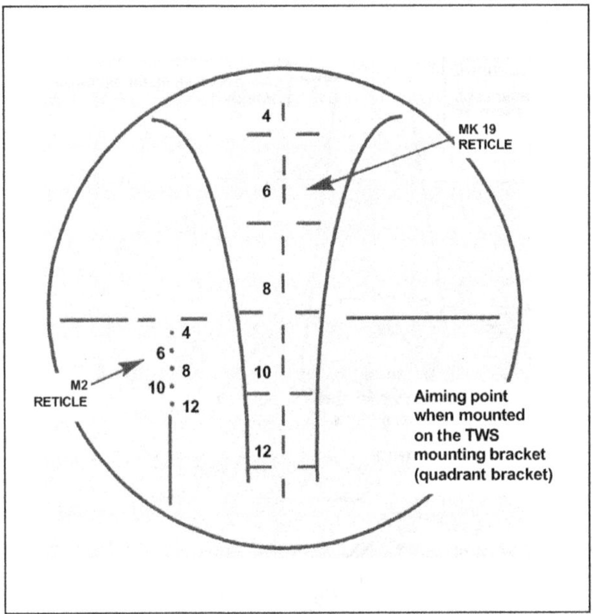

Figure G-7. Aiming points for the M2 (new reticle).

a. **Scoring**. For point target engagements (lightly armored vehicle targets, such as BRDMs and BTRs), give full credit (GO) when the gunner hits the target.

b. **Range Setup**. Set up targets and ranges as follows. Select target ranges IAW the scorecard for the applicable firing table. When choosing or placing targets, make sure no dead space falls within 100 meters of any of them. Any dead space near the target could keep the gunner from seeing the round impact, which he has to do in order to adjust fire:

(1) ***Hull Targets***. You need not modify hull targets on an impact range.

(2) ***Popup Silhouettes***. Set up a thermal source on popup silhouette targets so the gunner can acquire them with the HWTS. If the gunner will be using the AN/TVS-5, place a light source on each target.

- Targets between 400 and 600 meters--two chemical lights.
- Targets between 600 and 900 meters--three chemical lights.

c. **Grading**. Each firing point requires one grader.

(1) ***Grading Equipment***. While firing the day tables, the grader needs a set of binoculars. His other equipment needs vary according to the type of range used at night.

(a) *Night Vision Assistance*. When firing at hull targets, or when popup targets provide no downrange feedback, the grader must obtain an appropriate night vision device with which to observe the strike of the round.

- AN/PVS-14, 7B, with 3X magnifier.
- AN/TAS-4.
- AN/PAS-13 (heavy).

(b) *Target Scenario*. The grader needs this.

(c) *Range and Range Card*. At each particular firing point, the grader gives the gunner a target and a range at which the gunner must engage the target. The grader needs a range card for that firing range. The range card should include numbered targets and the ranges to them.

(2) ***Start and End Times***. Time starts as soon as the grader provides the target range and the target appears--which occur in this order, since graders must provide all required information before the target appears. Sometimes the targets consist of vehicle hulls in stationary positions. In this case, time starts as soon as the grader tells the gunner which target to engage and gives him the range to the target. Regardless of the type of target, time ends when the time designated for that task expires, or when the gunner successfully engages the target.

(3) ***Ammunition***. Ammunition breaks down by task. The gunner places each belt beside the assistant gunner in order of use. The number of rounds authorized for each task equals the number of rounds in each belt. For example, eight tasks require eight belts of ammunition, which should all sit within reach of the assistant gunner in the order the gunner plans to fire them.

(4) ***Fire Control***. Controlling M2 firing presents no problem when the range setup includes a firing lane for each firing point. However, this ideal situation seldom occurs with hull targets. In fact, some ranges must use the same target for each lane. This can cause confusion. When, inevitably, more than one gunner fires at the same target, no one can tell who hit it. The 400-meter target usually presents this problem. To prevent it, make sure that only one gunner at a time engages each target. If you want to let more than one gunner to fire at the same time, and then mix up the tasks so the gunners are firing at different targets. For example, have Point One fire at the 1,500-meter target while Point Two engages the 600-meter target, and so on. This requires careful coordination and communication between the graders.

d. **Firing Tables**. Figures G-8 and G-9 (page G-16), Figure G-10 (page G-17), and Figure G-11 (page G-18) show completed examples of the recommended day and night practice and qualification tables for the infantry M2 gunner and crew. Blank, reproducible forms are included at the back of the manual for users to copy onto 8 1/2 by 11-inch paper. They are also available on the Army Electronic Library (AEL) CD-ROM (EM0001) and at the USAPA website: http://www.usapa.army.mil.

(1) The first task allows evaluation of field zeroing. The gunner confirms the zero, even if he boresighted the weapon. If the gunner fails to zero within 14 rounds, graders remove him from the line and train him some more before letting him refire the table.

(2) All tasks have point targets. Graders should change one or two targets between the ranges of 600 and 900 meters to area targets, but should not change the round count or the time. They base their changes on range resources and the commander's guidance.

(3) Graders match the correct table to each target and weapon-sight configuration.

(4) Graders use range finders to predetermine the range to each target from each firing point.

(5) Graders set up M2s mounted or dismounted (in the tripod configuration), based on the range constraints and the commander's guidance.

(6) Leaders can have the unit fire the tasks in any order, and should do so to avoid redundancy in practice and qualification.

(7) For NBC qualification, the leader uses the practice tables.

e. **Ammunition Requirements for Night Training Strategy**. Table G-1 shows the ammunition requirements for a night training strategy.

G-9. SPECIAL MOUNTING PROCEDURES

This paragraph explains the special mounting and dismounting procedures required to mount a night vision device on the M2 heavy barrel .50 caliber machine gun.

a. **Mounting Procedure**. Before mounting the sight, the gunner ensures the bolt is forward and the rear sight is in the down position.

(1) Release and raise the top cover assembly to the upright position.

(2) Place the mounting bracket over the breach and slide it rearward until it stops.

(3) Push the three locking cams rearward until the bracket is secured (lock the side-locking cam first), and close the top cover assembly.

M2 CALIBER .50 HEAVY BARREL MACHINE GUN FIRING TABLE I
DAY PRACTICE SCORECARD

For use of this form, see FM 23-65, Change 1; the proponent agency is TRADOC.

DATA REQUIRED BY THE PRIVACY ACT OF 1974

AUTHORITY: 10 USC 3012(g)/Executive order 9397.
PRINCIPAL PURPOSE: To aid individual training on targets at various ranges.
ROUTINE USES: To evaluate individual proficiency. SSN is used for positive identification purposes only.
DISCLOSURE: Voluntary. However, mass rating and scoring require some tracking method. Therefore, only those soldiers who provide their SSNs can receive scores or qualify on weapons or equipment.

1a. LAST NAME	1b. FIRST NAME	1c. MI	2. DATE *(YYYYMMDD)*
DOE	JOHN	A	20021201

3. SSN	4. UNIT	5. RANGE	6. LANE
222-22-2222	2 / 29 TH INF	RUTH	4

7a. ENGAGEMENT	7b. STANDARDS	7c. AMMO	7d. TIME *(Minutes)*	7e. RANGE *(Meters)*	7f. GO	7g. NO GO
ZERO	1 BURST HIT	14	NA	400	NA	NA
2	1 BURST HIT	14	1.5	1,100 (±200)	X	
3	1 BURST HIT	14	1.5	1,500 (±200)	X	
4	1 BURST HIT	14	1.5	600 (±100)	X	
5	1 BURST HIT	14	1.5	800 (±100)	X	
6	1 BURST HIT	14	1.5	400	X	
7h. MULTIPLE TARGETS						
7	1 BURST HIT	28	2.5	1,100 (±200)	X	
	1 BURST HIT			600 (±100)	X	
8	1 BURST HIT	28	2.5	800 (±100)	X	
	1 BURST HIT			1,500 (±200)	X	

EXAMPLE

9. SCORING *(Check one)*

EXPERT - 9	X	SHARPSHOOTER - 8	
MARKSMAN - 6		UNQUALIFIED - 5 OR BELOW	

10. GUNNER'S SIGNATURE

PFC John A. Doe

11a. GRADER'S PRINTED OR TYPED NAME	11b. GRADER'S SIGNATURE
SSG DAVID R. JONES	SSG David R. Jones

DA FORM 7448-R, JAN 2002 USAPA V1.00

Figure G-8. Example completed DA Form 7448-R (Day Practice Scorecard).

M2 CALIBER .50 HEAVY BARREL MACHINE GUN FIRING TABLE II
DAY QUALIFICATION SCORECARD

For use of this form, see FM 23-65, Change 1; the proponent agency is TRADOC.

DATA REQUIRED BY THE PRIVACY ACT OF 1974

AUTHORITY: 10 USC 3012(g)/Executive order 9397.
PRINCIPAL PURPOSE: To aid individual training on targets at various ranges.
ROUTINE USES: To evaluate individual proficiency. SSN is used for positive identification purposes only.
DISCLOSURE: Voluntary. However, mass rating and scoring require some tracking method. Therefore, only those soldiers who provide their SSNs can receive scores or qualify on weapons or equipment.

1a. LAST NAME	1b. FIRST NAME	1c. MI	2. DATE *(YYYYMMDD)*
DOE	JOHN	A	20021201

3. SSN	4. UNIT	5. RANGE	6. LANE
222-22-2222	2 / 29 TH INF	RUTH	4

7a. ENGAGEMENT	7b. STANDARDS	7c. AMMO	7d. TIME *(Minutes)*	7e. RANGE *(Meters)*	7f. GO	7g. NO GO
ZERO	1 BURST HIT	14	NA	400	NA	NA
2	1 BURST HIT	14	1	1,100 (±200)		X
3	1 BURST HIT	14	1	1,500 (±200)	X	
4	1 BURST HIT	14	1	600 (±100)	X	
5	1 BURST HIT	14	1	800 (±100)	X	
6	1 BURST HIT	14	1	400	X	
8. MULTIPLE TARGETS						
7	1 BURST HIT	28	2	1,100 (±200)	X	
	1 BURST HIT			600 (±100)	X	
8	1 BURST HIT	28	2	800 (±100)	X	
	1 BURST HIT			1,500 (±200)	X	

EXAMPLE

9. SCORING *(Check one)*

EXPERT - 9 [ ]	SHARPSHOOTER - 8 [X]
MARKSMAN - 6 [ ]	UNQUALIFIED - 5 OR BELOW [ ]

10. GUNNER'S SIGNATURE
PFC John A. Doe

11a. GRADER'S PRINTED OR TYPED NAME	11b. GRADER'S SIGNATURE
SSG DAVID R. JONES	SSG David R. Jones

DA FORM 7449-R, JAN 2002 USAPA V1.00ES

Figure G-9. Example completed DA Form 7449-R (Day Qualification Scorecard).

M2 CALIBER .50 HEAVY BARREL MACHINE GUN FIRING TABLES III (A) AND III (B)
NIGHT PRACTICE SCORECARDS

For use of this form, see FM 23-65, Change 2; the proponent agency is TRADOC.

DATA REQUIRED BY THE PRIVACY ACT OF 1974

AUTHORITY:	10 USC 3012(g)/Executive order 9397.
PRINCIPAL PURPOSE:	To aid individual training on targets at various ranges.
ROUTINE USES:	To evaluate individual proficiency. SSN is used for positive identification purposes only.
DISCLOSURE:	Voluntary. However, mass rating and scoring require some tracking method. Therefore, only those soldiers who provide their SSNs can receive scores or qualify on weapons or equipment.

1a. LAST NAME	1b. FIRST NAME	1c. MI	1d. RANK	2. SSN
Doe	John	A	PFC	222-22-2222

EXAMPLE

TABLE III (A). NIGHT PRACTICE SCORECARD

NOTE: Use this table if the M2 has the TWS, or if it has an AN/TVS-5 that has been upgraded with the third-generation tube.

a. RANGE	b. UNIT	c. LANE	d. DATE *(YYYYMMDD)*
Ruth	2 / 29 th	4	20021120

e. ENGAGEMENT	f. STANDARDS	g. AMMO	h. TIME *(Minutes)*	i. RANGE *(Meters)*	j. GO	k. NO GO
ZERO	1 BURST HIT	14	NA	400	NA	NA
2	1 BURST HIT	14	1.5	1,100 (±200)	X	
3	1 BURST HIT	14	1.5	1,500 (±200)	X	
4	1 BURST HIT	14	1.5	600 (±100)		X
5	1 BURST HIT	14	1.5	800 (±100)	X	
6	1 BURST HIT	14	1.5	400	X	
l. MULTIPLE TARGETS						
7	1 BURST HIT	28	2.5	1,100 (±200)	X	
	1 BURST HIT			600 (±100)	X	
8	1 BURST HIT	28	2.5	800 (±100)	X	
	1 BURST HIT			1,600 (±200)	X	

4. SCORING *(Check one)*
- EXPERT - 9 [ ]
- SHARPSHOOTER - 8 [X]
- MARKSMAN - 6 [ ]
- UNQUALIFIED - 5 OR BELOW [ ]

5. GUNNER'S SIGNATURE: PFC John A. Doe
6a. GRADER'S PRINTED OR TYPED NAME: SSG David R. Jones
6b. GRADER'S SIGNATURE: SSG David R. Jones

7. TABLE III (B). NIGHT PRACTICE SCORECARD

NOTE: Use this table if the M2 has an AN/TVS-5 that has not been upgraded.

a. RANGE	b. UNIT	c. LANE	d. DATE *(YYYYMMDD)*
Ruth	2 / 29 th	4	20021120

e. ENGAGEMENT	f. STANDARDS	g. AMMO	h. TIME *(Minutes)*	i. RANGE *(Meters)*	j. GO	k. NO GO
ZERO	1 BURST HIT	14	NA	400	NA	NA
2	1 BURST HIT	14	1.5	1,100 (±200)	X	
3	1 BURST HIT	14	1.5	800 (±100)	X	
4	1 BURST HIT	14	1.5	600 (±100)	X	
5	1 BURST HIT	14	1.5	800 (±100)	X	
6	1 BURST HIT	14	1.5	400	X	
l. MULTIPLE TARGETS						
7	1 BURST HIT	28	2.5	1,100 (±200)	X	
	1 BURST HIT			600 (±100)	X	
8	1 BURST HIT	28	2.5	800 (±100)	X	
	1 BURST HIT			600 (±100)	X	

8. SCORING *(Check one)*
- EXPERT - 9 [X]
- SHARPSHOOTER - 8 [ ]
- MARKSMAN - 6 [ ]
- UNQUALIFIED - 5 OR BELOW [ ]

9. GUNNER'S SIGNATURE: PFC John A. Doe
10a. GRADER'S PRINTED OR TYPED NAME: SSG David R. Jones
10b. GRADER'S SIGNATURE: SSG David R. Jones

DA FORM 7450-R, OCT 2002 — DA FORM 7450-R, JAN 2002, IS OBSOLETE. — USAPA V1.00ES

Figure G-10. Example completed DA Form 7450-R (Night Practice Scorecard).

M2 CALIBER .50 HEAVY BARREL MACHINE GUN FIRING TABLES IV (A) AND IV (B) NIGHT QUALIFICATION SCORECARDS

For use of this form, see FM 23-65, Change 2; the proponent agency is TRADOC.

DATA REQUIRED BY THE PRIVACY ACT OF 1974

AUTHORITY: 10 USC 3012(g)/Executive order 9397.
PRINCIPAL PURPOSE: To aid individual training on targets at various ranges.
ROUTINE USES: To evaluate individual proficiency. SSN is used for positive identification purposes only.
DISCLOSURE: Voluntary. However, mass rating and scoring require some tracking method. Therefore, only those soldiers who provide their SSNs can receive scores or qualify on weapons or equipment.

1a. LAST NAME	1b. FIRST NAME	1c. MI	1d. RANK	2. SSN
Doe	John	A	PFC	222-22-2222

3. TABLE IV (A). NIGHT QUALIFICATION SCORECARD

NOTE: Use this table if the M2 has the TWS, or if it has an AN/TVS-5 that has been upgraded with the third-generation tube.

a. RANGE	b. UNIT	c. LANE	d. DATE *(YYYYMMDD)*
Ruth	2 / 29	4	20021204

e. ENGAGEMENT	f. STANDARDS	g. AMMO	h. TIME *(Minutes)*	i. RANGE *(Meters)*	j. GO	k. NO GO
ZERO	1 BURST HIT	14	NA	400	NA	NA
2	1 BURST HIT	14	1	1,100 (± 200)	X	
3	1 BURST HIT	14	1	1,500 (± 200)	X	
4	1 BURST HIT	14	1	600 (± 100)		X
5	1 BURST HIT	14	1	800 (± 100)	X	
6	1 BURST HIT	14	1	400	X	
l. MULTIPLE TARGETS						
7	1 BURST HIT	28	2	1,100 (± 200)	X	
	1 BURST HIT			600 (± 100)		X
8	1 BURST HIT	28	2	800 (± 100)	X	
	1 BURST HIT			1,500 (± 200)	X	

4. SCORING *(Check one)*
EXPERT - 9 [ ]
SHARPSHOOTER - 8 [ ]
MARKSMAN - 6 [X]
UNQUALIFIED - 5 OR BELOW [ ]

5. GUNNER'S SIGNATURE: PFC John A. Doe
6a. GRADER'S PRINTED OR TYPED NAME: SSG David R. Jones
6b. GRADER'S SIGNATURE: SSG David R. Jones

7. TABLE IV (B). NIGHT QUALIFICATION SCORECARD

NOTE: Use this table if the M2 has an AN/TVS-5 that has not been upgraded.

a. RANGE	b. UNIT	c. LANE	d. DATE *(YYYYMMDD)*
Ruth	2 / 29	4	20021204

e. ENGAGEMENT	f. STANDARDS	g. AMMO	h. TIME *(Minutes)*	i. RANGE *(Meters)*	j. GO	k. NO GO
ZERO	1 BURST HIT	14	NA	400	NA	NA
2	1 BURST HIT	14	1	1,100 (± 200)	X	
3	1 BURST HIT	14	1	800 (± 100)	X	
4	1 BURST HIT	14	1	600 (± 100)	X	
5	1 BURST HIT	14	1	800 (± 100)	X	
6	1 BURST HIT	14	1	400	X	
l. MULTIPLE TARGETS						
7	1 BURST HIT	28	2	1,100 (± 200)		X
	1 BURST HIT			600 (± 100)	X	
8	1 BURST HIT	28	2	800 (± 100)	X	
	1 BURST HIT			600 (± 100)	X	

8. SCORING *(Check one)*
EXPERT - 9 [ ]
SHARPSHOOTER - 8 [X]
MARKSMAN - 6 [ ]
UNQUALIFIED - 5 OR BELOW [ ]

9. GUNNER'S SIGNATURE: PFC John A. Doe
10a. GRADER'S PRINTED OR TYPED NAME: SSG David R. Jones
10b. GRADER'S SIGNATURE: SSG David R. Jones

EXAMPLE

DA FORM 7451-R, OCT 2002 — DA FORM 7451-R, JAN 2002, IS OBSOLETE. — USAPA V1.00ES

Figure G-11. Example completed DA Form 7451-R (Night Qualification Scorecard).

DANGER

BEFORE INSTALLING THE BRACKET, MAKE SURE THE WEAPON IS *UNLOADED* AND THE SAFETY IS SET ON *SAFE*. A LOADED WEAPON CAN DISCHARGE ACCIDENTALLY, INJURING OR KILLING SOMEONE.

NOTE: If the bracket sticks before reaching its correct position, rock the bracket up and down while sliding it.

b. **Dismounting Procedure**. Ensure the weapon is clear and the bolt is forward.

(1) Raise the top cover assembly to the upright position.

(2) Push the three locking cams forward until the bracket is free.

(3) Slide the mounting bracket forward over the breach until the mounting bracket clears the weapon.

Section IV. BORE-LIGHTING PROCESS AND BORESIGHT TARGET OFFSETS

The bore light allows gunners to accurately zero weapons and most aided vision equipment without the use of bullets. Table G-2 shows different combinations of weapon and aided-vision device that gunners can use to zero the bore light.

DANGER

- ***DO NOT STARE*** **INTO THE VISIBLE LASER BEAM.**
- ***DO NOT LOOK*** **INTO THE VISIBLE LASER BEAM THROUGH A TELESCOPE OR A PAIR OF BINOCULARS.**
- ***DO NOT POINT*** **THE VISIBLE LASER BEAM AT A MIRROR-LIKE SURFACE.**
- ***DO NOT SHINE*** **THE VISIBLE LASER BEAM INTO OTHER PEOPLE'S EYES.**

WARNING

- **Make sure the weapon is CLEAR and on SAFE before using the bore light.**
- **When rotating the bore light to zero it, ensure the mandrel turns. *Do not* tighten the bore light down on the mandrel; doing so could strip or break the mandrel's stud.**

	M16A2	M4/MWS	M249	M240B/M60	M2	M203	AT4
AN/PAQ-4C	X	X	X	X		X	X
AN/PEQ-2A	X	X	X	X	X	X	X
AN/PAS-13	X	X	X	X	X		X
M68 CCO	X	X					
AN/PVS-4	X	X	X	X		X	
AN/TVS-5					X		

Table G-2. Possible weapon-aided vision-device combinations.

G-10. BORESIGHTING PROCESS

Properly boresighting a weapon requires two gunners--a firer and a target-holder.

a. **Personnel.**

(1) *Firer.* The firer zeroes the bore light and adjusts the aided vision device. Before boresighting, he locks the M2 down in a cradle or some other device to keep the weapon stable during the boresighting process. In the absence of such a device, he assumes the most stable supported firing position possible and makes sure the weapon does not cant during boresighting.

(2) *Target holder.* The target holder keeps the target straight up and down at the appropriate distance from the firer and directs the firer in making necessary sight adjustments. The target holder *must* wear night vision goggles when viewing laser aiming light offsets.

b. **Steps**.

(1) *Align Bore Light.* Checking the alignment of the bore light requires the following actions:

(a) Place the appropriate mandrel, with bore light attached, in the muzzle of the weapon.

(b) Turn the bore light on so that the laser beam strikes the target (offset) 10 meters away.

(c) Slowly rotate the bore light one-half turn (180 degrees) while watching the beam on the target area, noting any circular patterns made.

(d) If the beam remains stationary, you have boresighted the M2. Move forward to the next paragraph and use the appropriate boresight target for the given weapon.

(e) If the beam rotates in a circle, adjust the windage, elevation, or both, until the beam remains stationary or rotates on itself no more than 1 centimeter.

(2) *Adjust Bore Light*. If necessary, the firer and target holder must adjust the bore light.

(a) Move the target 2 meters away.

(b) Mark the location of the laser beam.

(c) Slowly rotate the bore light one-half turn.

(d) Note the new location of the laser beam.

(e) Adjust the windage and elevation until the laser beam moves halfway back to its original location.

(3) *Adjust elevation adjustment screw*. One click at 25 meters equals 4 millimeters; clockwise equals down.

(4) *Adjust windage adjustment screw*. One click at 25 meters equals 4 millimeters; clockwise equals left.

(5) *Continue adjusting the bore light*. Keep adjusting it until the laser beam either stops moving or spins upon itself within 1 centimeter.

(6) *Move the target, and recheck the boresight*. Move it 10 meters away and recheck the boresight. Repeat this process every 10 meters if necessary.

NOTE: Because the M2 requires such a large offset, place the offset for the boresight target 5 meters away.

(7) *Boresight the Aided Vision Device to the Weapon*. Select the boresight target offset for the appropriate aided vision device. For laser aiming lights, position the weapon so the bore light strikes the black dot on the boresight target.

(a) Adjust the *laser* until it centers on the cross on the boresight target offset. For aided vision optics, position the weapon so the *reticle* centers on the cross on the boresight target offset (the firer must physically aim the weapon).

(b) Adjust the aided vision optic until the bore light strikes the black dot on the boresight target. Refer to specific instructions on each boresight-offset target.

(c) You have boresighted the weapon when both the laser bore light and the aiming point or laser move to the appropriate position on the boresight target offset.

G-11. TARGET OFFSETS

Figure G-12 (page G-23), Figure G-13 (page G-24), and Figure G-14 (page G-25) show example boresight target offsets. The gunner measures boresight offsets in 1-centimeter squares. For targets, he copies M16A2 25-meter zero targets, with the silhouette representing the point of aim.

a. **Aligning the M2 with the AN/PEQ-2A**. Using the 10-meter boresight target, align the bore light on the dot and adjust the AN/PEQ-2A to the cross. This gives an 800-meter zero (Figure G-12, page G-23).

b. **Aligning the M2 with the AN/TVS-5**. Using the 10-meter boresight target, place the target at a distance of 10 meters. Aim the 400-meter AN/TVS-5 reticle at the cross and adjust the sight so that the bore light strikes the dot (Figure G-13, page G-24).

c. **Aligning the M2 with the AN/PAS-13**. Using the 10-meter boresight target, place the tips of the index fingers on the gray circles. Aim between the hot spots provided by the fingers. Adjust the TWS so that the bore light strikes the dot when you aim between the hot spots at 10 meters. If using a spacer, increase the offset by the height of the spacer (Figure G-14, page G-25).

10-meter boresight target AN/PEQ-2A mounted on an AN/TVS-5 feed tray cover bracket and plastic spacer on M2 MG

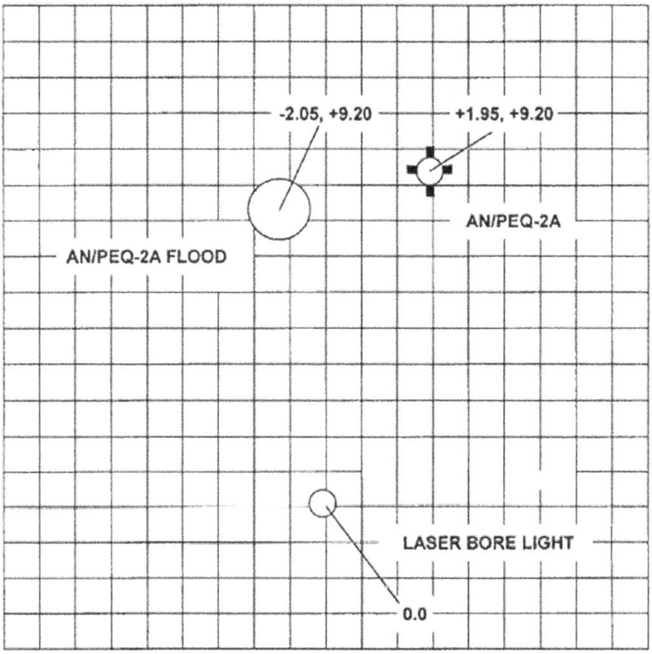

This is the 10-meter boresight target for 500-meter zero.
Grids are 1-centimeter square.
Units manufacture targets locally.

Zero target data for M2 MG:

1. **Stabilize the weapon.**
2. **Align laser borelight on its dot.**
3. **Adjust the AN/PEQ-2A until aiming laser centers on the dot cross hair.**

Figure G-12. Alignment of M2 with the AN/PEQ-2A.

10-meter boresight target
AN/TVS-5 mounted on AN/TVS-5 feed tray cover bracket on the M2 MG.

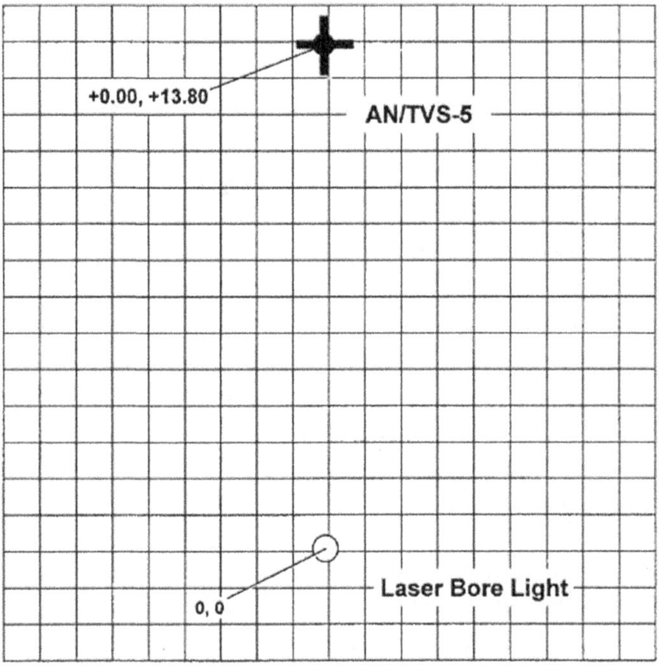

This is the 10-meter boresight target for the 500-meter zero. Grids measure 1 cm square. Units manufacture targets locally.

Zero target data for M2 MG:

1. Stabilize the weapon.
2. Aim the AN/TVS-5 at the cross hair and adjust until the borelight strikes the laser's borelight dot.

Figure G-13. Alignment of M2 with the AN/TVS-5.

10-meter boresight target AN/PAS-13 mounted on AN/PAS-13 feed tray cover bracket on the M2 MG.

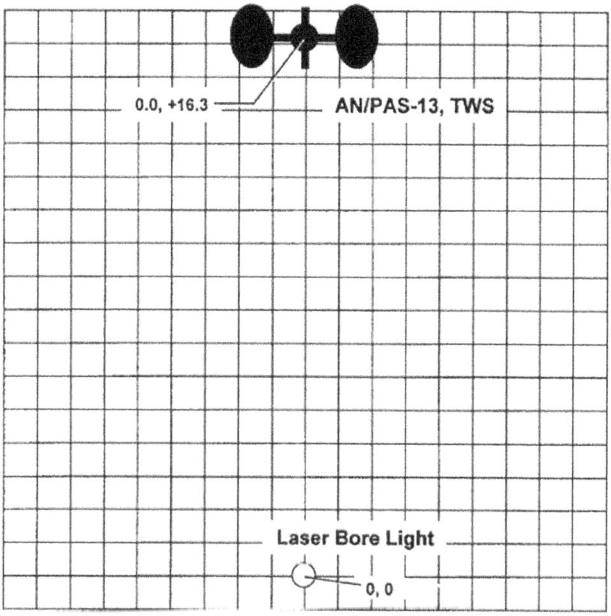

This is the 10-meter boresight target for the 500-meter zero. Grids measure 1 cm square. Units manufacture targets locally.

Zero target data for M2 MG:

1. Stabilize the weapon.
2. Place the thermal signature material or the tips of fingers on the gray circles.
3. Aim between the gray circles at the cross hairs.
4. Adjust the AN/PAS-13 until the borelight laser centers on its dot.
5. Zero the reticle in *both* wide and narrow fields of view.

Figure G-14. Alignment of M2 with the AN/PAS-13.

Section V. T&E MANIPULATION

The T&E mechanism allows engagement of preselected target areas at night or during limited visibility. The gunner records (in mils) the direction and elevation readings from the traversing bar and T&E mechanism.

G-12. ZEROING THE T&E MECHANISM

Before a gunner can effectively engage targets with the T&E, he must zero it to the weapon:

a. **Zero Traversing Handwheel.** Hold the T&E so the traversing handwheel sits on the left as you look at it.

(1) Turn the traversing handwheel toward you until it stops.

(2) Loosen the locking nut slightly.

(3) Align the zero on the scale with the zero on the elevating screw yoke.

(4) Hold the scale (with the zeroes aligned) and tighten the locking nut. Make sure the zeroes remain aligned.

(5) Turn the traversing handwheel two complete revolutions away from you. If doing this at night, count 50 "clicks" away from you.

b. **Zero Elevating Handwheel to Upper Elevating Screw.** Align the two zeroes.

(1) Rotate the elevating handwheel up or down until you can see a zero with a line below it on the upper elevating screw.

(2) Position the elevating handwheel so the indicator points to the zero on the handwheel.

c. **Zero Elevating Mechanism Sleeve to Lower Elevating Screw.**

(1) Rotate the elevating mechanism sleeve all the way up, and then rotate it down until it stops. Note the number of complete turns down.

(2) Rotate the elevating mechanism sleeve up half that many turns.

(3) Position the slide lock lever to face you.

(4) Attach the T&E to the tripod and gun.

G-13. LAYING THE GUN FOR DIRECTION

After the gunner receives an assigned sector of fire, he should--

a. Pick up the rear legs of the tripod.

b. Shift the tripod until the muzzle of the weapon points to the center of the sector of fire.

c. After laying the weapon for direction, firmly stamp in the tripod shoes and place sandbags on the legs. This aids stability and may prevent accidental movement.

d. Obtain and record directional readings to all targets within the sector and perform the following:

(1) Loosen the traversing slide lock lever and move the slide along the traversing bar until the weapon lays either on the center of a point target or on the flank of a linear target.

(2) Lock the traversing bar. Read the direction from the scale on the traversing bar. If the left edge of the traversing slide falls anywhere

outside the 5-mil tick mark, the gunner moves the left edge of the traversing bar slide back to the next smaller mil reading. Then, the gunner uses the traversing handwheel to complete the initial lay.

e. Obtain a reading to the target by the direction of the weapon barrel. If the barrel is moved to the right, the gunner records a right heading. He reads the number on the traversing bar from the left side of the traversing slide lock. If the barrel is moved to the left and the traversing slide lock is on the right side of the zero, he records a left reading.

f. After taking a directional reading for a target, measure the width of the target in mils. Then, he uses the traversing handwheel to move the barrel across the target--one click equals one mil.

g. Before moving to another target, reposition the traversing mechanism.

h. Obtain elevation readings. The gunner verifies that the weapon is laid on the center base of the target. Read the elevation from two scales.

i. Get the first, or major, part of the elevation from the elevating screw plate scale.

j. Get the second, or minor, part of the elevation from the elevating handwheel.

k. Separate the two parts of the elevation reading with a slash (/). For example, the gunner could write "-50/3." An elevation reading might apply only to the T&E where he reads it.

l. Note that, if the number of threads increases or decreases after he records the data, he cannot place accurate fire on the target. For example, rotating the base of the T&E mechanism to engage a secondary sector makes the data incorrect--unless the same number of threads appear both before and after the move.

* GLOSSARY

AP	armor piercing
APC	armored personnel carrier
approx	approximately
attn	attention
ba	battery
BFA	blank firing attachment
BFV	Bradley fighting vehicle
BMP	Soviet fighting vehicle
cal	caliber
cdr	commander
CLP	cleaner, lubricant, and preservative
CVKI	combat vehicle kill indicator
DA	Department of the Army
FEBA	forward edge of the battle area
FM	field manual
FOV	field of view
FPF	final protective fires
FPL	final protective line
FSN	federal stock number
FTX	field training exercise
GA	Georgia
HB	heavy barrel
HMMWV	high-mobility, multi-purpose, wheeled vehicle
HWTS	heavy weapon thermal sight
IAW	in accordance with
IN	infantry
IR	infrared
LED	light-emitting diode
LFX	live-fire exercise
LSA	semifluid lubricating oil
LTA	local training area
m	meter
METL	mission-essential task list
MG	machine gun
MILES	multiple integrated laser engagement system
mm	millimeter
MOPP	mission-oriented protection posture
mph	miles per hour
MOUT	military operations on urbanized terrain
MTA	major training area
MTP	mission training plan
NBC	nuclear, biological, and chemical
NCO	noncommissioned officer

NCOES	noncommissioned officer education system
NCOIC	noncommissioned officer in charge
NSN	national stock number
NVD	night vision device
OIC	officer in charge
PDF	principal direction of fire
PL-M	lubricating oil, general purpose
PL-S	lubricating oil, special purpose
POI	program of instruction
RABA	recoil amplifier barrel assembly
RBC	rifle bore cleaner
SM	soldiers manual
SMCT	soldiers manual of common tasks
SOP	standing operating procedure
SRTA	short-range training ammunition
SRTA-T	short-range training ammunition-tracer
STP	soldiers training publication
STRAC	Standards in Training Commission
STX	situational training exercise
T&E	traversing and elevating
TM	technical manual
TRADOC	Training and Doctrine Command
TRP	target reference point
US	United States
v	volt

* REFERENCES

SOURCES USED

These are the Army publications quoted or paraphrased in this manual.

DA Pam 350-38	Standards in Weapons Training. 3 July 1997
FM 3-5	NBC Decontamination. 17 November 1993.
FM 44-80	Visual Aircraft Recognition. 30 September 1996.
STP 21-24-SMCT	Soldier's Manual of Common Tasks, Skill Levels 2/3/4. 1 October 1992.
TC 25-8	Training Ranges. 25 February 1992, with Change 1, 16 November 1992.
TM 9-1005-213-10	Operator's Manual for Machine Guns, Caliber .50; Browning, M2, Heavy Barrel Flexible, 6 August 1992, with Changes 1 and 2, 1 August 1995 and 6 August 1996.
TM 9-1300-200	Ammunition General. 3 October 1969; with changes 1 through 5, dated 22 January 1971 through 30 September 1993.
TM 11-5855-214-10	Operator's Manual for Night Vision Sight, Crew Served Weapon, AN/TVS-5. 15 February 1989.
TM 11-5855-302-12&P	Operator's and Unit Maintenance Manual (including Repair Parts and Special Tools List), Sight, Thermal, AN/PAS-13(V)2, AN/PAS-13(V)3. To be published.
TM 11-5855-312-10	Operator's Manual, Sight, Thermal, AN/PAS-13B(V)2, AN/PAS-13B(V)3. To be published.

DOCUMENTS NEEDED

These Army publications must be available to the intended users of this manual.

Army Forms

These forms can be locally reproduced on 81/2 x 11-inch paper. The reproducible form is located at the back of this manual. The forms are also available on the Army Electronic Library (AEL) CD-ROM (EM0001) and at the USAPA website located at (www.usapa.army.mil):

DA Form 2404	Equipment Inspection and Maintenance Worksheet. 1 April 1979.
DA Form 5517-R	Standard Range Cards (LRA). February 1986.
DA Form 7007-R	Machine Gun Scorecard for M2. August 2001.
DA Form 7448-R	M2 Caliber .50 Heavy Barrel Machine Gun, Firing Table I, Day Practice Scorecard. January 2002.
DA Form 7449-R	M2 Caliber .50 Heavy Barrel Machine Gun, Firing Table II, Day Qualification Scorecard. January 2002.
* DA Form 7450-R	M2 Caliber .50 Heavy Barrel Machine Gun, Firing Tables III(A) and III(B), Night Practice Scorecard. October 2002.
* DA Form 7451-R	M2 Caliber .50 Heavy Barrel Machine Gun, Firing Tables IV(A), Night Qualification Scorecard; and IV(B), Day Practice Scorecard. October 2002.

Technical Manuals

TM 11-5855-306-10	Operator's Manual Monocular Night Vision Device (MNVD), AN/PVS-14. June 2000.
TM 11-5855-306-23&P	Unit and Direct Support Maintenance Manual (Including Repair Parts and Special Tools List), Monocular Night Vision Device (MNVD), AN/PVS-14. June 2000.
TM 11-5855-309-12&P	Operator's and Unit Maintenance Manual (including Repair Parts and Special Tools List), Sight, Thermal, AN/PAS-13A(V)2, AN/PAS-13A(V)3. To be published.

INTERNET WEB SITES

U.S. Army Publishing Agency (USAPA):
www.usapa.army.mil

Reimer Digital Library (RDL):
www.adtdl.army.mil

Air Force Publishing:
afpubs.hq.af.mil/

* INDEX

MACHINE GUN SCORECARD FOR M2

For use of this form, see FM 23-65, Change 1, or the back of this form. The proponent agency is TRADOC.

DATA REQUIRED BY THE PRIVACY ACT OF 1974

AUTHORITY: 10 USC 3012(g)/Executive order 9397.

PRINCIPAL PURPOSE: To aid individual training on targets at various ranges.

ROUTINE USE: To evaluate individual proficiency. SSN is used for positive identification purposes only.

DISCLOSURE: Voluntary. However, mass rating and scoring require some tracking method. Therefore, only those soldiers who provide their SSNs can receive scores or qualify on weapons or equipment.

1a. LAST NAME	1b. FIRST NAME	1c. MI	2. DATE *(YYYYMMDD)*

3. SSN	4. UNIT	5. LANE

	6a. FIRING TABLE I			7a. FIRING TABLE II					8a. FIRING TABLE III					9a. FIRING TABLE IV				
TSK	RANGE (M)	HIT	PTS	RANGE (M)	TIME	HIT	PTS	BON	RANGE (M)	TIME	HIT	PTS	BON	RANGE (M)	TIME	HIT	PTS	BON
1	10	NA	NA	550	None	NA	NA	NA	550	None	NA	NA	NA	550	None	NA	NA	NA
2	10	NA	NA	800	20 Sec				800	20 Sec				800	20 Sec			
3	10	NA	NA	400	20 Sec				400	20 Sec				400	20 Sec			
4	10	NA	NA	700	25 Sec				700	25 Sec				700	25 Sec			
5	10			1,000	25 Sec				1,000	25 Sec				1,000	25 Sec			
6	10			400	35 Sec				400	35 Sec				400	35 Sec			
				700					700					700				
7	10			550	35 Sec				550	35 Sec				550	35 Sec			
				800					800					800				
8	NA	NA	NA	400	45 Sec				400	45 Sec				400	45 Sec			
				550					550					550				
				1,000					1,000					1,000				

6b. SUBTOTAL FIRING TABLE I	7b. SUBTOTAL FIRING TABLE II	8b. SUBTOTAL FIRING TABLE III	9b. SUBTOTAL FIRING TABLE IV

10. TOTAL SCORE	11. RATING
12a. OIC's PRINTED OR TYPED NAME	13a. GRADER'S PRINTED OR TYPED NAME
12b. OIC'S SIGNATURE	13b. GRADER'S SIGNATURE

COMPLETE THE BLOCKS ON THIS SCORECARD AS FOLLOWS:

1. NAME. Enter last name, first name, middle initial.

2. DATE. Enter today's date in YYYYMMDD format.

3. SSN. Enter gunner's social security number.

4. UNIT. Enter gunner's unit.

5. LANE. Enter gunner's firing lane.

6a - 9a. FIRING TABLE I thru IV

HIT column. As applicable, mark block with an "X" if gunner hit target.

POINTS column. As applicable, enter points earned for each target hit.

BON *(Bonus)* column *(Firing Tables I thru III only)*. As applicable, enter bonus points earned for each target hit.

6b-9b. SUBTOTALS. Subtotal all of the points earned for each firing table.

10. TOTAL SCORE. Add firing table subtotals *(Blocks 6b - 9b)* to obtain this score.

11. RATING. Compare total score *(Block 10)* to rating chart to determine the soldier's rating.

12a. OIC's PRINTED OR TYPED NAME. Print or type OIC's name.
12b. OIC's SIGNATURE. *(OIC must sign the actual form)*.

13a. GRADER's PRINTED OR TYPED NAME. Print or type grader's name.
13b. GRADER's SIGNATURE. *(Grader must sign the actual form)*.

COMPUTE RATING:

Maximum..............................	218 points
Expert..................................	196 to 217 points
Gunner first class.....................	174 to 195 points
Gunner second class................	153 to 173 points
Unqualified..............................	152 points or less

M2 CALIBER .50 HEAVY BARREL MACHINE GUN FIRING TABLE I
DAY PRACTICE SCORECARD

For use of this form, see FM 23-65, Change 1; the proponent agency is TRADOC.

DATA REQUIRED BY THE PRIVACY ACT OF 1974

AUTHORITY: 10 USC 3012(g)/Executive order 9397.

PRINCIPAL PURPOSE: To aid individual training on targets at various ranges.

ROUTINE USES: To evaluate individual proficiency. SSN is used for positive identification purposes only.

DISCLOSURE: Voluntary. However, mass rating and scoring require some tracking method. Therefore, only those soldiers who provide their SSNs can receive scores or qualify on weapons or equipment.

1a. LAST NAME	1b. FIRST NAME	1c. MI	2. DATE *(YYYYMMDD)*

3. SSN	4. UNIT	5. RANGE	6. LANE

7a. ENGAGEMENT	7b. STANDARDS	7c. AMMO	7d. TIME *(Minutes)*	7e. RANGE *(Meters)*	7f. GO	7g. NO GO
ZERO	1 BURST HIT	14	NA	400	NA	NA
2	1 BURST HIT	14	1.5	1,100 (±200)		
3	1 BURST HIT	14	1.5	1,500 (±200)		
4	1 BURST HIT	14	1.5	600 (±100)		
5	1 BURST HIT	14	1.5	800 (±100)		
6	1 BURST HIT	14	1.5	400		
8. MULTIPLE TARGETS						
7	1 BURST HIT	28	2.5	1,100 (±200)		
	1 BURST HIT			600 (±100)		
8	1 BURST HIT	28	2.5	800 (±100)		
	1 BURST HIT			1,500 (±200)		

9. SCORING *(Check one)*

EXPERT - 9 ☐ SHARPSHOOTER - 8 ☐

MARKSMAN - 6 ☐ UNQUALIFIED - 5 OR BELOW ☐

10. GUNNER'S SIGNATURE

11a. GRADER'S PRINTED OR TYPED NAME	11b. GRADER'S SIGNATURE

DA FORM 7448-R, JAN 2002 USAPA V1.00

M2 CALIBER .50 HEAVY BARREL MACHINE GUN FIRING TABLE II
DAY QUALIFICATION SCORECARD

For use of this form, see FM 23-65, Change 1; the proponent agency is TRADOC.

DATA REQUIRED BY THE PRIVACY ACT OF 1974

AUTHORITY: 10 USC 3012(g)/Executive order 9397.
PRINCIPAL PURPOSE: To aid individual training on targets at various ranges.
ROUTINE USES: To evaluate individual proficiency. SSN is used for positive identification purposes only.
DISCLOSURE: Voluntary. However, mass rating and scoring require some tracking method. Therefore, only those soldiers who provide their SSNs can receive scores or qualify on weapons or equipment.

1a. LAST NAME	1b. FIRST NAME	1c. MI	2. DATE *(YYYYMMDD)*

3. SSN	4. UNIT	5. RANGE	6. LANE

7a. ENGAGEMENT	7b. STANDARDS	7c. AMMO	7d. TIME *(Minutes)*	7e. RANGE *(Meters)*	7f. GO	7g. NO GO
ZERO	1 BURST HIT	14	NA	400	NA	NA
2	1 BURST HIT	14	1	1,100 (±200)		
3	1 BURST HIT	14	1	1,500 (±200)		
4	1 BURST HIT	14	1	600 (±100)		
5	1 BURST HIT	14	1	800 (±100)		
6	1 BURST HIT	14	1	400		
8. MULTIPLE TARGETS						
7	1 BURST HIT	28	2	1,100 (±200)		
	1 BURST HIT			600 (±100)		
8	1 BURST HIT	28	2	800 (±100)		
	1 BURST HIT			1,500 (±200)		

9. SCORING *(Check one)*

EXPERT - 9 ☐ SHARPSHOOTER - 8 ☐

MARKSMAN - 6 ☐ UNQUALIFIED - 5 OR BELOW ☐

10. GUNNER's SIGNATURE

11a. GRADER'S PRINTED OR TYPED NAME	11b. GRADER'S SIGNATURE

DA FORM 7449-R, JAN 2002 USAPA V1.00

M2 CALIBER .50 HEAVY BARREL MACHINE GUN FIRING TABLES III (A) AND III (B) NIGHT PRACTICE SCORECARDS

For use of this form, see FM 23-65, Change 2; the proponent agency is TRADOC.

DATA REQUIRED BY THE PRIVACY ACT OF 1974

AUTHORITY: 10 USC 3012(g)/Executive order 9397.
PRINCIPAL PURPOSE: To aid individual training on targets at various ranges.
ROUTINE USES: To evaluate individual proficiency. SSN is used for positive identification purposes only.
DISCLOSURE: Voluntary. However, mass rating and scoring require some tracking method. Therefore, only those soldiers who provide their SSNs can receive scores or qualify on weapons or equipment.

1a. LAST NAME	1b. FIRST NAME	1c. MI	1d. RANK	2. SSN

TABLE III (A). NIGHT PRACTICE SCORECARD

NOTE: Use this table if the M2 has the TWS, or if it has an AN/TVS-5 that has been upgraded with the third-generation tube.

a. RANGE	b. UNIT	c. LANE	d. DATE *(YYYYMMDD)*

e. ENGAGEMENT	f. STANDARDS	g. AMMO	h. TIME *(Minutes)*	i. RANGE *(Meters)*	j. GO	k. NO GO
ZERO	1 BURST HIT	14	NA	400	NA	NA
2	1 BURST HIT	14	1.5	1,100 (±200)		
3	1 BURST HIT	14	1.5	1,500 (±200)		
4	1 BURST HIT	14	1.5	600 (±100)		
5	1 BURST HIT	14	1.5	800 (±100)		
6	1 BURST HIT	14	1.5	400		
			l. MULTIPLE TARGETS			
7	1 BURST HIT	28	2.5	1,100 (±200)		
	1 BURST HIT			600 (±100)		
8	1 BURST HIT	28	2.5	800 (±100)		
	1 BURST HIT			1,500 (±200)		

4. SCORING *(Check one)*
EXPERT - 9 ☐
SHARPSHOOTER - 8 ☐
MARKSMAN - 6 ☐
UNQUALIFIED - 5 OR BELOW ☐

5. GUNNER'S SIGNATURE

6a. GRADER'S PRINTED OR TYPED NAME

6b. GRADER'S SIGNATURE

7. TABLE III (B). NIGHT PRACTICE SCORECARD

NOTE: Use this table if the M2 has an AN/TVS-5 that has not been upgraded.

a. RANGE	b. UNIT	c. LANE	d. DATE *(YYYYMMDD)*

e. ENGAGEMENT	f. STANDARDS	g. AMMO	h. TIME *(Minutes)*	i. RANGE *(Meters)*	j. GO	k. NO GO
ZERO	1 BURST HIT	14	NA	400	NA	NA
2	1 BURST HIT	14	1.5	1,100 (±200)		
3	1 BURST HIT	14	1.5	800 (±100)		
4	1 BURST HIT	14	1.5	600 (±100)		
5	1 BURST HIT	14	1.5	800 (±100)		
6	1 BURST HIT	14	1.5	400		
			l. MULTIPLE TARGETS			
7	1 BURST HIT	28	2.5	1,100 (±200)		
	1 BURST HIT			600 (±100)		
8	1 BURST HIT	28	2.5	800 (±100)		
	1 BURST HIT			600 (±100)		

8. SCORING *(Check one)*
EXPERT - 9 ☐
SHARPSHOOTER - 8 ☐
MARKSMAN - 6 ☐
UNQUALIFIED - 5 OR BELOW ☐

9. GUNNER'S SIGNATURE

10a. GRADER'S PRINTED OR TYPED NAME

10b. GRADER'S SIGNATURE

M2 CALIBER .50 HEAVY BARREL MACHINE GUN FIRING TABLES IV (A) AND IV (B) NIGHT QUALIFICATION SCORECARDS

For use of this form, see FM 23-65, Change 2; the proponent agency is TRADOC.

DATA REQUIRED BY THE PRIVACY ACT OF 1974

AUTHORITY: 10 USC 3012(g)/Executive order 9397.
PRINCIPAL PURPOSE: To aid individual training on targets at various ranges.
ROUTINE USES: To evaluate individual proficiency. SSN is used for positive identification purposes only.
DISCLOSURE: Voluntary. However, mass rating and scoring require some tracking method. Therefore, only those soldiers who provide their SSNs can receive scores or qualify on weapons or equipment.

1a. LAST NAME	1b. FIRST NAME	1c. MI	1d. RANK	2. SSN

3. TABLE IV (A). NIGHT QUALIFICATION SCORECARD

NOTE: Use this table if the M2 has the TWS, or if it has an AN/TVS-5 that has been upgraded with the third-generation tube.

a. RANGE	b. UNIT	c. LANE	d. DATE *(YYYYMMDD)*

e. ENGAGEMENT	f. STANDARDS	g. AMMO	h. TIME *(Minutes)*	i. RANGE *(Meters)*	j. GO	k. NO GO
ZERO	1 BURST HIT	14	NA	400	NA	NA
2	1 BURST HIT	14	1	1,100 (±200)		
3	1 BURST HIT	14	1	1,500 (±200)		
4	1 BURST HIT	14	1	600 (±100)		
5	1 BURST HIT	14	1	800 (±100)		
6	1 BURST HIT	14	1	400		
l. MULTIPLE TARGETS						
7	1 BURST HIT	28	2	1,100 (±200)		
	1 BURST HIT			600 (±100)		
8	1 BURST HIT	28	2	800 (±100)		
	1 BURST HIT			1,500 (±200)		

4. SCORING *(Check one)*
EXPERT - 9 ☐
SHARPSHOOTER - 8 ☐
MARKSMAN - 6 ☐
UNQUALIFIED - 5 OR BELOW ☐

5. GUNNER'S SIGNATURE

6a. GRADER'S PRINTED OR TYPED NAME

6b. GRADER'S SIGNATURE

7. TABLE IV (B). NIGHT QUALIFICATION SCORECARD

NOTE: Use this table if the M2 has an AN/TVS-5 that has not been upgraded.

a. RANGE	b. UNIT	c. LANE	d. DATE *(YYYYMMDD)*

e. ENGAGEMENT	f. STANDARDS	g. AMMO	h. TIME *(Minutes)*	i. RANGE *(Meters)*	j. GO	k. NO GO
ZERO	1 BURST HIT	14	NA	400	NA	NA
2	1 BURST HIT	14	1	1,100 (±200)		
3	1 BURST HIT	14	1	800 (±100)		
4	1 BURST HIT	14	1	600 (±100)		
5	1 BURST HIT	14	1	800 (±100)		
6	1 BURST HIT	14	1	400		
l. MULTIPLE TARGETS						
7	1 BURST HIT	28	2	1,100 (±200)		
	1 BURST HIT			600 (±100)		
8	1 BURST HIT	28	2	800 (±100)		
	1 BURST HIT			600 (±100)		

8. SCORING *(Check one)*
EXPERT - 9 ☐
SHARPSHOOTER - 8 ☐
MARKSMAN - 6 ☐
UNQUALIFIED - 5 OR BELOW ☐

9. GUNNER'S SIGNATURE

10a. GRADER'S PRINTED OR TYPED NAME

10b. GRADER'S SIGNATURE

FIELD MANUAL
NO. 3-22.27

HEADQUARTERS
DEPARTMENTS OF THE ARMY
WASHINGTON, DC, 28 November 2003

MK 19, 40-mm GRENADE MACHINE GUN, MOD 3

CONTENTS

DISTRIBUTION RESTRICTION: Approved for public release; distribution is unlimited.

*This publication supersedes FM 23-27, dated 27 December 1988.

PREFACE

This manual discusses how to train the unit to use the MK 19, 40-mm grenade machine gun, model (MOD) 3, referred to in this manual as the MK 19. This manual highlights mechanical training, weapon capabilities, and gunnery principles, methods, techniques, and standards that apply to the MOD 3. It also includes preliminary gunnery, a gunnery skills test, gunnery tables, and qualification tables. If this information conflicts with an applicable technical manual (TM), revised TM, or TM update, the crew will follow the guidance in the most recently published document.

The primary audiences for this manual are soldiers, trainers, and staff officers. Units can modify the gunnery program to meet local training restraints. Although the gunnery tables are intended for use with live fire, trainers can use the tactical engagement simulator system (TESS), a video disc trainer (VDT), a multipurpose arcade combat simulator (MACS), or other training device, except on qualification tables.

In all cases, units must evaluate their training to ensure that it follows the building-block principle and adheres to sound training policy. Only a crew that is trained and does well in preliminary gunnery exercises is likely to do well in live-fire exercises (LFX) and in combat situations.

The proponent for this publication is the U.S. Army Infantry School. Send comments and recommendations to doctrine@benning.army.mil or on a DA Form 2028 (Recommended Changes to Publications and Blank Forms) directly to Commandant, U.S. Army Infantry School, ATTN: ATSH-ATD, Fort Benning, Georgia 31905-5593.

Unless this manual states otherwise, masculine nouns and pronouns do not refer exclusively to men.

CHAPTER 1
INTRODUCTION

Although the MK 19 is a recent entry into the Army's inventory, development began in 1963. The first version was a hand-cranked, multiple grenade launcher called the MK 18. In 1966 the need for more firepower inspired the development of a self-powered 40-mm machine gun called the MK 19, MOD 0. This model was neither reliable nor safe enough for use as a military weapon system. Product improvements begun in 1971 resulted in the 1972 MOD 1, of which only six were produced. The MOD 1 performed effectively in Navy riverine patrol craft and broader applications for the MK 19 were found. In 1973 the Navy developed the MOD 2, which featured improved reliability, safety, and maintainability. In 1976 a complete redesign resulted in the MK 19, MOD 3, which the Army adopted in 1983. The Army now uses the MK 19 within the tactical environment for defense, retrograde, patrolling, rear area security, urban operations, and special operations.

This chapter provides applications, training strategies, and descriptive, technical, and operational data for the MK 19 (Figure 1-1).

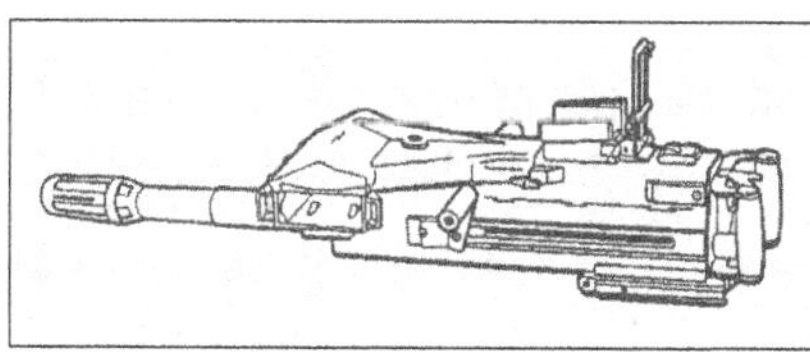

Figure 1-1. MK 19, 40-mm grenade machine gun, MOD 3.

1-1. APPLICATIONS

The MK 19 supports the soldier in both the offense and defense. It gives the unit the capability of laying down a heavy volume of close, accurate, and continuous fire. The MK 19 can also:

- Protect motor movements, assembly areas, and supply trains in a bivouac.
- Defend against hovering rotary aircraft.
- Destroy lightly-armored vehicles.
- Fire on suspected enemy positions.
- Provide high volumes of fire into an engagement area (EA).
- Cover obstacles.
- Provide indirect fires from defilade positions.

1-2. DESCRIPTION

The MK 19 is an air-cooled, blowback-operated machine gun with five major assemblies (Figure 1-2). A disintegrating metallic link belt feeds ammunition through the left side of the weapon. Tables 1-1 and 1-2 provide MK 19 technical and operational data, respectively.

a. **Receiver Assembly.** Holds the barrel and other parts of the gun. Ammunition is fed into the left side of the receiver through the feed throat assembly. The MK 19's barrel will not overheat, even after prolonged firing.

b. **Feed Slide Assembly and Tray.** Holds the rounds in the feeder and indexes the ammunition into position for delinking.

c. **Top Cover Assembly.** Holds the feed slide assembly and tray. It is opened by a latch (left side) for loading or to clean and inspect feeder area. A blade-type front sight is attached to the top cover assembly (Figure 1-3).

d. **Sear Assembly.** Holds the receiver sear. Trigger action releases the sear and allows the bolt to go forward. The safety is attached to the sear assembly.

e. **Bolt and Backplate Assembly.** The bolt fires the round when the sear is depressed by trigger action. The recoil springs drive the bolt forward on the receiver rails. The guide rods hold the springs in position. Trigger and handgrips are located on the backplate assembly.

f. **Feed Throat Assembly.** Allows smooth feeding of 40-mm ammunition. It attaches to the forward left side of the receiver by two sets of spring-loaded retaining pins. Without a feed throat, machine gun stoppages may occur because of twisted or misaligned rounds.

g. **Leaf-Type Rear Sight (with adjustable range plate).** Is marked in 100-meter intervals from 300 to 1,500 meters. The sight is mounted on a spring dovetail base to the receiver assembly (Figure 1-4). Before moving the weapon, the gunner folds the sight forward to a horizontal position. The rear sight can be adjusted for range and windage.

(1) ***Range.*** Different adjustments can be made to the range. Use the rear sight slide release to make *major adjustments* to the range. Use the elevation wheel to make *fine adjustments* to the range.

(2) ***Windage.*** Use the rear sight to adjust for windage. One click equals a 1-mil change. To move the sight to the *right*, turn the windage screw clockwise. To move the sight to the *left*, turn the windage screw counterclockwise.

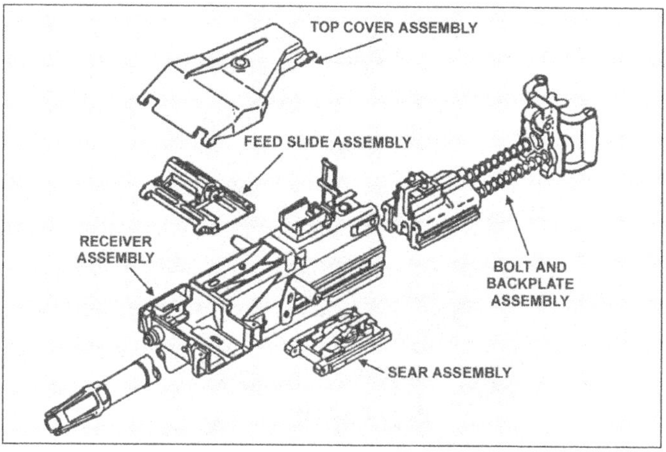

Figure 1-2. Five major assemblies.

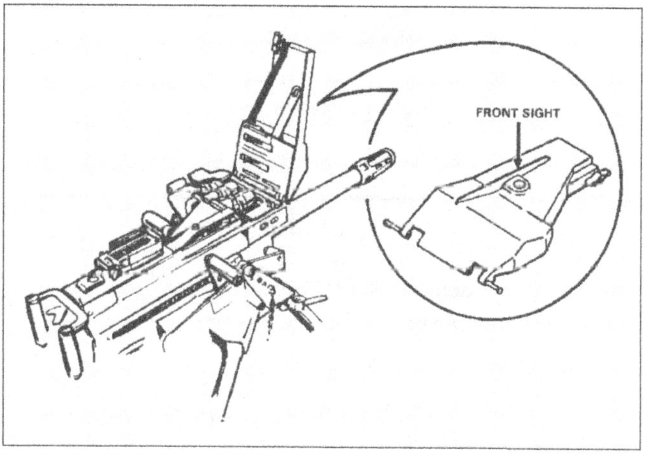

Figure 1-3. Front sight on top cover assembly.

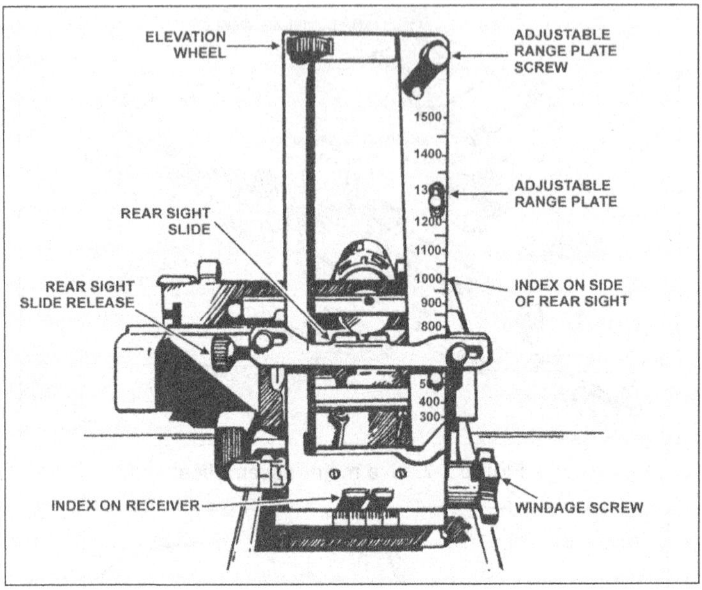

Figure 1-4. Rear sight.

> **WARNING**
>
> **Use only prescribed ammunition. Mixing types of ammunition could result in injury.**

h. The MK 19 uses the following 40-mm cartridges (Figure 1-5, page 1-8).

(1) ***High-Explosive, Dual-Purpose M430 Cartridge.*** The high-explosive, dual-purpose (HEDP) M430 cartridge is the standard round for the MK 19 (Department of Defense Identification Code [DODIC] B542). They are linked with M16A2 links. The HEDP round, the top-curved portion of the projectile, is olive drab with a yellow ogive and yellow markings. It is packed in M548 (48 rounds) or PA120 (32 rounds) ammunition containers. The HEDP, an impact-type round, can penetrate 2 inches of steel armor at 0-degree obliquity and inflict personnel casualties out to 15 meters from impact. It arms within 18 to 30 meters of the gun muzzle and has a point-initiating, base-detonating (PIBD) fuze.

(2) ***High-Explosive M383 Cartridge.*** The high-explosive (HE) M383 cartridge round is olive drab with a yellow ogive and yellow markings. It is packed in a metal ammunition container (48 rounds, linked, in each container). The HE round has a wound radius of 15

meters. It lacks the armor-penetrating ability of the HEDP M430 round. The HE arms between 18 to 36 meters of the gun muzzle fuze.

(3) ***M922 Dummy Cartridges.*** Each MK 19 is issued with one 10-round of inert dummy rounds belt (DODIC B472). M16A2 links join the dummy rounds into a 10-round belt packed in an M2A1 metal box. Trainers use dummy rounds to check weapon function and to train crews.

(4) ***M918 Cartridge.*** The M918 is a training practice cartridge that has the same muzzle velocity of 790 feet per second (fps), signature, and sound as the HE round (DODIC B584).

MK 19 TECHNICAL DATA		
MK 19 (MOD 3):		
Weight without feed throat	77.6 pounds	
Weight with feed throat	78.0 pounds	
Length	43.1 inches	
Width	14.0 inches	
Height	8.8 inches	
MK 64 (MOD 7) gun cradle:		
Weight	21.0 pounds	
Length	17.5 inches	
Height	9.5 inches	
Tripod (M3) weight:	44.00 pounds	
Gun and cradle:		
Weight without feed throat	98.6 pounds	
Weight with feed throat	99.0 pounds	
Gun, cradle, and tripod:		
Weight without feed throat	142.6 pounds	
Weight with feed throat	143.0 pounds	
Mounts:	**Ammunition:**	
M3 tripod		M430 (HEDP)
M4 pedestal		M383 (HE)
M66 ring		M918 (TP)
HMMWV weapon platform		M922 (dummy)
M113 APC commander's cupola		
NOTE: Keep the feed throat attached to the weapon.		

Table 1-1. Technical data.

MK 19 OPERATIONAL DATA		
Maximum range:	2,212	meters
Maximum effective range:	1,500	meters (point target)
	2,212	meters (area target)
Rates of fire:		
Sustained	40	rounds per minute
Rapid	60	rounds per minute
Cyclic	325 to 375	rounds per minute
Ammunition:		
M430 HEDP	2	inch armor
	15	meter casualty radius
M383 HE	15	meter casualty radius
Service frequency:	50,000	rounds
Elevation, tripod controlled:	100	mils
Depression, tripod controlled:	258	mils
Traverse, tripod controlled:	800	mils (400 left plus 400 right)
Muzzle velocity (average):	798	feet per second
Recoil forces (average):	500	pounds
Angle of automatic fire:	0 to 70	degrees elevation (automatic fire), based on mounting arrangements
Weights:		
Rounds	62	pounds (48 rounds in M548 metal container)
	42	pounds (32 rounds in PA120 metal container)
Planned operating load	400	prescribed by local commanders

Table 1-2. Operational data.

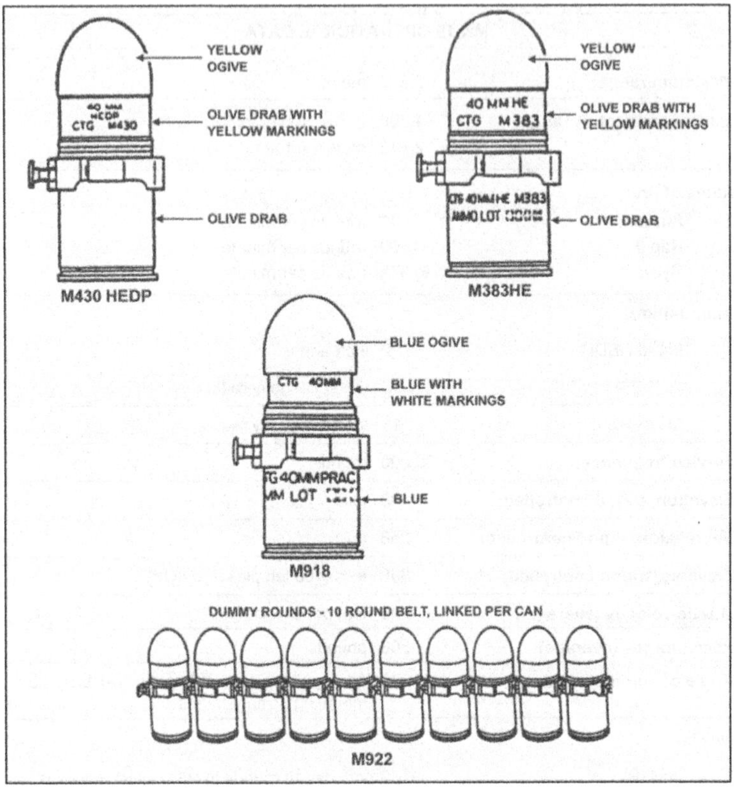

Figure 1-5. 40-mm cartridges.

1-3. TRAINING STRATEGY

A training strategy that integrates resources with the instruction and training of individual and collective skills is necessary to carry out a unit's wartime mission. TRADOC institutions, such as the noncommissioned officers education system (NCOES), and units implement overall training strategies for the MK 19. This multifaceted approach includes specific plans for using resources such as publications, ranges, ammunition, training aids, devices, simulators, and simulations. These specific strategies develop critical soldier skills and training tasks, as well as leader skills needed to support the intended outcome (see Appendix A, Training Strategy, for more details).

CHAPTER 2

OPERATION AND FUNCTION

This chapter includes the cycle of operation; operating precautions; clearing; disassembly and assembly; loading; malfunctions, stoppages, and corrections; operations under adverse conditions; and destruction of the MK 19. Refer to Appendix B, Operations Checklist, for procedures to be used before, during, and after operation of the MK 19 and Appendix C, Stoppages, Malfunctions, Immediate Actions, and Remedial Actions, for procedures to correct stoppages and malfunctions.

2-1. CYCLE OF OPERATION

The MK 19's cycle of operation includes six steps: charging, extracting (delinking), cocking, firing, blowback, and automatic feeding. More than one step may be done at the same time.

a. **Charging.** The charger handles are used to pull the bolt (1) to the rear, aligning the round with the bolt extractors. The rearward movement of the bolt causes the primary drive lever (2) to move to the left, which moves the secondary drive lever (3) to the right. The forked end of the secondary drive lever, which rests on the feed slide pin (4), moves the feed slide (5) to the right. The feed pawls (6) on the feed slide move the linked rounds (7) over one place in the ammunition-feed area of the receiver. The leading round (8) lines up with the bolt extractor (9). See Figure 2-1.

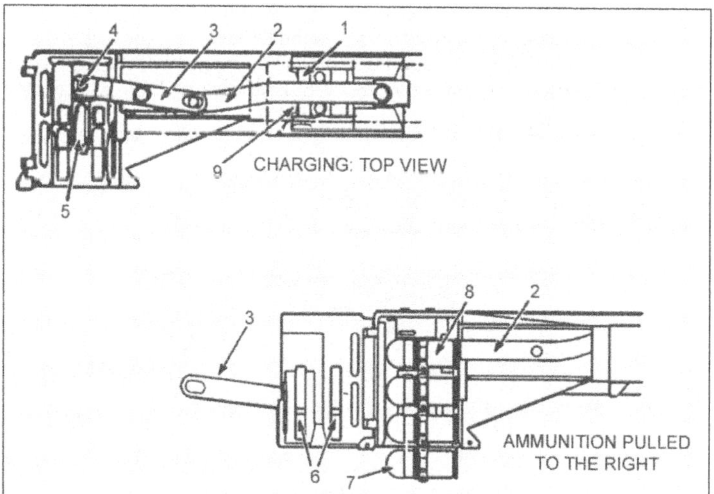

Figure 2-1. Charging cycle.

b. **Extracting (Delinking).** When a round is stripped from the belt, it is extracted or "delinked." This happens, after the MK 19 has been charged once, when the trigger (10) is pressed. The bolt slams forward and the bolt's extractors (11) snap over the rim of the cartridge case. When the MK 19 is charged again, the extractor pulls the leading round to the rear and separates the male and female links. The curved edge of the vertical cam (12) forces the lead round out of the extractors and into the bolt fingers (13). With the bolt completely to the rear, the round lines up with the chamber (14), and is ready to fire. As the original leading round chambers, the next round aligns with the bolt extractors. See Figure 2-2.

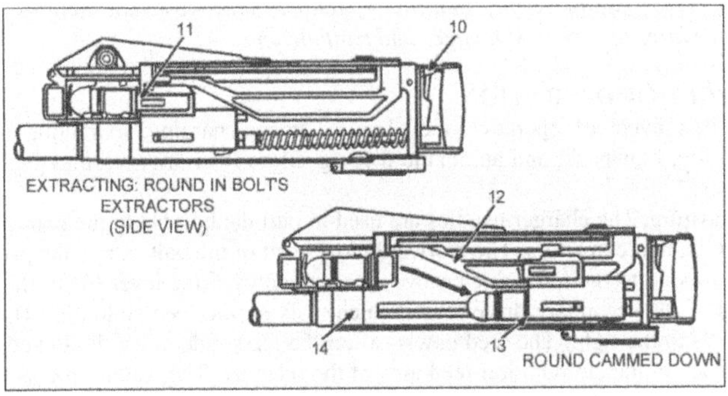

Figure 2-2. Extracting (delinking) round.

c. **Cocking.** The rearward movement of the bolt causes the cocking lever (15) to retract the firing pin (16). When the cocking lever hits the rear end of the left receiver rail slot (17), the cocking lever is forced forward. When the cocking lever retracts the firing pin, the firing pin sear holds the firing pin to the rear (18). See Figure 2-3.

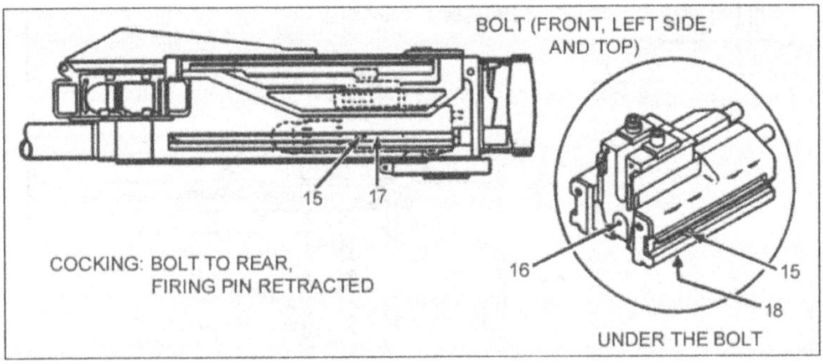

Figure 2-3. Cocking the MK 19.

d. **Firing Cycle.** The safety switch is on F (FIRE). Pressing the trigger (20) depresses the tip of the receiver sear (21). The receiver sear disengages the bolt sear (22), which releases the bolt forward under spring pressure with a round in the bolt fingers. The cocking lever hits the forward end of the left receiver rail slot, forcing the lever to the rear. The bolt sear hits a plate in the bottom of the receiver, which pushes the firing pin sear up to release the firing pin. A combination of the bolt's inertia and pressure from the firing pin spring drive the firing pin forward. The tip of the firing pin detonates the primer. The round is not completely inside the chamber at the moment the weapon is fired. The cartridge case, held by the bolt fingers, protrudes from the chamber (23). The explosion forces the projectile down the bore. See Figure 2-4.

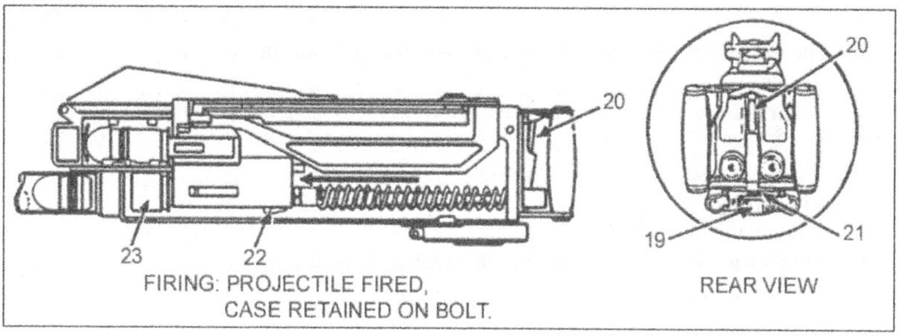

Figure 2-4. Firing cycle.

e. **Blowback and Automatic Feeding Cycle.** The gases from the burning powder force the bolt, with a new round in its extractors, to the rear. During this blowback, several things happen at once. First, the curved rail of the vertical cam delinks and forces down the new round on top of the spent case, which forces the spent case out of the bolt fingers and ejects it out the bottom of the gun. Next, the feed slide assembly pulls the rounds to the right in the receiver ammunition feed area, where a new round is ready to pick up (automatic feed). During the bolt's travel to the rear, the cocking lever is pushed forward, which cocks the firing pin. When the bolt reaches the limit of its rearward travel, the recoil springs (24) are completely compressed. The bolt buffers (25) absorb over-travel, reducing trunnion load (recoil force) at the gun-mount attaching points. The bolt sear will not engage the receiver sear if the trigger is still depressed, and another firing cycle occurs. Release of the trigger causes the bolt sear to engage the receiver sear, which prevents the bolt from going forward, and firing stops. See Figure 2-5.

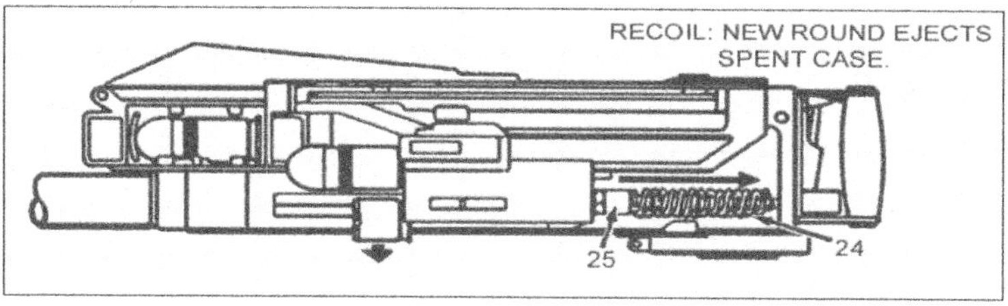

Figure 2-5. Blowback and automatic feed.

2-2. OPERATING PRECAUTIONS

Precautions are taken before, during, and after firing of the MK 19.

a. Before firing the weapon ensure that:

(1) The correct ammunition is on hand.

(2) The ammunition is clean and dent-free.

(3) All ogives on the cartridges are tight.

(4) The line of fire is clear of obstructions.

b. While firing the weapon ensure that:

(1) The top cover is closed.

(2) The muzzle of the MK 19 is pointed downrange at all times.

NOTE: In the event of a runaway gun, lower one of the charging handles.

(3) Use the suggested three to five-round bursts.

c. After firing the weapon:

(1) Unload and clear the MK 19.

(2) Note weapon discrepancies and report them to the armorer.

(3) Clean and lubricate the MK 19 before storage.

2-3. CLEARANCE PROCEDURES

The MK 19 is cleared differently in a firing situation than in a nonfiring situation.

a. **Firing Situation.** In a firing situation, use the following procedures to clear the MK 19:

(1) Move the safety switch to S (SAFE) (Figure 2-6).

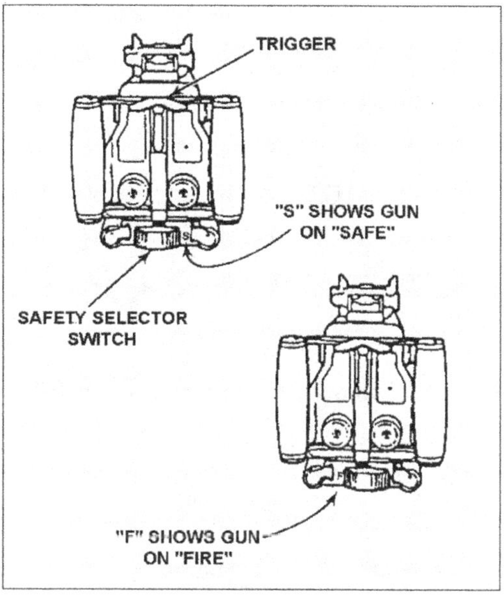

Figure 2-6. S (SAFE) position.

(2) Open the top cover assembly (Figure 2-7). If all the ammunition has NOT been fired, the bolt is to the rear and a round is on the bolt face. If the bolt is forward, lock it to the rear.

Figure 2-7. Opening the top cover assembly.

(3) Take the ammunition from the feed tray by reaching beneath the feed tray and pressing the primary and secondary positioning pawls. While pressing the position pawls, slide the linked rounds out of the MK 19 through the feed throat (Figure 2-8).

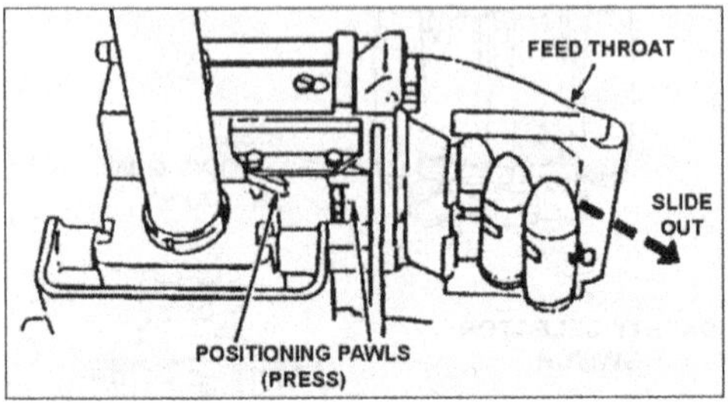

Figure 2-8. Removal of linked rounds from feeder.

(4) Insert a section of the cleaning rod or bayonet through either side of the receiver rail. Place it on top of the live round or cartridge case, as close to the bolt face as possible, and push down. This action forces the round out of the MK 19.

(5) Lower and pull both charger handles to the rear (Figure 2-9).

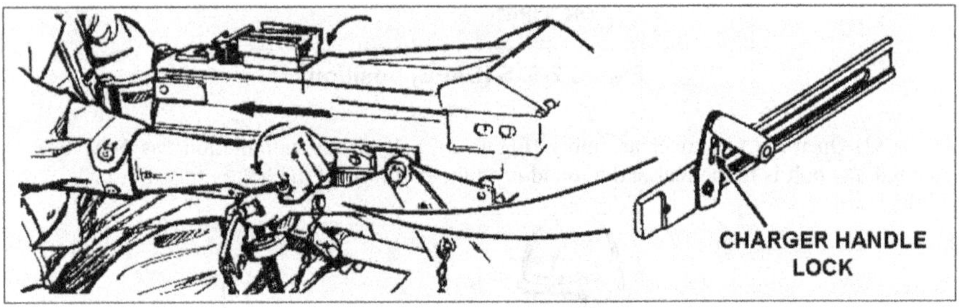

Figure 2-9. Charger handles.

(6) Inspect the chamber and bolt face to ensure that no live rounds are in the weapon.

(7) Place the safety switch on F (FIRE) (Figure 2-6).

(8) Maintain rearward pressure on the charging handle, press the trigger and ease the bolt forward.

(9) Place the safety switch on S (SAFE).

b. **Nonfiring Situation.** In a nonfiring situation, use the following procedures to clear the MK 19:

(1) Place the safety switch on S (SAFE) (Figure 2-6).

(2) Open the top cover assembly.

(3) Lower one or both charger handles.

(4) Pull the charger handle slightly to the rear.

(5) Allow sufficient space between the face of the bolt and the chamber to see both.

(6) Check for live ammunition.

(7) Ride the bolt forward.

(8) Return the charger handle to its original upright position.

2-4. DISASSEMBLY AND ASSEMBLY

Only qualified ordnance personnel should disassemble the MK 19 beyond the steps detailed in this paragraph (Figure 2-10 and Table 2-1).

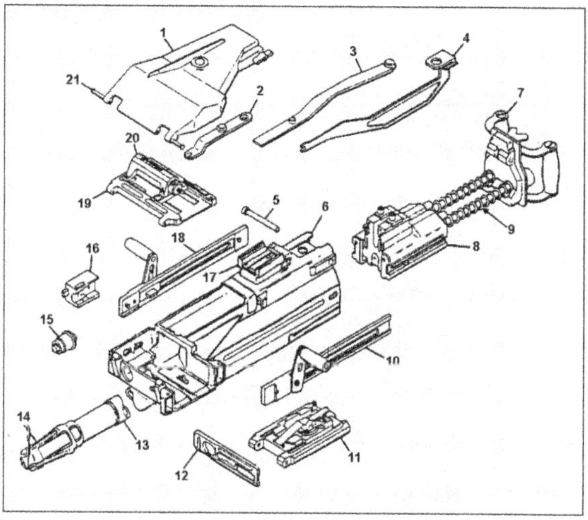

Figure 2-10. MK 19 disassembled.

ITEM	DESCRIPTION
1	Top Cover
2	Secondary Drive Lever
3	Primary Drive Lever
4	Vertical Cam Assembly
5	Backplate Pin
6	Receiver Assembly
7	Control Grip Assembly
8	Bolt
9	Guide Rods and Springs
10	Charger Assembly, Left
11	Sear Assemble
12	Alignment Guide Assembly
13	Barrel
14	Flash Suppressor
15	Ogive Plunger Assembly
16	Round-Positioning Block
17	Rear Sight Assembly
18	Charger Assembly, Right
19	Feed Tray
20	Feed Slide
21	Cover Pins

Table 2-1. Names of the MK 19 major components.

NOTE: High friction areas — the under side of the primary drive lever (3), the vertical cam assembly (4), the bolt (8), the area where the feed slide (20) and feed slide tray (19) touch have been treated to prolong the life of the MK 19 — should be handled carefully during cleaning and handling.

CAUTION

The sear assembly should be removed only in a clean and well-lit area because a small pin could possibly fall out and become lost, which makes the gun inoperable.

a. **Disassembly.** Disassembly includes removal of parts to the extent explained in this chapter. To ensure that parts are not lost and are replaced properly, place them (in the order in which they are taken off) on a clean flat surface. A spent cartridge casing, a section of the cleaning rod, and a cartridge link may be used as removal and replacement tools. To disassemble:

(1) Clear the weapon.

(2) Take out the secondary drive lever by following these procedures:

(a) Raise the top cover assembly and push the secondary drive lever pivot post from the outside of the top cover assembly.

(b) Separate the secondary drive lever from the top cover assembly (Figure 2-11).

(c) Tap lightly on the secondary lever pivot post with any available tool, if necessary.

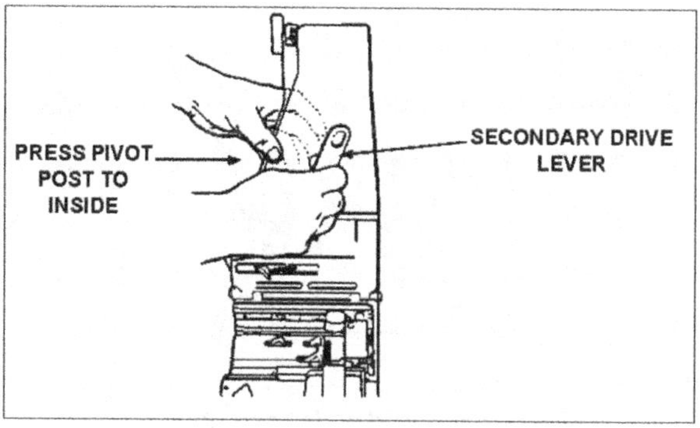

Figure 2-11. Secondary drive lever removal.

(d) Take the secondary drive lever from the slide assembly, and allow the feed slide and tray assembly to close.

(3) Take off the top cover assembly by following these procedures:

(a) Hold the top cover straight up with one hand.

(b) Pull the top cover pins from both sides (Figure 2-12).

(c) Lift the top cover assembly straight up and off.

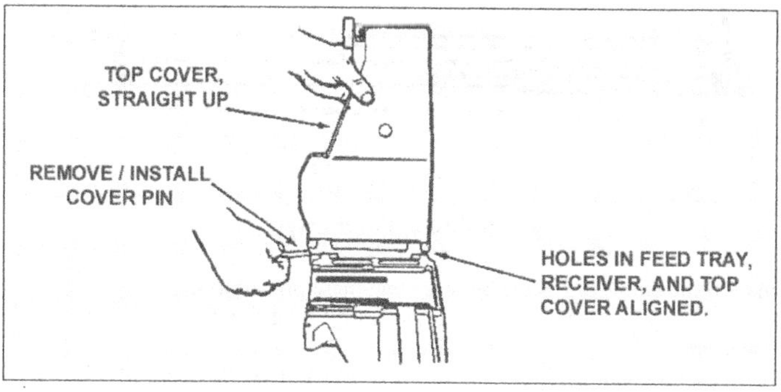

Figure 2-12. Top cover assembly.

(4) Take out the feed slide assembly and separate the feed slide from the feed tray by following these procedures:

(a) Align the tabs on the feed slide with the slots in the feed tray and lift them straight up (Figure 2-13).

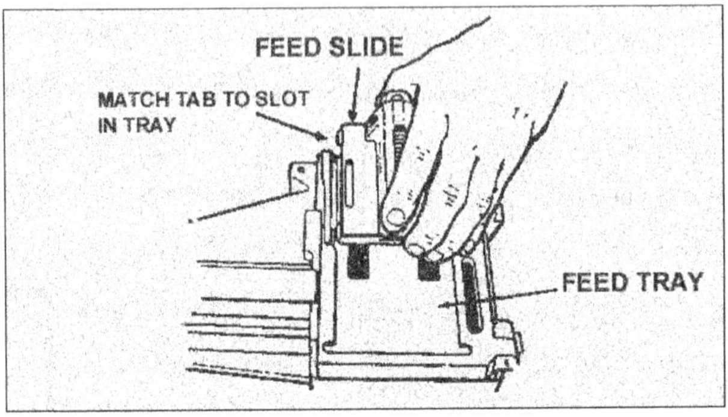

Figure 2-13. Feed slide assembly.

(b) Take out the feed tray by lifting it straight up (Figure 2-14).

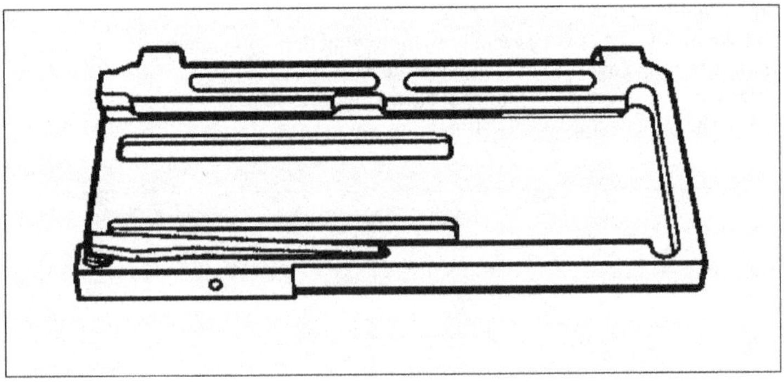

Figure 2-14. Feed tray.

(5) Take off the alignment guide by following these procedures:

(a) Depress the tip of the alignment guide spring with the secondary drive lever or a cleaning rod section.

(b) Slide the alignment guide out of the receiver by pulling the assembly slightly rearward (Figure 2-15).

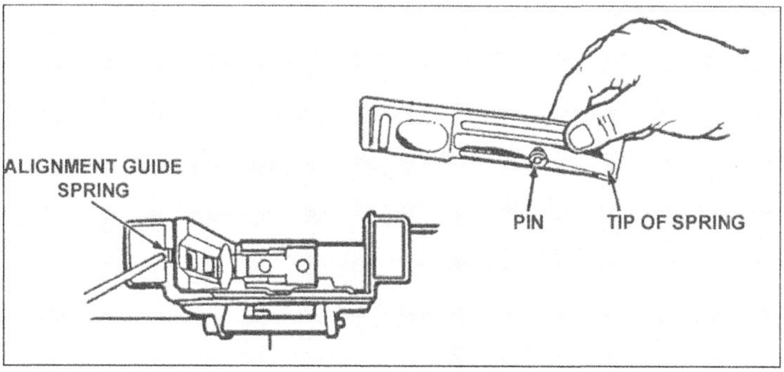

Figure 2-15. Alignment guide.

(6) Take out the ogive plunger by pulling it out through the inside wall of the receiver (Figure 2-16).

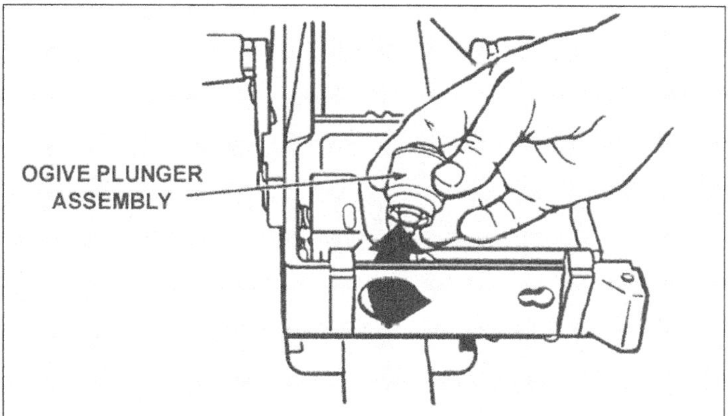

Figure 2-16. Ogive plunger.

(7) Take off the round-positioning block by following these procedures:

(a) Push the round-positioning block into the side of the gun.

(b) Slide the round-positioning block forward.

(c) Release the round-positioning block forward from the keyslots in the receiver wall (Figure 2-17).

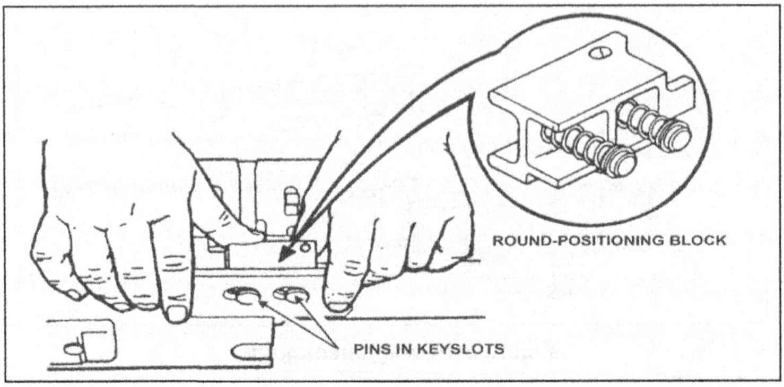

Figure 2-17. Round-positioning block.

(8) Take out the bolt and backplate assembly by following these procedures:

WARNING
The backplate is under extreme pressure when the bolt is in the rear position. To avoid serious injury, ensure the bolt is forward before removing the backplate pin.

(a) Place the safety switch in the F (FIRE) position.

(b) Take out the backplate pin using the rim of a spent cartridge case or metal link.

(c) Pry outward on the pin lip and remove the pin with the fingers (Figure 2-18). If the pin cannot be removed with the fingers, hit the small end of the pin with the secondary drive lever until the pin comes loose.

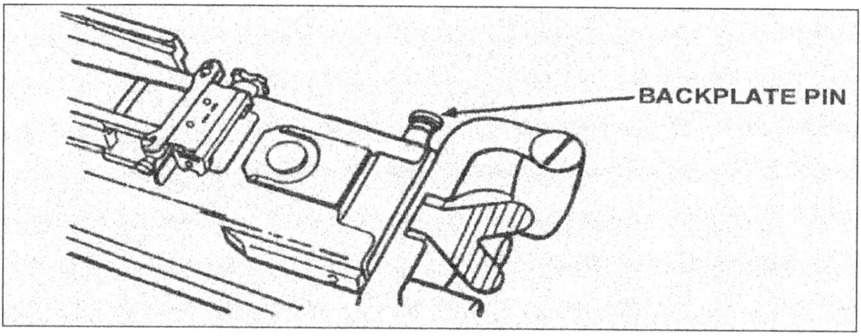

Figure 2-18. Backplate pin.

(d) Grasp the control grips with both hands and lift up slightly to disengage the backplate from the locking lugs in the receiver.

(e) Pull the bolt and backplate assembly to the rear.

(f) Once the bolt clears the sear, catch the bolt in one hand to prevent damage to the backplate assembly (Figure 2-19).

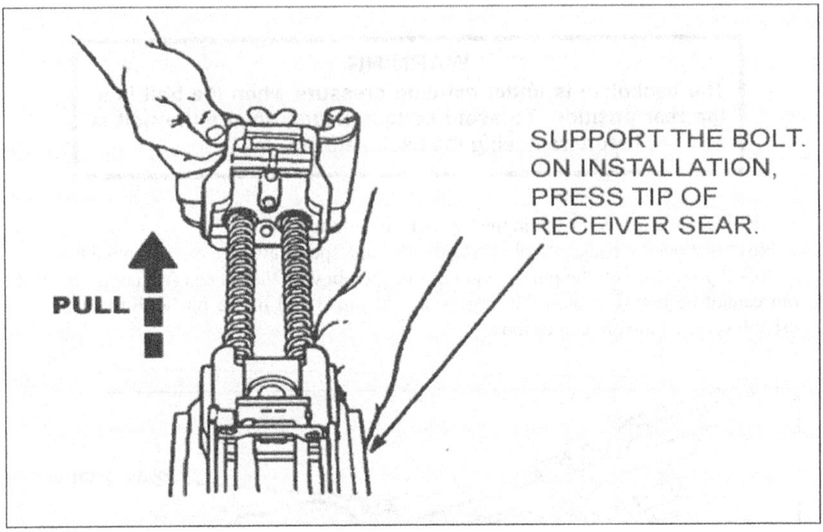

Figure 2-19. Removal of bolt and backplate assembly.

(9) Take off the primary drive lever and vertical cam.

(a) Grasp the primary drive lever and vertical cam with one hand and ensure they do not fall when released.

(b) Reach under the top of the receiver and locate the drive lever lock.

(c) Slide the lock one-quarter of an inch to the rear.

(d) Press down on the primary drive lever pivot post, which releases both the primary drive and vertical cam.

(e) Pull the primary drive lever from the front of the weapon and the vertical cam from the back (Figure 2-20).

CAUTION

Ensure you do not scratch the shiny portion of the vertical cam and primary drive lever. These areas have been specially treated to reduce friction; any scratches or use of abrasives will take off the treatment and increase wear.

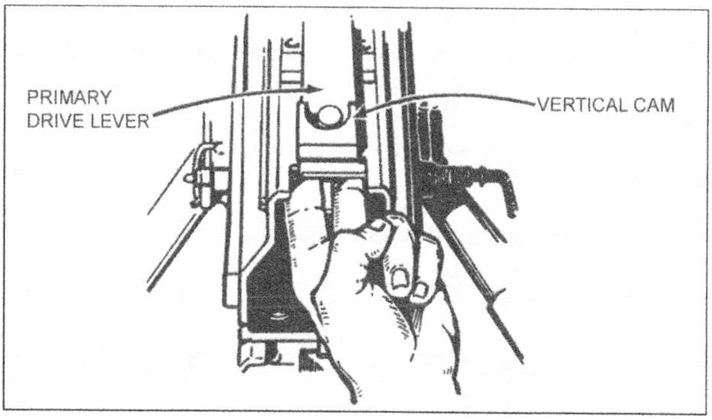

Figure 2-20. Primary drive lever and vertical cam.

(10) Take out the charger assemblies from both sides by following these procedures.

(a) Place the charger assemblies in the upright position.

(b) Using a metal link or spent cartridge case, retract the lock plunger at the base of the charger arm.

(c) Slide the charger housing rearward to disengage the lugs from the keyslots in the receiver.

(d) Lift the charger assembly away from the receiver (Figure 2-21).

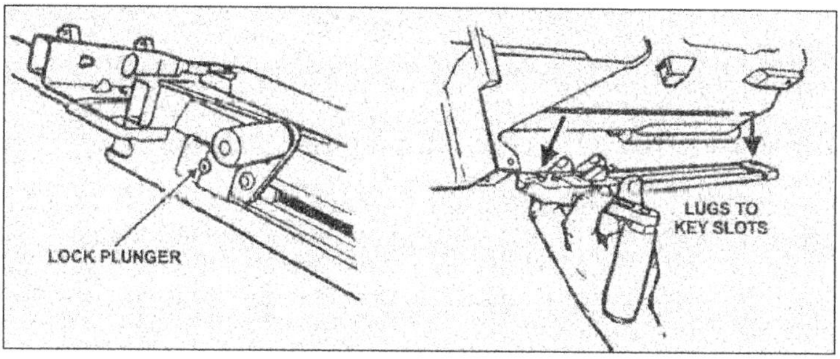

Figure 2-21. Charger assemblies.

(11) Take off the sear assembly by following these procedures:

(a) Turn the MK 19 on its side or upside down.

(b) Move the safety selector switch to F (FIRE), and use the rim of a spent cartridge case to lift up the sear lock plunger.

(c) At the same time, squeeze the sear and rotate the sear assembly (left or right) until you can no longer hold the sear (Figure 2-22).

(d) Release the plunger, and apply downward pressure to the sear assembly. (A spring in the sear will push the sear away from the receiver preventing further turning of the sear.)

(e) With the sear pressed flush to the receiver, continue to turn the sear until you hear a metallic click.

(f) Place the safety selector switch on S (SAFE) and continue to rotate the sear until it is at a 90-degree angle to the receiver.

(g) Lift the sear straight off the receiver.

(h) Place the safety switch in the S (SAFE) position.

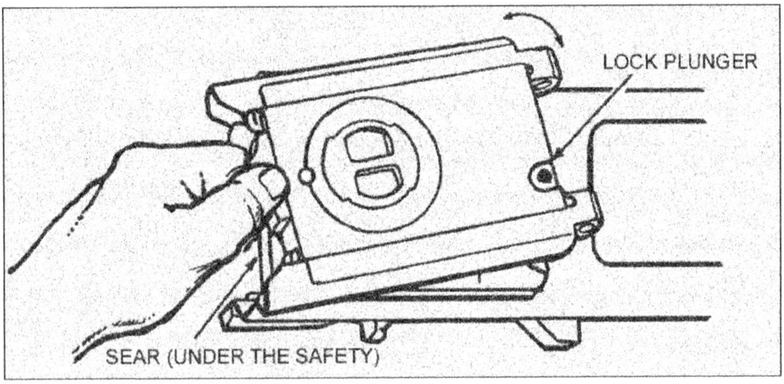

Figure 2-22. Sear assembly.

b. **Assembly.** To assemble the gun, replace the groups in the exact reverse order in which they were taken off. Be sure the components are lubricated prior to reassembly.

(1) Install left and right hand charger assemblies by following these procedures:

(a) Turn receiver upright.

(b) Rotate charger handle to the forward straight-up position.

(c) Line up lugs on charger with slots in receiver rail. Insert charger lugs into slots.

(d) Hold charger tightly against rail. Slide charger forward until it locks in place.

(2) Install round-positioning block. Insert pins into the slots on the receiver and slide the block forward until the pins lock into the small sections of the holes.

(3) Install the ogive plunger assembly. Insert ogive plunger as shown in Figure 2-16.

(4) Install alignment guide assembly by following these procedures:

(a) Position the alignment guide assembly so that the pin is lined up with the slot in the feeder wall.

(b) Hold the alignment guide against the front wall and slide the alignment guide into the receiver until it locks or 'clicks.'

(5) Install feed tray and feed slide assembly by following these procedures (Figure 2-23):

(a) Place tray into top of feeder, recessed side up. The tabs on the feed slide should align with the slots on the feed slide tray.

(b) The hinge pinholes on the tray should line up with the hinge pinholes on the receiver.

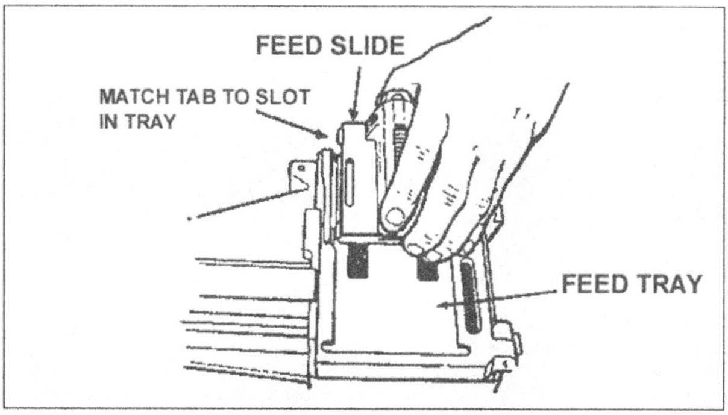

Figure 2-23. Feed slide assembly.

(6) Install top cover assembly by following these procedures:

> **CAUTION**
> To avoid breaking the crosspin, be sure it is fully inserted into the receiver before closing the top cover.

(a) Feed tray should be in place resting on the receiver.

(b) Place the top cover on the receiver with the hinge pinholes in line with the receiver lug end feed tray hinge pinholes.

(c) Hold top cover at a right angle to the receiver so that the key slots are aligned. Insert top cover pins on both sides. Ensure the crosspin is fully inserted then rotate the top cover fully open.

(7) Install secondary drive lever by following these procedures (Figure 2-24):

> **CAUTION**
> If the secondary drive lever is not properly engaged with the feedslide pin, the gun will not fire.

(a) Rotate the feed slide assembly and tray upward.

(b) Engage forked end of secondary drive lever with the feed slide pin.

(c) Press raised pivot post through hole in top cover.

(d) Press secondary drive lever against top cover until it locks in place.

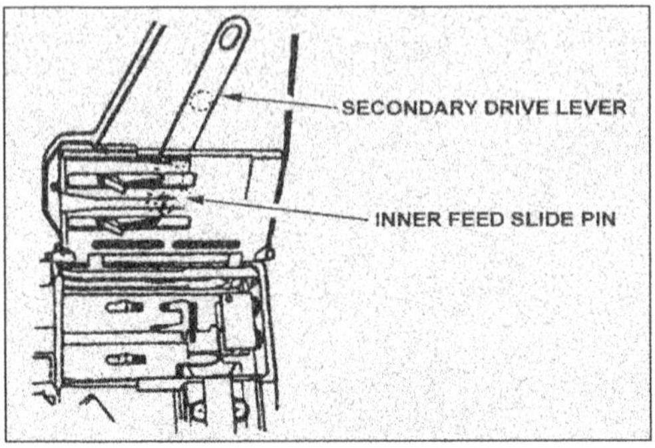

Figure 2-24. Installation of the secondary drive lever.

(8) Install vertical cam assembly by following these procedures:

> **CAUTION**
> Do not damage vertical cam by hitting it on the inside of the receiver.

(a) Slide vertical cam assembly through rear of receiver. Raised portion should slide over the top of the receiver. Drive lever lock should be underneath.

(b) Engage forked end in the notch.

(9) Engage primary drive lever by following these procedures:

(a) Hold vertical cam assembly in place and slide primary drive lever into receiver and rest the front end on the feed area.

(b) Slide drive lever lock rearward and engage pivot post of lever through holes in receiver and vertical cam.

(c) Slide drive lever lock (on the vertical cam just beneath top of receiver) forward.

(10) Install bolt and backplate assembly. With the sear assembly on the gun, assemble bolt and backplate using the following procedure (Figure 2-25):

> **CAUTION**
> Before inserting assembly, put cocking lever in forward position.

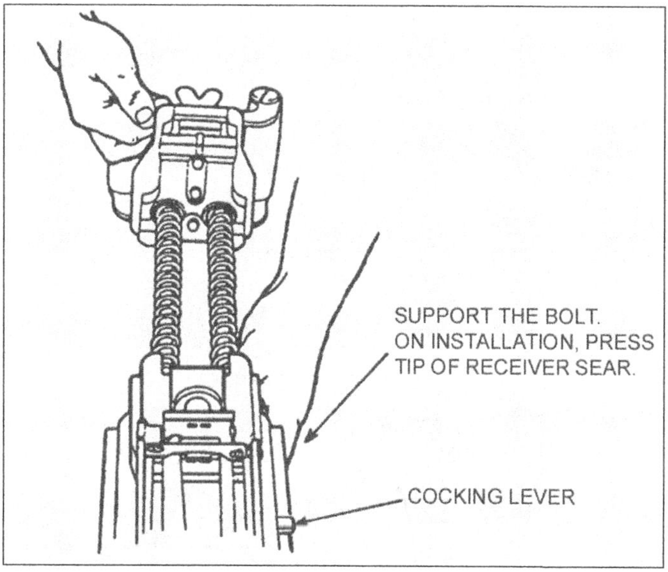

Figure 2-25. Installation of the bolt and backplate assembly.

(a) Place safety in F (FIRE) position.

(b) Ensure that the cocking lever is forward and align the bolt at the rear of the receiver and slide the receiver forward until it stops.

(c) Press down on the bolt release, slide the bolt slightly forward, and remove hands from the bolt release.

(d) Continue to slide the bolt forward until it stops again.

(e) Again, press down on the bolt release, slide the bolt slightly forward, and remove hands from the bolt release.

(f) Continue to slide the bolt forward until it reaches its most forward position.

(g) Seat the bottom of the backplate into the slots on the bottom of the receiver.

(h) Insert backplate pin to lock assembly in place.

(11) Install sear assembly by following these procedures:

(a) Turn receiver over on its top.

(b) Place sear housing on the receiver and line up sear housing assembly at a right angle to the barrel centerline.

(c) Put safety on F (FIRE) position.

(d) Press down and rotate the sear assembly until it locks in place.

(12) Install feed throat assembly by following these procedures:

(a) Squeeze the plungers, and align the pins with the holes in the receiver.

(b) Release plunger to reattach feed throat.

NOTE: Ensure the safety switch is in the F (FIRE) position so the sear can be easily depressed.

2-5. LOADING PROCEDURES

Before loading, the gunner should ensure the MK 19 is on S (SAFE) and the bolt is in the forward position.

a. Before loading, use the following procedures:

(1) Attach the feed throat by squeezing the spring-loaded pins on the feed throat (Figure 2-26).

(2) Insert the feed throat on both sides of the feeder.

(3) Ensure that the feed throat points down.

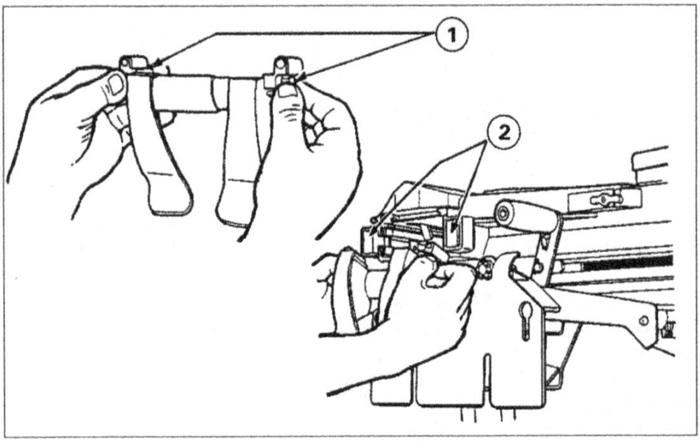

Figure 2-26. Attachment of feed throat.

b. When loading, use the following procedures:

(1) Insert the first round into the feeder (female link first) (Figure 2-27).

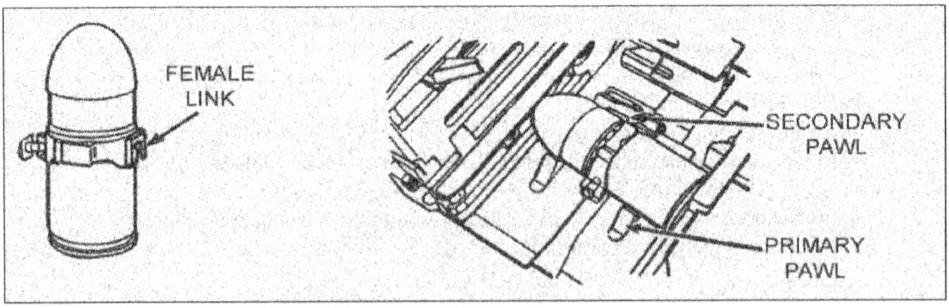

Figure 2-27. Loading of first round.

(2) Push the round across the secondary feed pawl. To move the feed slide to the left, push the secondary drive lever to the right (Figure 2-28).

(3) Close the cover.

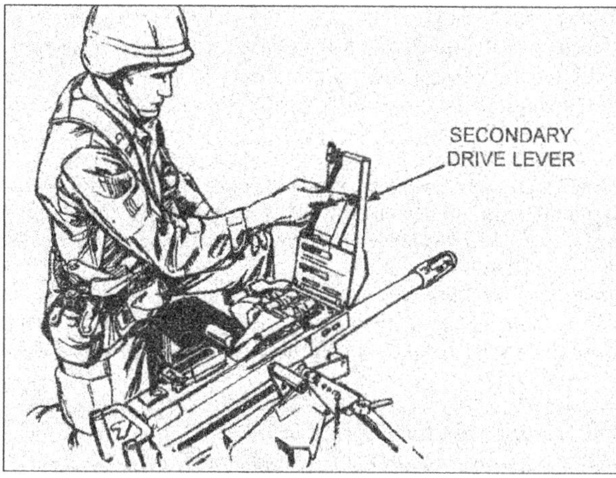

Figure 2-28. Positioning of feed slide assembly.

(4) Grasp the charger handles with the palms down.

(5) Press the charger handle locks in.

(6) Rotate the handles down and pull them sharply to the rear.

(7) Return the charger handles forward to their original upright position, after locking the bolt to the rear.

CAUTION

Failure to completely pull the bolt to the rear may result in the misalignment of the M16A2 links on the round, which causes the round to feed improperly.

(8) Place the safety switch on F (FIRE) and press the trigger. The bolt slams forward and grasps the first round in the bolt extractors.

(9) Grasp, unlock, and turn charger handles down, and lock the bolt to the rear again.

(10) Ensure the safety switch is on S (SAFE).

(11) Return the charger handles to their original upright position. The MK 19 is now ready to fire.

2-6. MALFUNCTIONS, STOPPAGES, AND CORRECTIONS

A malfunction is a failure of the weapon to function properly. Neither defective ammunition nor improper operation of the gun by a crewmember is considered a malfunction of the MK 19. A stoppage is any interruption in the cycle of operation caused by faulty action of the weapon or ammunition. Appendix C, Stoppages, Malfunctions, Immediate Actions, and Remedial Actions, provides inspections, corrective actions, and emergency actions of malfunctions and stoppages. For more information on troubleshooting common malfunctions and stoppages, refer to TM 9-1010-230-10.

2-7. CARE AND CLEANING

The MK 19 requires preventive maintenance checks and services (PMCS) and lubrication before, during, and after firing. It should be checked and cleaned daily when used and not fired, and weekly if not used. For more detailed guidance on the care, cleaning, and PMCS for the MK 19, refer to TM 9-1010-230-10.

a. **Cleaning and Lubrication.** Several types of lubricants can be used on the MK19. High friction areas of the gun have a protective coating and care must be taken to prevent its removal. Listed below are authorized lubricants, and other information about lubricating and cleaning the MK19.

(1) ***Authorized Lubricants.*** These are the authorized lubricants used on the MK 19:

(a) Lubricant, artic weather, automatic weapons (LAW).

(b) Lubricating oil, semifluid, automatic weapons (LSA).

(c) Lubricating oil, semifluid, automatic weapons, temperate (LSAT).

(d) Rifle bore cleaner (RBC). RBC should be used sparingly only in the bore and when necessary.

(e) Grease, molybdenum disulfide (GMD).

(2) ***Use of Lubrications.*** The following is general guidance on the use of lubricants on the MK 19.

(a) Never mix lubricants on the MK19. If you change lubricants or do not know for sure which lubricant is currently applied to the MK19, completely clean off all old materials before you apply a new lubricant.

(b) Never use cleaner lubricant petroleum (CLP) on a MK19. It gums up moving parts instead of lubricating them.

(c) Apply a heavy coat of lubricant and do not wipe it off.

(d) In cold weather, 0 degrees F to -25 degrees F, use LSAT, GMD, or LAW.

(e) In extra cold weather, -25 degrees F and below, use LAW.

(3) ***High Friction Areas.*** The under side of the primary drive lever, the vertical cam assembly, the bolt (where it rides on the rails), and the area where the feed slide and feed slide tray connect have been treated to prolong the life of the MK 19. Special care is required during cleaning and maintenance to prevent damage to the coating. CLP, acids, and abrasives should never be used on the MK 19 because they will remove the protective coating.

b. **General Cleaning and Lubrication Instructions.** The following is general guidance for the cleaning of the MK 19.

(1) Wipe or brush off dirt and grime.

(2) When lubricating the weapon, give extra attention to the feed pawls, the cocking lever rails, the pivot posts on the primary drive lever, and the bolt assembly.

(3) Lubricate each part. Give special care to those hard-to-get spots.

(4) Work in the lubricant by moving the parts.

c. **When to Clean, Lubricate and Inspect the MK 19.** Clean and inspect the gun thoroughly. Report all worn, burred, defective, or missing parts to your armorer or support maintenance. In general, lubricate and clean the MK19 as follows:

(1) Always clean and lubricate the gun after firing.

(2) Clean and lubricate the gun daily if it is used but not fired.

(3) Clean and lubricate the gun weekly if the weapon is stored.

d. **Inspections.** Several parts of the MK 19 should be checked whenever possible. These include:

(1) The vertical cam. Check closely for smoothness.

(2) The gap between the charger assemblies and the receiver.

(3) All parts for wear, cracks, missing items, and abnormal shapes.

2-8. OPERATION UNDER ADVERSE CONDITIONS

The MK 19 is a robust weapon and designed to operate over a wide range of weather conditions but additional care is required when the temperatures are hot or cold, and the conditions are wet, sandy, or icy.

a. During *hot, wet, or salty air conditions* follow these procedures:

(1) Inspect the weapon more frequently for signs of rust.

(2) Keep gun as moisture-free as possible.

(3) Field strip, clean, and lubricate more often to preserve metal and prevent rust.

(4) Use a generous second coat of lubrication for extra protection.

b. During *hot, dry, sandy, or dusty conditions* follow these procedures:

(1) Do not lubricate the entire exposed metal surfaces as this will only collect dust and sand.

(2) Clean the weapon thoroughly and lubricate only the moving components.

(3) Extreme heat dries up lubricant. Clean and lightly lubricate the moving components of the weapon more frequently.

c. During *cold, icy, or snowy conditions* follow these procedures:

(1) Prior to using cold-weather lubricant, be sure to clean off all existing lubrication. From 0 degrees F to –25 degrees F, use either LSAT, GMD, or LAW. At temperatures –25 degrees F and below, use LAW.

(2) If the weapon 'sweats,' dry and lube the parts before taking the weapon outdoors and apply a light second coat to provide protection.

(3) Cover weapon if left outside.

(4) If possible, thoroughly clean, dry, and lubricate the weapon in a warm place.

(5) If the weapon is brought indoors, keep it away from direct heat.

(6) Perform functional checks and lubricate daily to help prevent corrosion.

2-9. DESTRUCTION

Only the commander can direct the destruction of the MK 19. There are several ways to destroy it. Methods include destruction by mechanical means, burning, gunfire, demolition or disposal. Use the same methods to destroy equipment. However, ammunition must be handled differently. The best way to destroy ammunition is to fire it. Otherwise, dispose of it by burial, dumping it in a body of water, or using the demolition method.

WARNING
Ammunition must never be scattered (this may result in a hazard to friendly soldiers), burned (HE and HEDP rounds may explode), or destroyed mechanically.

a. **Mechanical Method.** If possible, use the mechanical method to destroy MK 19s and equipment. There are three ways to do this:

(1) Smash them. Using a sledgehammer or some similar heavy tool, damage the MK 19s so that they cannot be used.

(2) Bend the MK 19s' guide rods over the receivers.

(3) Take off (or out) the same part from each of the MK 19s. If a different part is taken from each gun, new MK 19s may be pieced together.

b. **Burning Method.** To burn equipment or MK 19s, use a fuel source that burns hot enough to take the temper from the metal. For example, use jet propulsion fuel (JP-4), oil, or gas. All friendly soldiers in the area are warned before equipment or MK 19s are bare burned, because a fire will reveal the positions of friendly forces.

c. **Gunfire Method.** This is the least reliable way to destroy MK 19s or equipment. Pile up all of the equipment and MK 19s and use MK 19s, grenades, or rifles to fire on them. They will scatter so check after shooting to see that they are all destroyed.

d. **Demolition Method.** Use an explosive charge such as composition 4 (C4) or any type of HE round. Place the charges so they will completely destroy all the MK 19s and equipment.

e. **Disposal Method.** To dispose of MK 19s and equipment, break them down and either scatter the parts into swamps or marshes, or bury them.

(2) Push the round across the secondary feed pawl. To move the feed slide to the left, push the secondary drive lever to the right (Figure 2-28).

(3) Close the cover.

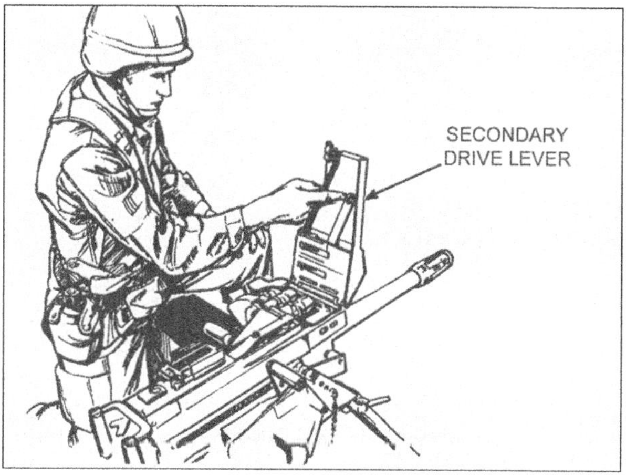

Figure 2-28. Positioning of feed slide assembly.

(4) Grasp the charger handles with the palms down.

(5) Press the charger handle locks in.

(6) Rotate the handles down and pull them sharply to the rear.

(7) Return the charger handles forward to their original upright position, after locking the bolt to the rear.

CAUTION

Failure to completely pull the bolt to the rear may result in the misalignment of the M16A2 links on the round, which causes the round to feed improperly.

(8) Place the safety switch on F (FIRE) and press the trigger. The bolt slams forward and grasps the first round in the bolt extractors.

(9) Grasp, unlock, and turn charger handles down, and lock the bolt to the rear again.

(10) Ensure the safety switch is on S (SAFE).

(11) Return the charger handles to their original upright position. The MK 19 is now ready to fire.

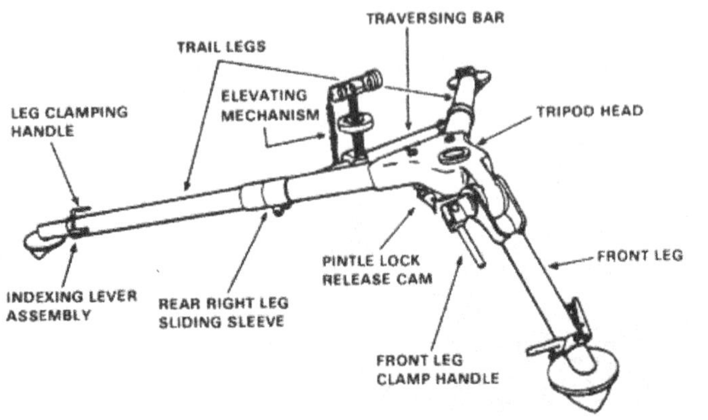

Figure 3-2. M3 tripod.

(1) Unscrew the leg-clamping handle; press down on the indexing lever, and extend the leg to the desired length.

(2) Align the indexing lever stud with one of the holes in the tripod leg extension.

(3) Release pressure on the indexing lever, allowing the stud to fit the desired hole. Tighten the leg-clamping handle.

b. Set the front leg of the tripod by following these procedures:

(1) Turn the front leg clamp handle counterclockwise to loosen the front leg.

(2) Adjust the leg to the desired angle, ensuring the tripod head is level, and tighten the front leg clamp.

c. Secure the tripod legs by following these procedures:

(1) Stamp the metal shoe on each tripod leg into the ground.

(2) Sandbag each leg to stabilize the MK 19 for firing.

d. Mount the M64 gun cradle onto the M3 tripod by following these procedures:

(1) Unlock the tripod pintle lock release cam.

(2) Insert the gun cradle's pintle into the tripod pintle bushing (Figure 3-3).

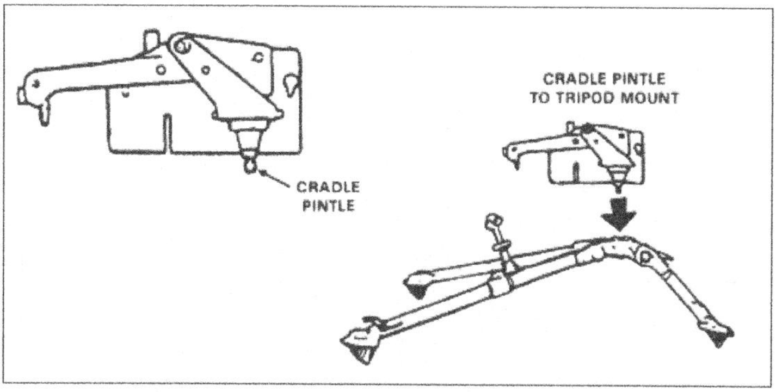

Figure 3-3. Insertion of pintle into pintle bushing.

(3) Lock the pintle lock release cam to secure the gun cradle. Check the gun cradle, by pulling up on it slightly, to ensure that it is seated and locked.

e. Attach the (traverse and elevating) T&E mechanism by following these procedures (Figure 3-4):

(1) Zero the T&E mechanism by zeroing the traversing handwheel, elevating handwheel, and the elevating mechanism sleeve to the lower elevating screw.

(a) To zero the traversing handwheel, hold the T&E mechanism so that the traversing handwheel is on the left when looking at it, then turn the traversing handwheel toward you until it stops. Loosen the locking nut slightly. Align the zero on the scale with the zero on the elevating screw yoke. Hold the scale with the zeros aligned, and tighten the locking nut. Make sure the zeros stay aligned. Turn the traversing handwheel two complete revolutions away from you. If doing this at night, count 50 "clicks" away from you.

(b) To zero the elevating handwheel to the upper elevating screw, align the two zeros. Rotate the elevating handwheel up or down until a zero with a line below it is visible on the upper elevating screw. Position the elevating handwheel so the indicator is pointing at the zero on the handwheel.

(c) Zero the elevating mechanism sleeve to the lower elevating screw. Rotate the elevating mechanism sleeve all the way up; rotate it down until it stops and note the number of complete turns. Rotate the elevating mechanism sleeve up half that number of turns. Position the slide lock lever to face you.

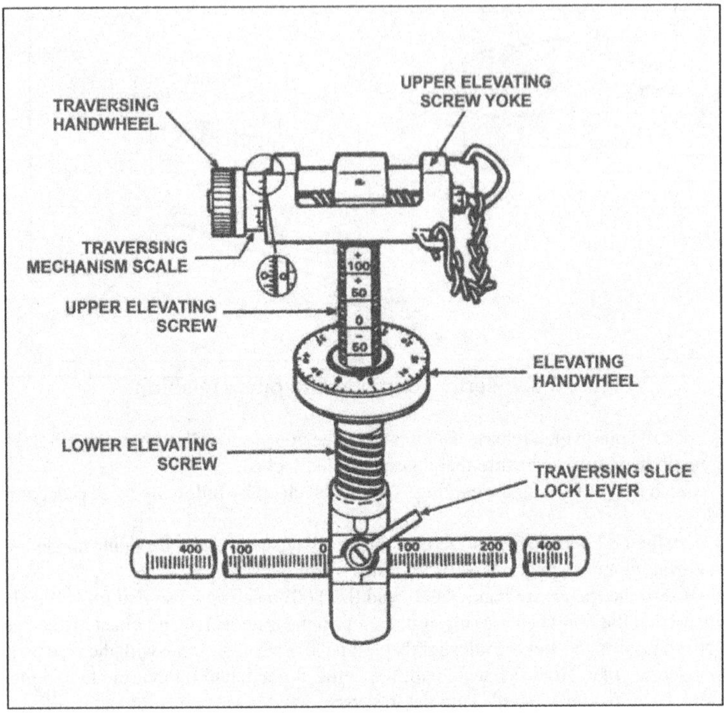

Figure 3-4. T&E mechanism.

(2) Remove the stow pin from the gun cradle (Figure 3-5).

(3) Align holes in the upper elevating screw yoke of the T&E mechanism with the rear holes in the gun cradle (Figure 3-6).

NOTE: The stow pin locks the cradle in a horizontal position, preventing it from depressing or elevating.

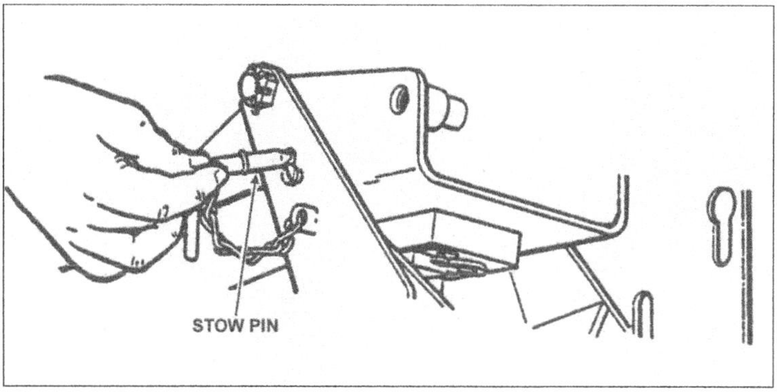

Figure 3-5. Stow pin.

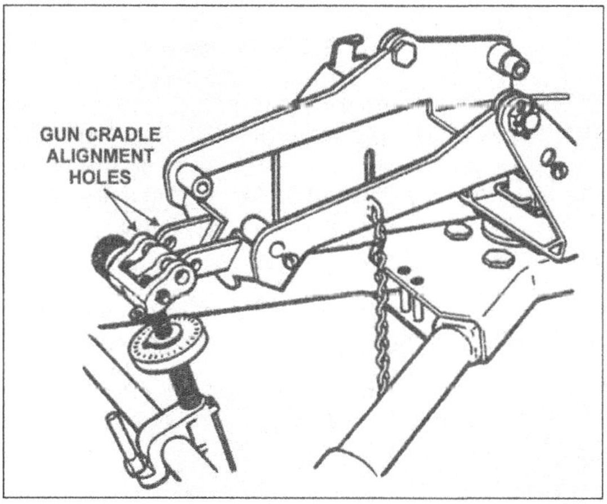

Figure 3-6. Alignment of gun cradle to T&E mechanism.

(4) Lock the elevating sleeve mechanism onto the center of the traversing bar.

(5) Insert the quick-release pin from the right (Figure 3-7).

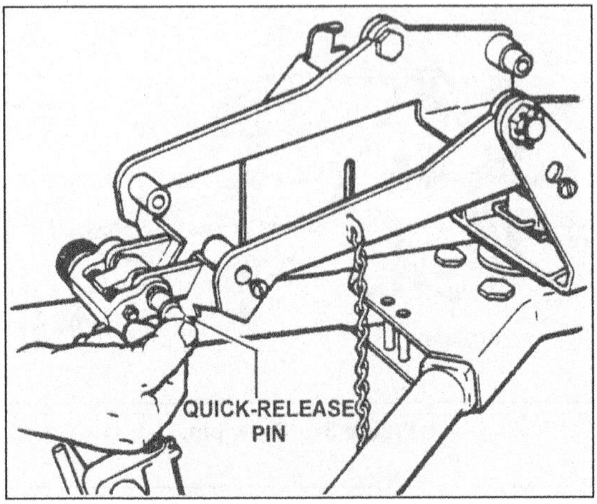

Figure 3-7. Quick-release pin.

f. Mount the MK 19 by following these procedures:

(1) Lift the MK 19 into the gun cradle. One soldier holds the barrel while another holds the control grips. The one holding the grips stands between the trail legs of the tripod; the other straddles the front leg. The soldiers lift the MK 19, barrel first, onto the cradle.

(2) Align the grooves on the receiver with the lugs in the gun cradle, and slide the receiver forward (Figure 3-8).

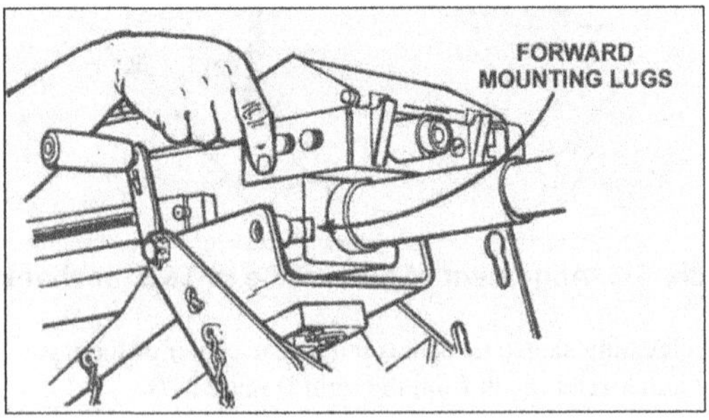

Figure 3-8. Mounting lugs.

(3) Align the sear mounting holes with the gun cradle mounting holes (Figure 3-9). Secure the rear of the receiver by inserting the retaining pin through the cradle and sear assembly and rotate it until it locks in place (Figure 3-10). If a safety clip is attached, use it to secure the retaining pin.

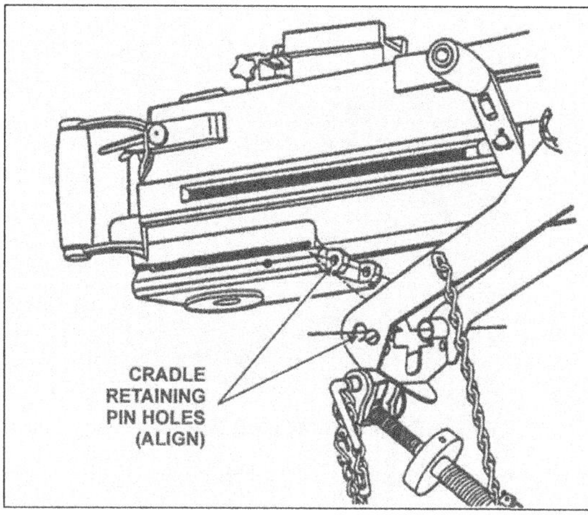

Figure 3-9. Alignment of the sear assembly and pinholes.

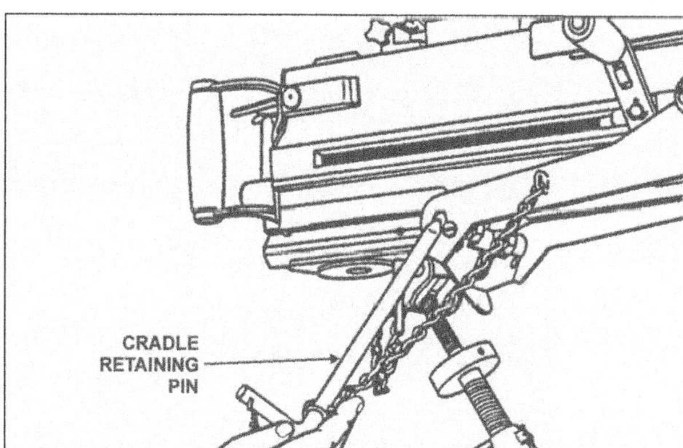

Figure 3-10. Insertion of the cradle retaining pin.

g. Attach the feed throat to the MK 19 by following these procedures:

(1) Squeeze together each set of grip pins (Figure 3-11).

(2) Attach the feed throat to the front left-hand side of the receiver assembly. The pins of the feed throat must line up with the pinholes in the receiver (Figure 3-12).

(3) Relax pressure on the spring-loaded grip pins so they will snap into place (Figure 3-13, page 3-9).

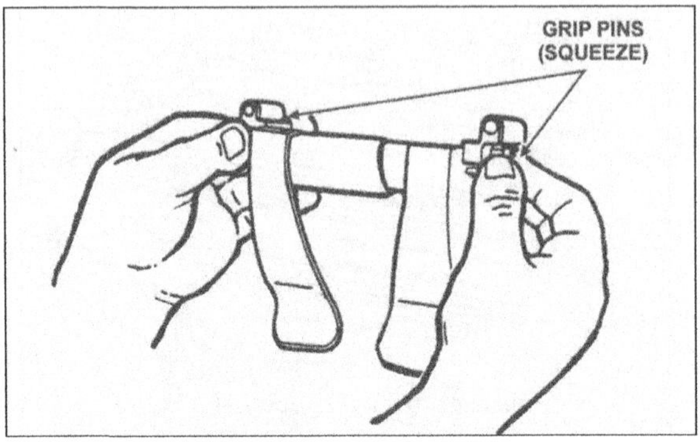

Figure 3-11. Grip pins.

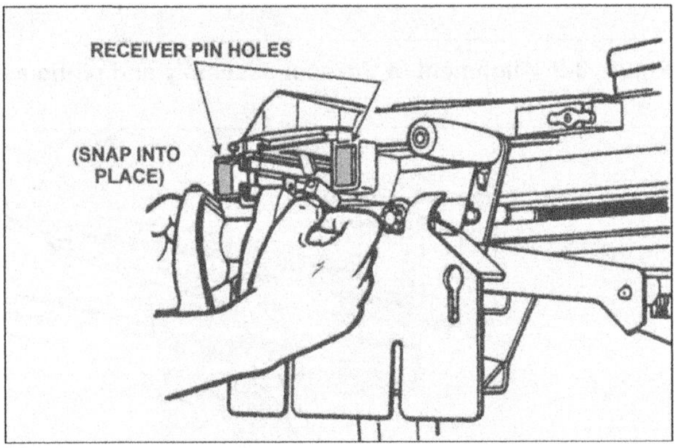

Figure 3-12. Feed throat alignment.

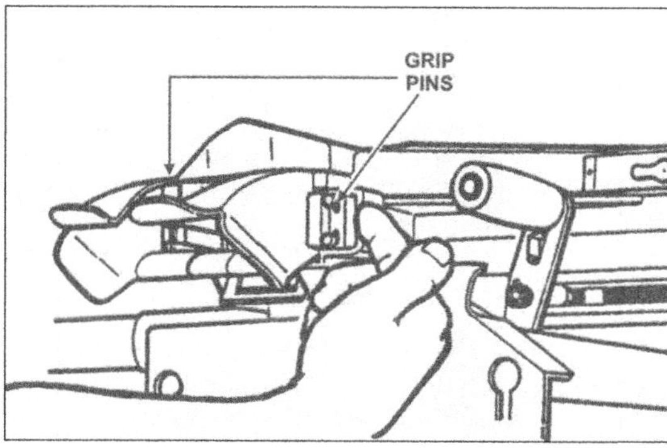

Figure 3-13. Feed throat installation.

NOTE: 1. When the MK 19 is mounted on the M3 tripod or any vehicle, it is called a weapon system.
2. The same procedure is used to mount the MK 19 to the gun cradle regardless of the type of mount used.

3-3. VEHICLE-MOUNT

This section lists the accessories used and outlines the steps to mount the MK 19 onto vehicles. Since mounting procedures for vehicles vary little, this section explains in detail how to mount the MK 19 on the HMMWV and highlights the differences in mounting procedures for other vehicles.

a. **Accessories.** Mounting the MK 19 on vehicles requires several accessories.

(1) ***Gun Cradle.*** The MK 64 gun cradle is used to mount the MK 19 onto any vehicle having a pedestal mount for the M2 machine gun.

(2) ***Pintle Adapter.*** The pintle adapter is used to mount the MK 19 in all vehicular modes (Figure 3-14). Other accessories are used to mount the MK 19 to specific vehicles. The upper end accepts the gun cradle's pintle, which is secured by a quick-release pin. The lower end of the adapter fits the mounting wells on the M4 pedestal, HMMWV weapon platform, M36A2 ring mount with M66 ring, and commander's cupola on the M113 APC.

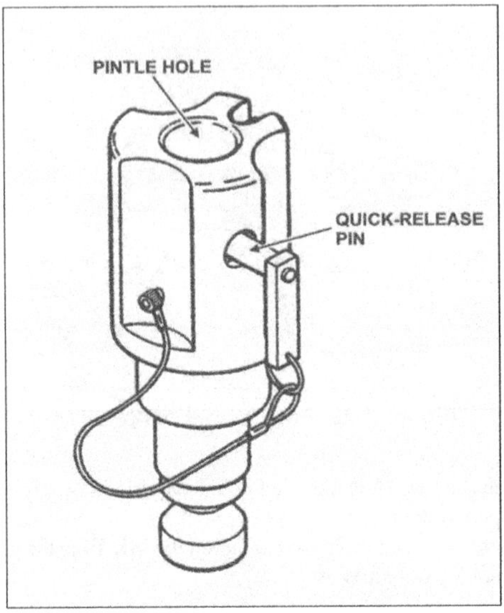

Figure 3-14. Pintle adapter.

(3) ***Train and Elevating Assembly.*** This assembly secures the MK 64 gun cradle to the M4 pedestal or HMMWV weapon platform pedestal and allows directional adjustments. The lower end of the train and elevating assembly is attached to the pedestal by a train lock clamp. The clamp is released or locked in position by a train lock handle. Two positioning clamps are supplied to prevent vertical movement of the train lock clamp on the M4 pedestal. When used on the HMMWV weapon platform pedestal, *only one clamp is needed above the train lock clamp.* The upper end of the train and elevating assembly is a standard caliber .50 T&E mechanism and it is attached to the lower rear holes in the M4 cradle by a retaining pin (Figure 3-15).

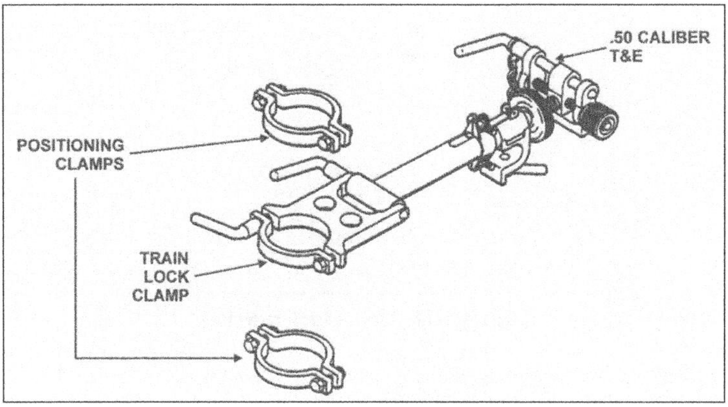

Figure 3-15. Train and elevating assembly.

(4) ***Travel Lock Adapter.*** This adapter is used to hold the gun cradle securely during travel. It is attached to the gun cradle by a retaining pin, which is inserted through the gun cradle train and elevating mechanism mounting holes (Figure 3-16).

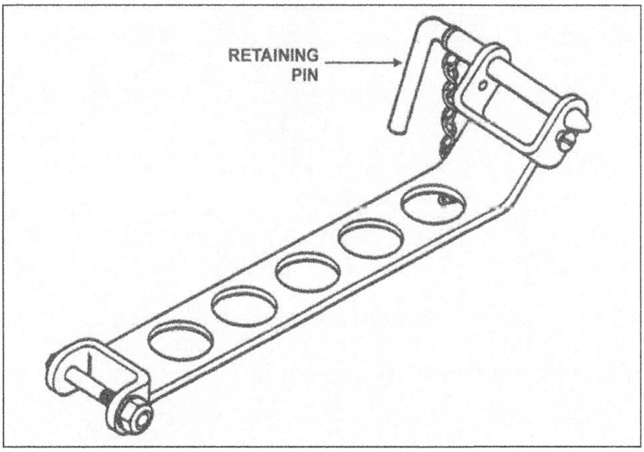

Figure 3-16. Travel lock adapter.

(5) ***Travel Lock Bracket.*** This bracket allows the travel lock adapter to be used with the M4 pedestal mount. It has two halves that are secured around the pedestal with nuts and bolts. The bracket is attached to the travel lock adapter by aligning the lower holes (Figure 3-17).

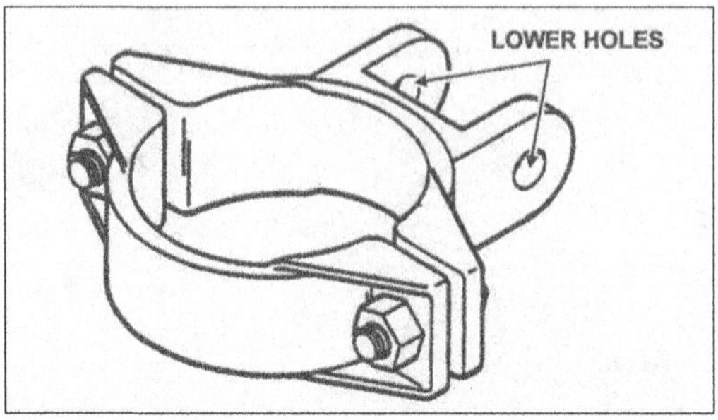

Figure 3-17. Travel lock bracket.

(6) ***Bracket Mounting Assembly.*** This assembly supplies a mount for the M548 metal ammunition container. It has a metal frame that attaches to the gun cradle, and a retaining pin that inserts through the top-inner end of the M548 ammunition container (Figure 3-18).

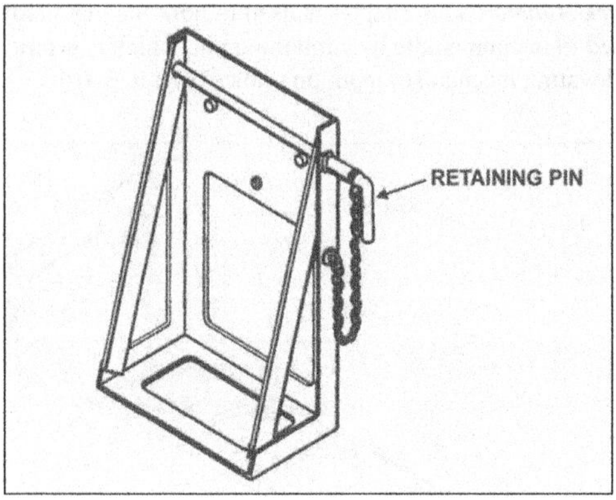

Figure 3-18. Bracket mounting assembly.

(7) ***Empty Case Catch Bag.*** This rubberized canvas bag catches empty cartridge cases as they are ejected. It is held in place by a metal rim, which attaches to the bottom of the cradle under the receiver by two hooks and a hanger (Figure 3-19).

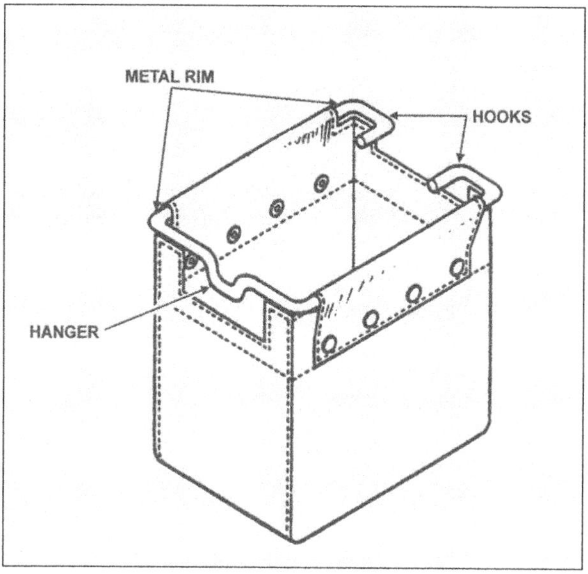

Figure 3-19. Empty case catch bag.

NOTE: The 9/16-, 7/16-, and 3/8-inch open end wrenches are necessary tools to mount the MK 19 on vehicles.

b. **Vehicle Mounts.** Various vehicle mounts can be used with the MK 19.

(1) ***HMMWV Weapon Platform.*** To mount the MK 19 on the HMMWV weapon platform (Figure 3-20):

(a) Insert the pintle adapter onto the HMMWV pedestal by following these procedures:

(1) Loosen the HMMWV pedestal lock screws by turning counterclockwise until the threaded ends are flushed with the pedestal's inner wall (Figure 3-20).

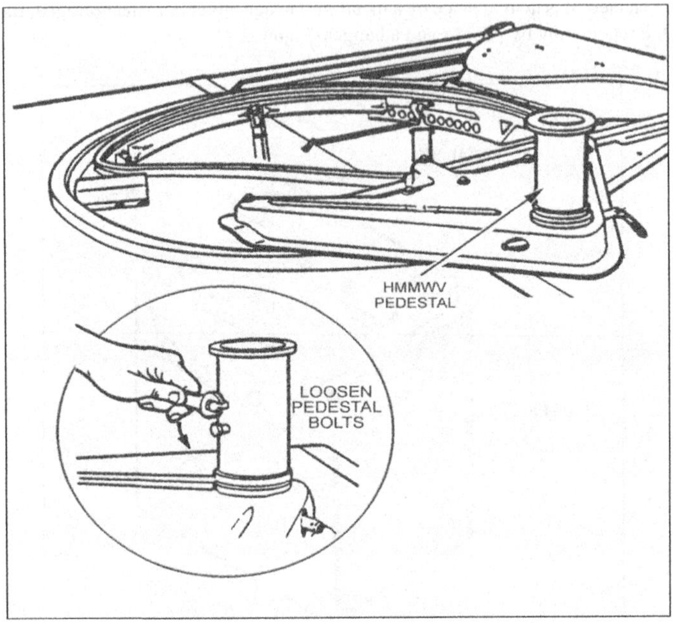

Figure 3-20. HMMWV pedestal.

(2) Insert the pintle adapter assembly by placing the lower end of the pintle adapter into the HMMWV pedestal (Figure 3-21).

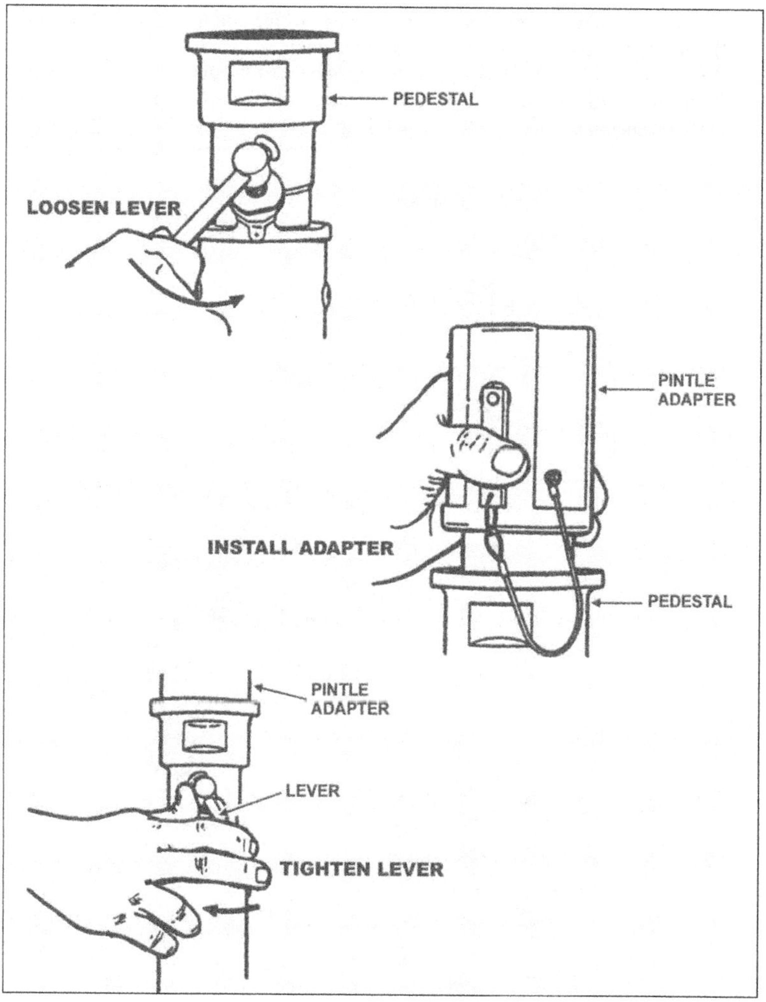

Figure 3-21. Pintle adapter assembly.

(3) Tighten the lock screws with a 3/8-inch open-end wrench. Pull upward on the pintle adapter to ensure it is secured in place.

(4) Remove the pintle adapter quick-release by pressing in on the pin's quick-release button and pulling the pin from the pintle adapter.

(b) Install the carriage and cradle assembly by following these procedures (Figure 3-22):

(1) Insert the front stow pin to prevent relative movement to the carriage and cradle.

(2) Insert the carriage pintle into the top of the pintle adapter assembly.

(3) Press in on the pin's quick-release button and insert the pin.

(4) Pull upward and twist the carriage and cradle assembly. It should lock into the pintle adapter, but it should traverse freely left and right.

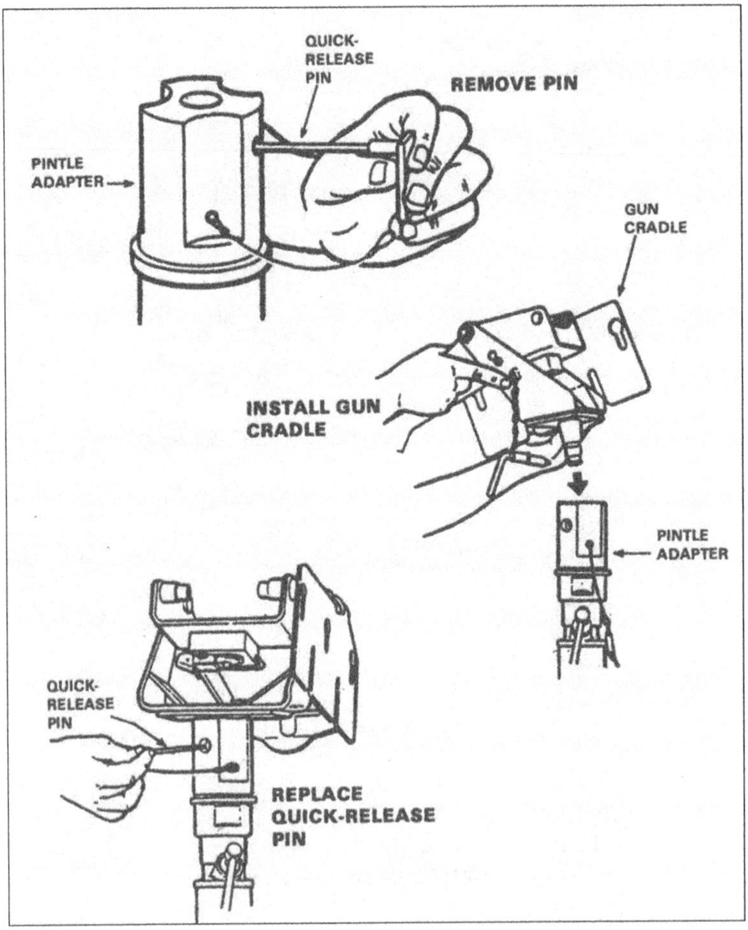

Figure 3-22. Install the carriage and cradle assembly.

(c) Install the T&E mechanism in the train and elevation assembly by following these procedures:

(1) Turn the elevating handwheel to read 250 mils.

(2) Loosen the pivot arm screws, using two 7/16-inch open-end wrenches.

(3) Insert the T&E mechanism all the way into the train and elevation assembly.

NOTE: The T&E mechanism is the upper portion of the training and elevation assembly; it must be installed in the train and elevation assembly *before* the training and elevation assembly can be installed on the vehicle.

(4) Position the traversing lock lever to the rear and the traversing knob to the left by tightening the pivot arm screws, using the two wrenches.

(d) Separate the middle clamp on the train and elevation assembly by following these procedures:

(1) Unscrew the train lock handle counterclockwise to remove.

(2) Using the 9/16-inch open-end wrench, remove the hex head screw on the other side of the middle clamp.

(e) Attach the train and elevation middle clamp to the HMMWV pedestal post by following these procedures (Figure 3-23):

(1) Assemble the middle clamp around the pedestal.

(2) Tighten the clamp by turning the train lock handle clockwise.

(3) Using the 9/16-inch open-end wrench, equally tighten the screw on the other side of the middle clamp.

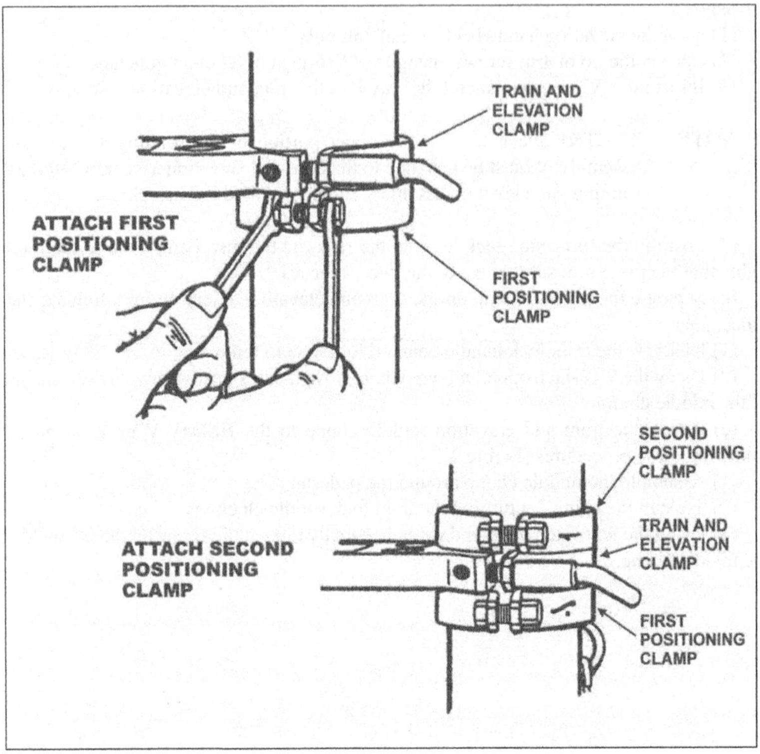

Figure 3-23. Train and elevation middle clamp.

(f) Install a support clamp above the middle clamp. To allow free left and right traverse of the carriage and cradle assembly, loosen the train lock handle. To lock in place, tighten the handle.

NOTE: The second or lower set of clamps is not used in this application since the train lock rests on the built-up base of the HMMWV pedestal. This will also allow for increased elevation. Using two 9/16-inch open-end wrenches, fasten one set of clamps above the middle clamp. Tighten each screw two turns until snug.

(g) Attach the train and elevation assembly to the carriage and cradle assembly by following these procedures:

(1) Pull out the front stow pin to allow the cradle to move.

(2) Pull out the train and elevation assembly's retaining pin.

(3) Ensure the elevating handwheel is set at 250 mils. Position the lock lever to the rear and ensure the traversing knob is on the left.

(4) Turn the traversing knob to center the elevating screw in the yoke.

NOTE: Ensure the elevating handwheel does not get jammed under the lip of the cradle.

(5) Turn the elevating handwheel to align the elevating mechanism holes with the lower rear holes in the cradle.

(6) Insert the retaining pin and rotate to the locked position.

(h) Mount the ammunition can bracket by following these procedures:

(1) Partly unscrew the wing nut on the threaded stud of the bracket mounting assembly. Align the stud with the forward groove in the gun cradle's side plate. Push the bracket mounting assembly up until the heads of the two mounting pins align with the two forward keyholes. Push the heads of the two mounting pins into the keyholes and allow the bracket mounting assembly to slide down. Tighten the wing nut behind the side plate of the cradle (Figure 3-24).

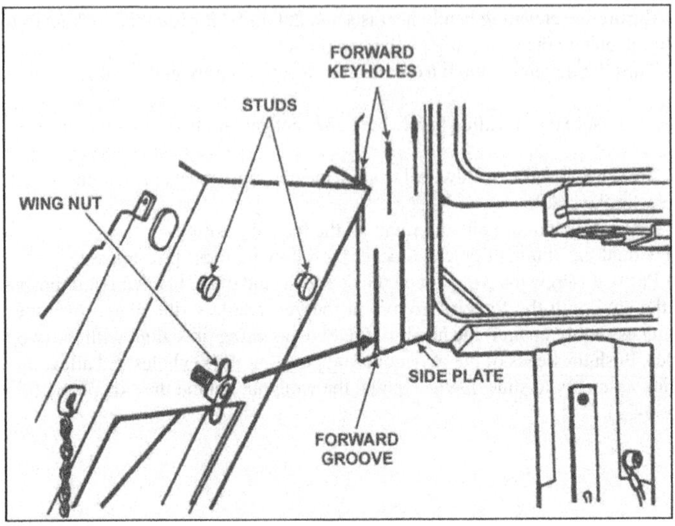

Figure 3-24. Attachment of bracket mounting assembly.

(2) Insert the two hooks on the empty case catch bag through the rear holes in the gun cradle. Engage the single front hanger on the catch bag with the hook on the gun cradle (Figure 3-25).

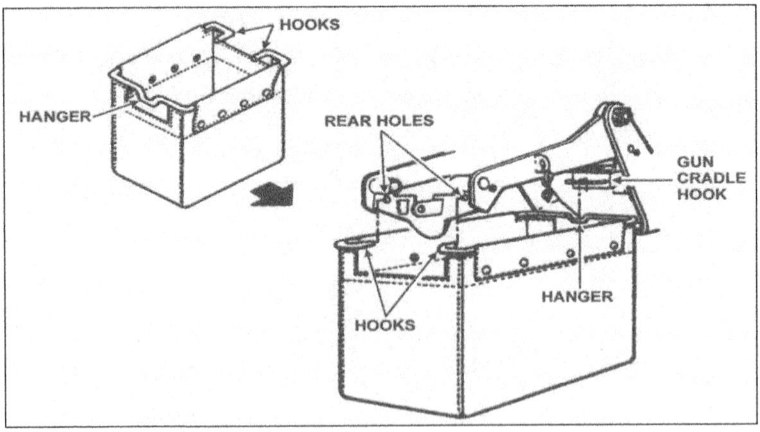

Figure 3-25. Attachment of empty case catch bag.

(2) ***M66 Ring Mount (2 1/2- to 5-ton cargo trucks).*** To mount the MK 19 on the M66 ring mount:

(a) Insert the pintle adapter into the M36A2 ring mount with the M66 ring receptacle.

(b) Install the gun cradle and mount the MK 19.

(c) Attach the bracket mounting assembly and the empty case catch bag.

(3) ***Commander's Cupola (M113 APC).*** To mount the MK 19 on the commander's cupola follow these procedures:

(a) Install the pintle adapter and the gun cradle.

(b) Mount the MK 19 and attach the bracket mounting assembly and the empty case catch bag.

(c) Install the machine gun pintle adapter in the commander's cupola mount by pushing the lock handle down.

(d) Pull the lock handle up, if it does not come up when the pintle is installed.

(e) Insert the front stow pin to prevent relative movement of the carriage and cradle.

(f) Insert the carriage pintle into the top of the pintle adapter assembly.

(g) Press in on the pin's quick-release button and insert the pin.

(h) Pull upward and twist the carriage and cradle assembly. It should be locked into the pintle adapter, but it should traverse freely left and right.

(i) To remove, reverse the steps.

(4) ***The M88 Recovery Vehicle Mount.*** To mount the MK 19 on the M88 recovery vehicle mount follow these procedures:

(a) Loosen the traverse locking screw.

(b) Install the machine gun pintle in the machine gun mount.

(c) Insert the front stow pin to prevent movement of the carriage and cradle. Insert the carriage pintle into the top of the pintle adapter assembly.

(d) Press in on the pin's quick-release button and insert the pin.

(e) Pull upward and twist the carriage and cradle assembly. It should be locked into the pintle adapter, but it should traverse freely left and right.

(f) To remove, reverse the steps.

(5) ***Weather Cover.*** To install the weather cover on the MK 19 follow these procedures:

(a) Unzip the zipper on the weather cover.

(b) Pull the weather cover over the barrel, ammunition can, and rear of the MK 19.

(c) Zip the zipper to secure.

NOTE: The weather cover for the MK 19 is an auxiliary component not common to all MK 64 mounts.

CHAPTER 4

MARKSMANSHIP

This chapter gives guidance for instruction and development of MK 19 gunnery skills. It provides data needed to develop training programs, plans, and lessons used to build the gunner's confidence and skill in firing the MK 19. Marksmanship training ensures that each soldier keeps his combat firing skills at a consistent level with the unit's mission. See Appendix D, Gunnery Tests, for more details concerning diagnostic and intermediate tests.

Section I. PREPARATORY MARKSMANSHIP TRAINING

Preparatory marksmanship training for the MK 19 covers the firing positions, the MK 19 fighting position, use of the T&E mechanism, and manipulation exercises. Thorough instruction during the preparatory training and exercises helps ensure efficient use of time and ammunition during range firing.

4-1. FIRING POSITIONS

Use the basic sitting, standing, and kneeling positions during training and range firing of the MK 19.

a. **Sitting.** When the tripod is used in the low or high position, sit directly behind the gun between the trail legs of the tripod. Extend your legs under the tripod, cross them, or brace your feet on the tripod (Figure 4-1). Place your elbows on the inside of your thighs for support when crossing your legs or bracing your feet on the tripod.

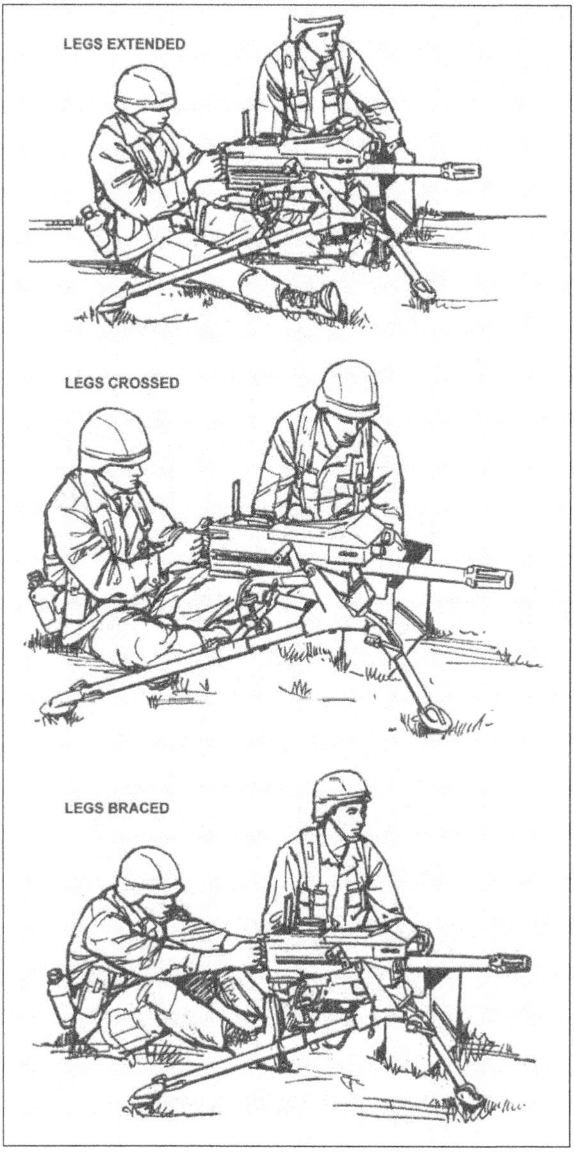

Figure 4-1. Seated firing positions.

b. **Standing.** When the MK 19 is mounted on a vehicle, stand with both of your hands on the control grips and your thumbs resting on the trigger. Keep your elbows against your body, your body forward, and your chest against your hands to brace the gun (Figure 4-2).

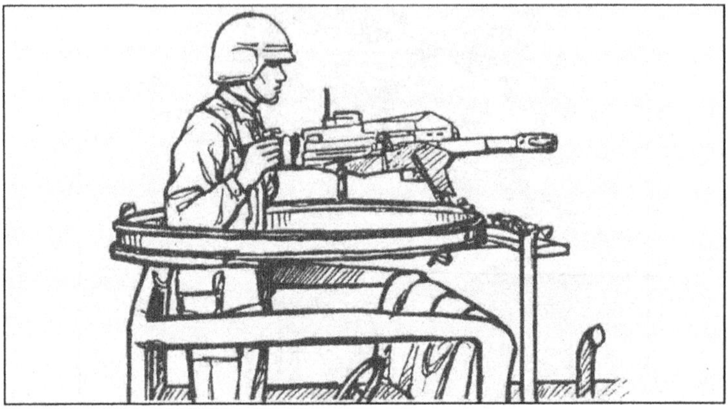

Figure 4-2. Standing position.

c. **Kneeling.** When the MK 19 is in a fighting or hasty tripod-mounted position, kneel and grasp the control grips with your thumbs on the trigger (Figure 4-3).

Figure 4-3. Hasty tripod position.

4-2. FIGHTING POSITION

A fighting position is dug when the unit has time to prepare it or is in a defensive position. Make the fighting position according to Figure 4-4.

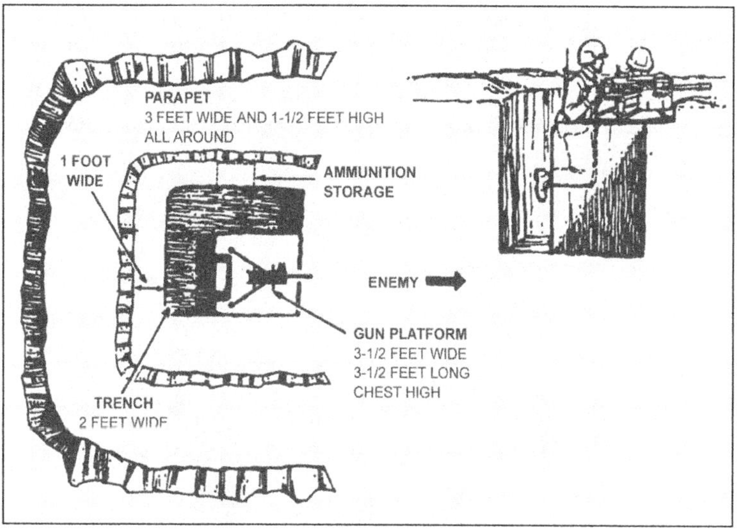

Figure 4-4. Dug-in position.

a. Position the MK 19 so it is oriented on the center of the assigned sector of fire. Mark the position of the tripod legs and trace the outline of the firing platform, which will be dug about 3 inches deep.

b. Dig the position in an "L" shape around the firing platform. The position should be chest-deep and wide enough to allow the MK 19 crew to load, operate, and place effective fire on an assigned sector of fire. The position can be made so that the gunner can fire from the kneeling position by digging the desired depth directly behind the gun.

c. Use the dirt from the hole to build flank parapets 3 feet wide and 1 ½ feet high. Dig the ammunition storage area into the left wall of the "L" and close to the ground.

d. Make sure the wall behind the firing platform is sloped to allow for entrance to and exit from the fighting position.

e. Follow these steps to build overhead cover:

(1) Put flank support logs, 4 to 6 inches in diameter, on top of each other along the entire length of the flank parapets.

(2) Put logs, 4 to 6 inches in diameter, side by side across the support logs as the base for the overhead cover.

(3) Put a waterproof layer over the base logs.

(4) Put 18 inches of dirt on top of the waterproofing material.

(5) Mold and camouflage the cover to blend with the terrain.

f. Camouflage the position with natural materials such as rocks, logs, live bushes, and grass.

4-3. TRAVERSE BAR AND T&E MECHANISM

Use the T&E mechanism to engage preselected target areas at night or during degraded light conditions. Record direction and elevation readings from the traversing bar and T&E mechanism. Record all readings in mils.

a. **Zero the T&E Mechanism.** Before the gunner can use the T&E mechanism to engage targets, he must zero it (Figure 4-5). Refer to section 3-2 for more information on zeroing the T&E mechanism.

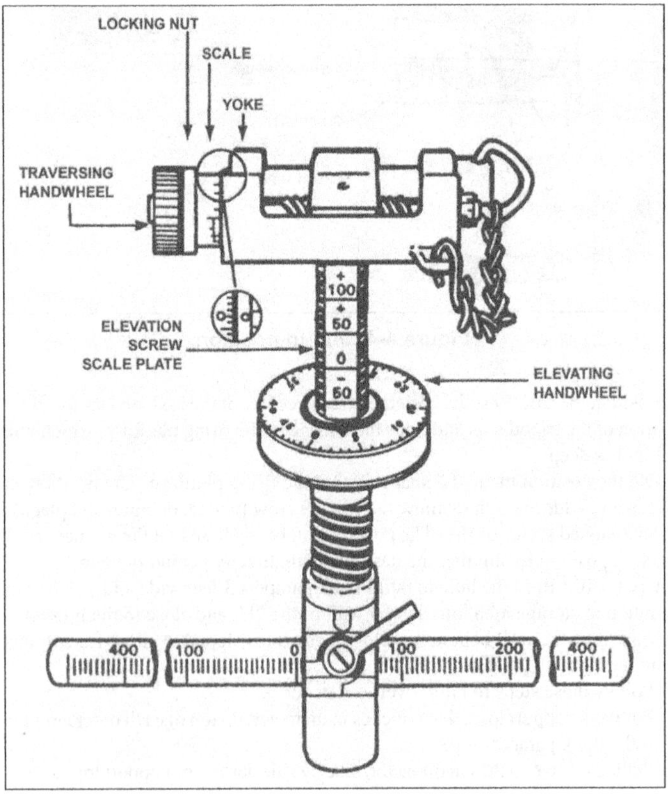

Figure 4-5. Zeroing the T&E mechanism.

b. **Lay the Gun for Direction.** When a sector of fire has been assigned, pick up the rear legs of the tripod and shift the tripod until the muzzle of the weapon points to the center of the sector of fire. Once the MK 19 is laid for direction, firmly stamp in the tripod shoes and place sandbags on the legs. This aids stability and may prevent accidental movement.

c. **Obtain and Record Direction Readings to all Targets Within the Sector of Fire** (Figure 4-6)**.** Loosen the traversing slide lock lever and move the slide along the traversing bar until the MK 19 is laid on the center of a point target or on either flank of a linear target. Lock the traversing bar and read the direction from the scale on the traversing bar. If the left edge of the traversing slide does not fall exactly on a 5-mil tickmark, move the left edge of the traversing bar slide back to the next smaller mil reading. Use the traversing handwheel to complete the initial lay. Obtain a reading to the target by the direction of the MK 19 barrel. If the barrel is moved to the right, record a right reading. Read the number on the traversing bar from the left side of the traversing slide lock. If the barrel is moved to the left and the traversing slide lock is on the right side of the zero, record a left reading. After taking a direction reading for a target, measure the width of the target in mils using the traversing handwheel to move across the target. Reposition the traversing mechanism before moving to another target.

d. **Obtain Elevation Reading.** Ensure the MK 19 is laid on the center base of the target. Read the elevation from two scales: the first, or major, part from the elevating screw plate scale; the second, or minor, part from the elevating handwheel. Separate the two parts of the elevation reading with a slash (for example, -50/3). An elevation reading may not be valid on any other T&E mechanism than the one from which it was read. If the number of threads is increased or decreased after the data is recorded, accurate fire cannot be placed on the target. For example, if a weapon is freed to engage a secondary sector, and the base of the T&E mechanism is rotated, the data is wrong unless the same number of threads is exposed before and after the move.

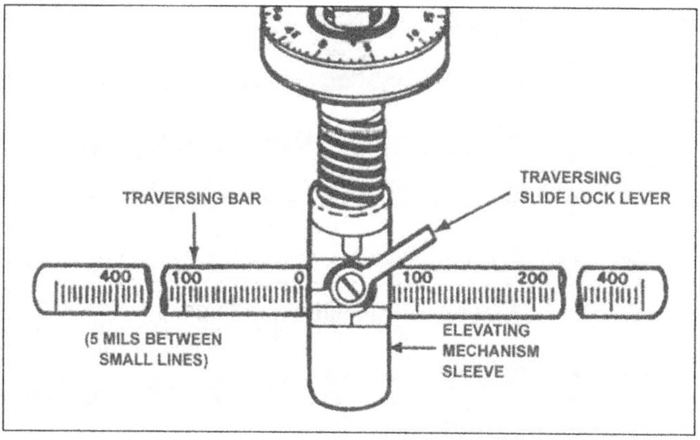

Figure 4-6. Direction reading.

(1) ***Major Reading.*** The elevating scale plate is graduated in 50-mil increments from -250 to +100. Except for the zero, each number has a plus or minus sign above it. All numbers, including zero, have an index line below them. To get the major elevation reading, hold the

T&E mechanism so that the eyes are even with the top of the elevating handwheel. The major reading (for example, -50) is the first number, with an index line, that is visible above the handwheel (Figure 4-7).

(2) ***Minor Reading.*** The elevating handwheel is marked in 50 increments of 1 mil each (Figure 4-7). The minor reading, which has no plus or minus sign, is the number that lines up with the pointer. The minor reading represents the number of mils the gun is laid below the major reading, for example, 3 mils. Verify a direction elevation reading by firing and adjusting on the target. Use the dry-fire method to get data to targets without live firing and adjusting. Set the rear sight on the range to the target, lay the gun on the center base of the target, and take direction and elevation readings. Use the dry-fire method only when firing is not feasible.

NOTE: Correct range setting is critical since any discrepancy causes an error in the elevation when the target is engaged.

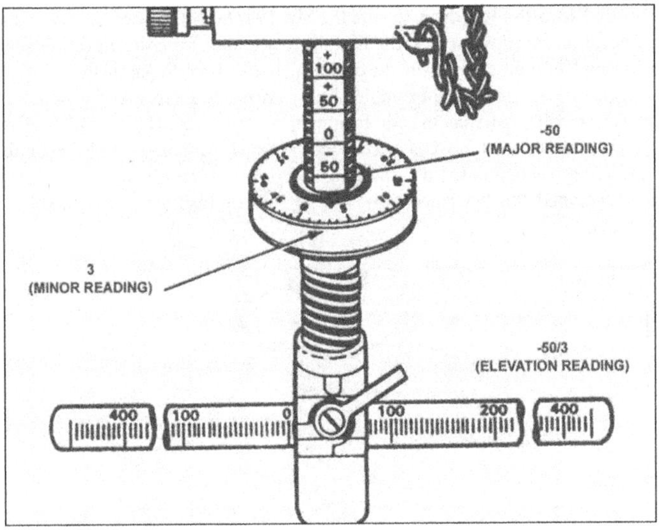

Figure 4-7. Elevation readings.

4-4. EXERCISES

The exercises include sighting and aiming; sight setting and laying; manipulating; determining the range; and preparing a range card.

a. **Sighting and Aiming Exercises.** There are two sighting and aiming exercises. The first calls for the use of the sighting bar.

(1) This exercise teaches how to obtain the correct sight picture, and also teaches alignment methods. Sight alignment means lining up the front and rear sights, with the top center of the front sight blade appearing in the center of the rear sight. Sight picture means lining up the front sight, the rear sight, and the target, with the tip of the front sight blade just touching the bottom center of the target. Gunners must show skill in aligning sights correctly and in getting the correct sight picture, using a sighting bar (Figure 4-8). Gunners must correctly show the point of aim, after sight alignment, 8 out of 10 times.

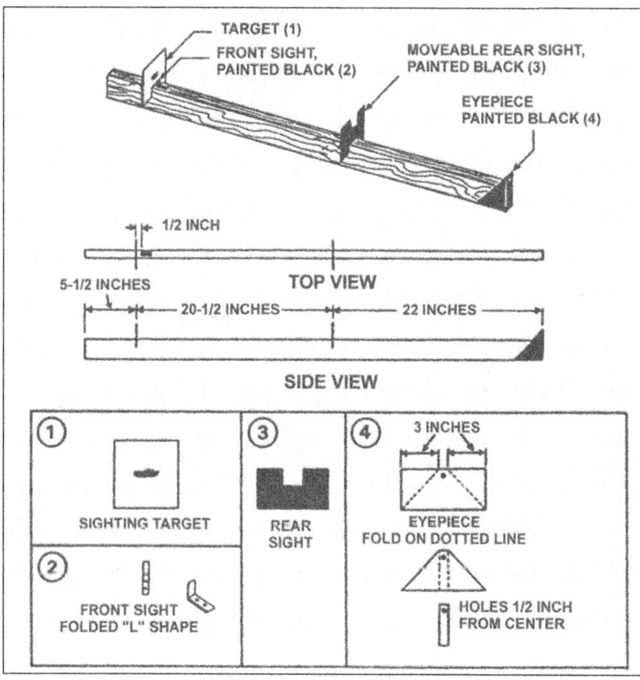

Figure 4-8. Sighting bar.

(2) The second sighting and aiming exercise, used with the MK 19 and target board, allows the gunner to apply lessons learned during the first exercise. Place the target board 10 meters from the muzzle of the MK 19 (Figure 4-9). Use the T&E mechanism to get the correct sight picture on each target that is called out. When the gunner feels he can sight and aim, the coach checks and critiques the gunner's ability to change from the start point on the target board to any new target called by the coach.

NOTE: Continue this exercise until the gunner is skilled in sighting and aiming.

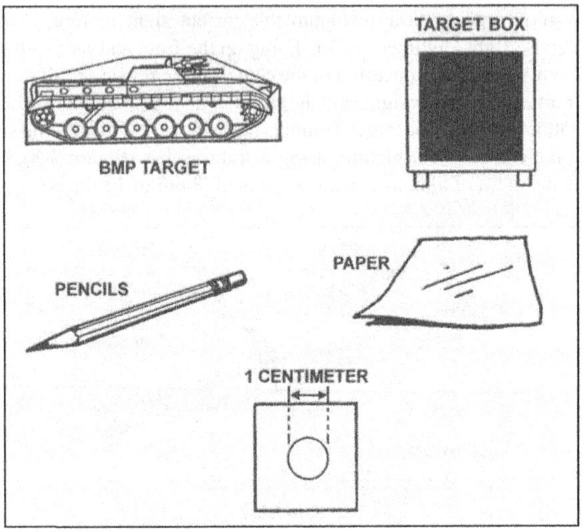

Figure 4-9. MK 19 target board.

b. **Sight Setting and Laying Exercises.** This first exercise teaches the soldier how to set the rear sight quickly and correctly (Figure 4-10). The second builds accuracy and speed in laying the gun on an aiming point, and allows extra practice in sight setting. Explain and show each exercise before beginning instruction.

(1) ***Sight Setting Exercise.*** Explain that, by training the elevating screw knob, the soldier can make minor adjustments in elevation. All major adjustments are made by using the slide release. Have one soldier act as gunner and another act as coach. Announce the range (for example, "EIGHT HUNDRED"). Have the soldier in the gunner's position conduct the following:

- Repeat the range.
- Set the sight at the announced range.
- Assume the correct gunner's position.
- Announce "Up."

The soldier, acting as coach, checks the setting of the slide and points out any errors. The soldiers work in pairs, change roles, and go through the exercise until each one can correctly and rapidly set the sight.

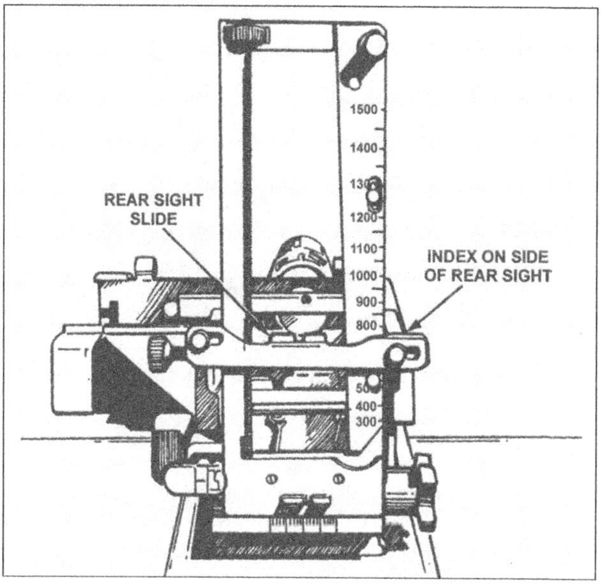

Figure 4-10. Rear sight adjustments.

(2) ***Laying Exercise.*** The MK 19 target board is used in this exercise. Explain that the exercise starts with the sight leaf down and the sight slide at 500 meters. Ranges less than 400 meters or greater than 1,500 meters are not announced. Have one soldier act as the gunner and another act as the coach at the gun. Announce an aiming point and range (for example, "TARGET NUMBER ONE," "ONE TWO HUNDRED"). Pause after each element to allow the gunner time to repeat it. After the gunner repeats each element, have the gunner:

- Set the sights
- Manipulate the gun by turning the T&E handwheels until sights are aligned on the chosen aiming point.
- Assume the correct gunner's position.
- Report "Up."

The soldier, acting as coach, checks the sight setting and lay of the gun. The soldiers work in pairs, changing roles until each is skilled in doing the exercise.

c. **Manipulation Exercise.** Manipulation means shifting the direction of the gun from one point to another. There are two manipulation exercises.

(1) The first exercise gives the gunner practice in manipulation. The coach stands 10 paces in front of the gun and uses hand signals to show the direction in which the gunner is to move the muzzle (Figure 4-11). The gunner works the handwheel to manipulate the gun. The coach observes and makes needed corrections. When the gunner reacts quickly and can manipulate the gun as he has been shown, his instruction may continue.

Figure 4-11. Coach and gunner exercise.

NOTE: Continue this exercise until the gunner is skilled in sighting and aiming.

(2) Once the soldier knows the basics of sighting and aiming, and can assume a good firing position, give him the second manipulation exercise. Show him how to manipulate the weapon to get a correct initial lay and how to skillfully shift the direction of the weapon to successive points. Ensure the soldier understands and knows how to perform the following instructions:

(a) Place the MK 19 target board 10 meters from the muzzle of the gun.

(b) Make large shifts in direction by releasing the traversing slide lock lever and moving the slide to the right or left. Make small changes in direction by turning the traversing handwheel with the left hand. One click on either T&E handwheel moves the strike 1 mil (1 cm on the target).

(c) Manipulate for elevation by rotating the elevating handwheel with the left hand (Figure 4-12).

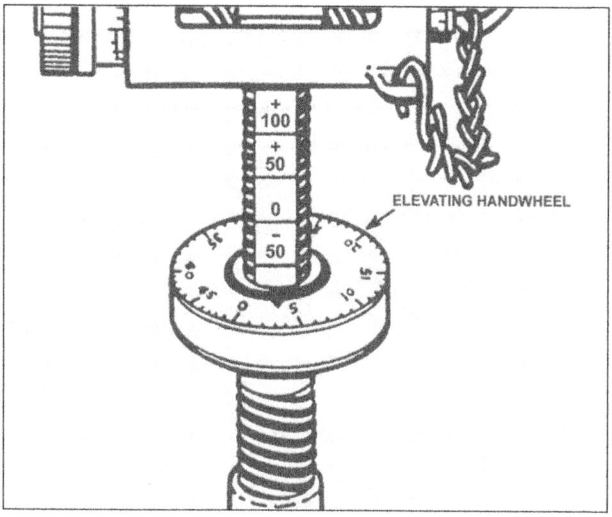

Figure 4-12. Elevating handwheel.

d. **Range Determination Exercise.** Range determination means finding the distance between the gunner's position and his target. The gunner's ability to engage a target depends on his ability to correctly find the range to the target. Under combat conditions, ranges may not be known in advance; the effectiveness of fire depends largely on the accuracy of the range used. There are several methods for finding range. Some of these methods include: estimating visually, firing the weapon, measuring range from a map or aerial photograph, and pacing the distance. Binoculars or laser range finders may be used (see Chapter 5, Techniques of Fire, for more on range determination).

e. **Range Card Exercise.** A range card is a record of the firing data needed to engage predetermined targets within a sector of fire at night or during degraded conditions. The range card may also be used to aid in target engagement during good visibility. It aids the leader in preparing his defense plan. Predetermined targets in the secondary sector are engaged by use of field expedients. A range card has two parts: a sketch section and a data section. Although the sketch is not drawn to scale, the data referring to the targets must be correct. DA Form 5517-R is the standard range card to be used (Figure 4-13).

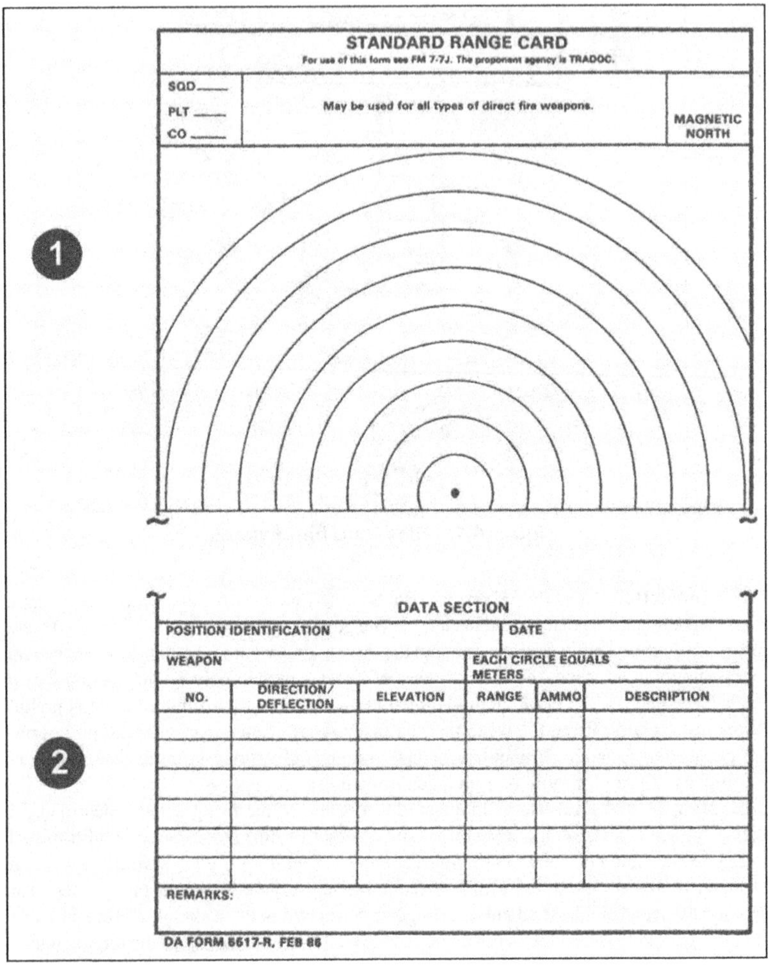
STANDARD RANGE CARD

For use of this form see FM 7-7J. The proponent agency is TRADOC.

SQD____ PLT ____ CO ____

May be used for all types of direct fire weapons.

MAGNETIC NORTH

1

DATA SECTION

POSITION IDENTIFICATION			DATE		
WEAPON			EACH CIRCLE EQUALS ____ METERS		
NO.	DIRECTION/ DEFLECTION	ELEVATION	RANGE	AMMO	DESCRIPTION

REMARKS:

DA FORM 5517-R, FEB 86

2

Figure 4-13. Range card sketch (1) and data sections (2).

(1) Range cards are prepared for primary, alternate, and supplementary positions. They are prepared immediately upon arrival in a position, regardless of the length of stay, and are updated as necessary. Range cards are prepared in duplicate. One copy stays at the gun position, and the other is sent to headquarters. Range cards are prepared on arrival in a position, regardless of the planned length of stay. Range cards are revised often during occupation of a position.

(2) Once the leader has decided where the firing position will be, the gunner prepares the range card.

(3) The range card is prepared as follows:

(a) Draw a sector sketch, covering the entire sector. Make the sketch as large as possible, not to exceed the largest circle. For a large area covered by trees or woods, draw only the outline and label the area "woods" or "orchard" (Figure 4-14).

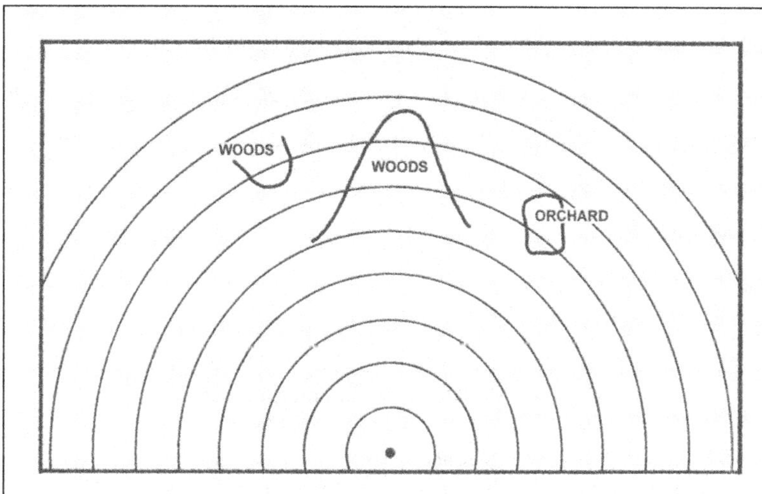

Figure 4-14. Sketch of area on range card.

(b) In the lower center of the sketch section, show the firing position by drawing the symbol for the weapon (Figure 4-15).

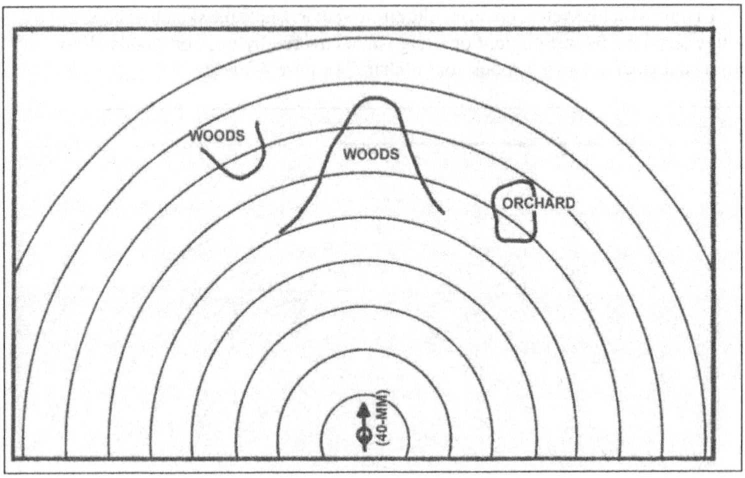

Figure 4-15. MK 19 symbol on range card.

(c) Show the location of the firing position by drawing a sketch of a nearby recognizable terrain feature. Label it and draw an arrow to the weapon symbol. Add the distance and azimuth from the terrain feature to the firing position (Figure 4-16).

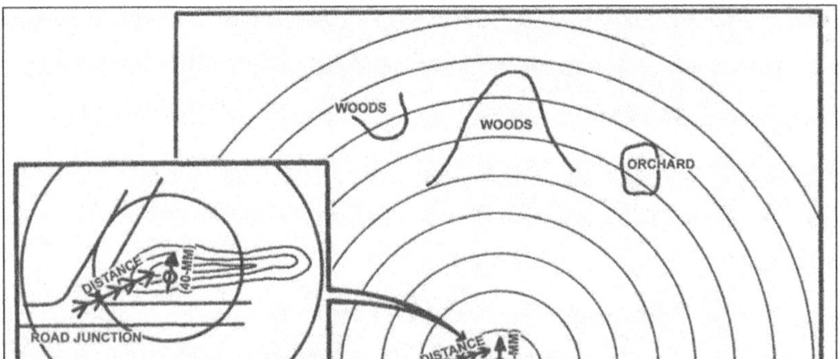

Figure 4-16. Distance and azimuth to terrain feature.

(d) Draw lines from the weapon symbol to reflect the left and right limits (Figure 4-17).

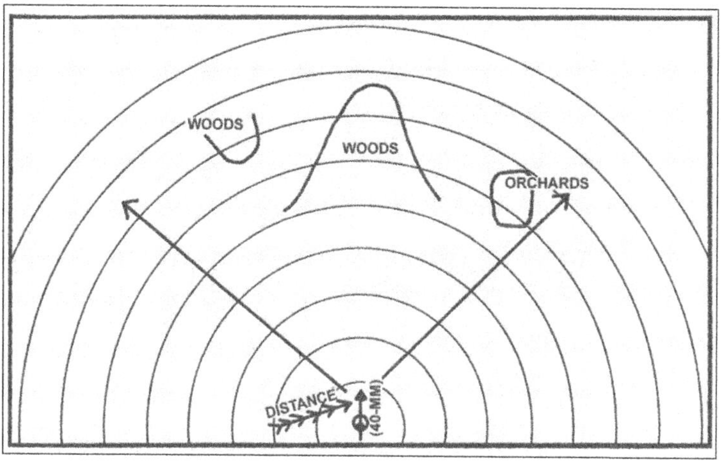

Figure 4-17. Sector of fire.

(e) Number the planned target engagement locations from left to right, and write them on the range card (Figure 4-18).

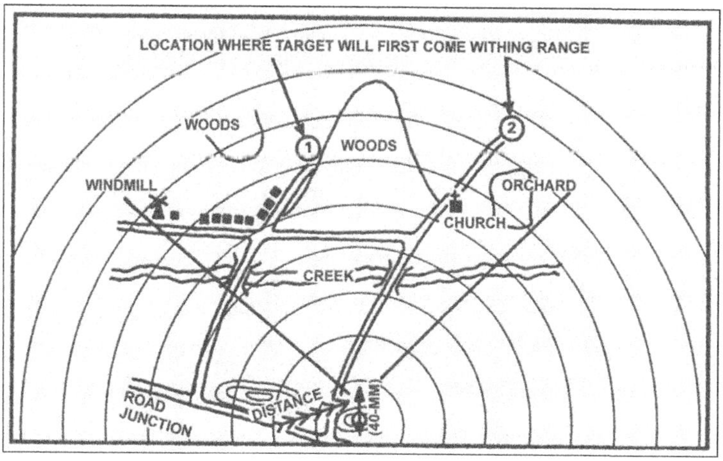

Figure 4-18. Planned target engagement areas.

(f) Number the target reference points (TRPs) and write them on the range card (Figure 4-19).

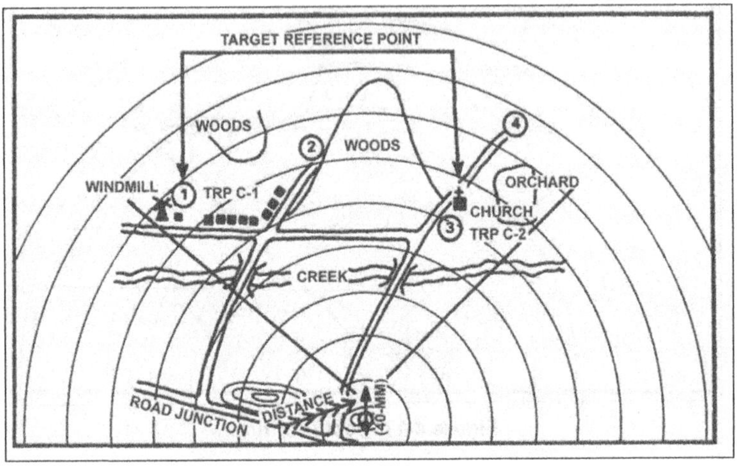

Figure 4-19. Target reference points.

(g) If no limiting factors exist, the maximum engagement line curves and joins the left and right sector-of-fire boundaries at the maximum engagement range (Figure 4-20). If limiting factors exist, draw the maximum engagement line in front of the limiting terrain features (Figure 4-21).

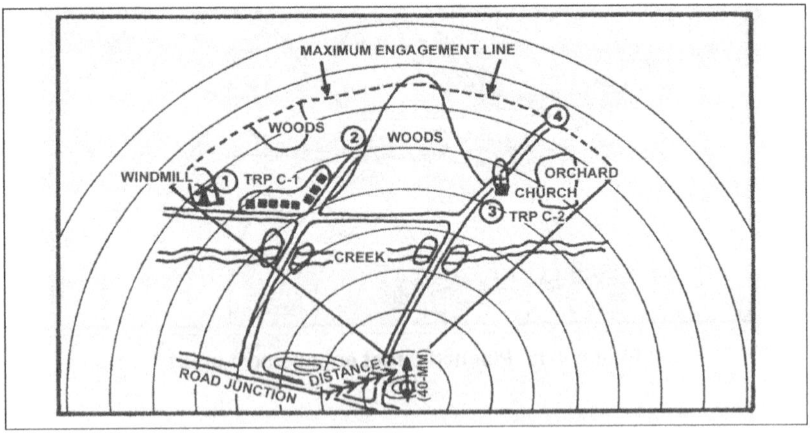

Figure 4-20. Maximum engagement line.

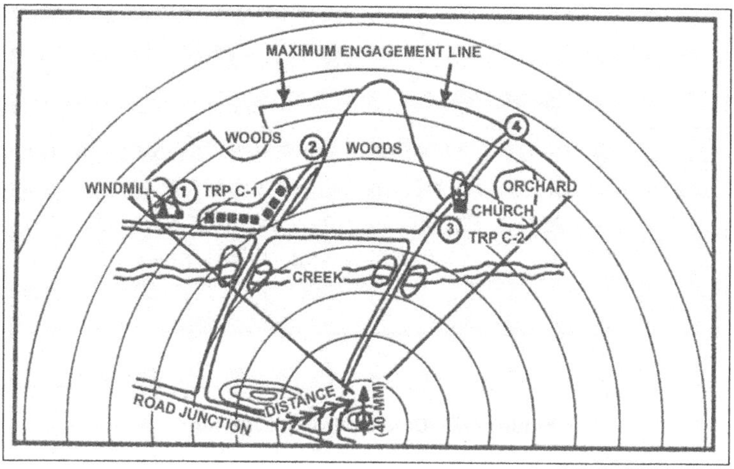

Figure 4-21. Limiting factors on maximum engagement line.

(h) Fill in the marginal information at the top of the card in the following manner (Figure 4-22):

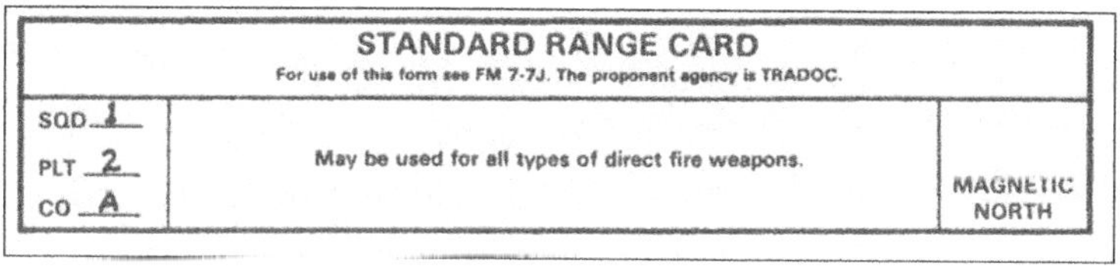

Figure 4-22. Marginal information.

- **Unit description.** Never show unit higher than company level.
- **Magnetic north.** Orient the range card with the terrain and find the direction of magnetic north with a compass. Draw a magnetic north arrow using the straight edge of the compass.

(i) Fill in the data section in the following manner (Figure 4-23):

- **Position identification.** Write primary, alternate, or secondary.
- **Weapon.** Write MK 19, grenade machine gun.
- **Date.** Write the day and month.

POSITION IDENTIFICATION: PRIMARY — DATE: 16 DEC

WEAPON: MK19 — EACH CIRCLE EQUALS ______ METERS

NO.	DIRECTION/ DEFLECTION	ELEVATION	RANGE	AMMO	DESCRIPTION
1	L200 MIL	+140/35	2050	M450	WINDMILL/TRP C-1
2	L30 mil	+160/35	1750	M430	ROAD
3	R25 mil	+140/35	1600	M430	CHURCH/TRP C-2
4	R100 mil	+150/10	1400	M430	ROAD

REMARKS:

Figure 4-23. Data section information.

- **Each circle equals 294 meters.** Write in the distance in meters between the circles. To find the distance, count the intervals from the weapon to the maximum engagement line. Divide the number of intervals into the range. This gives the distance between circles (Figure 4-24). For example:

$$\frac{2{,}212 \text{ meters}}{7.5 \text{ intervals}} = 294 \text{ meters between circles}$$

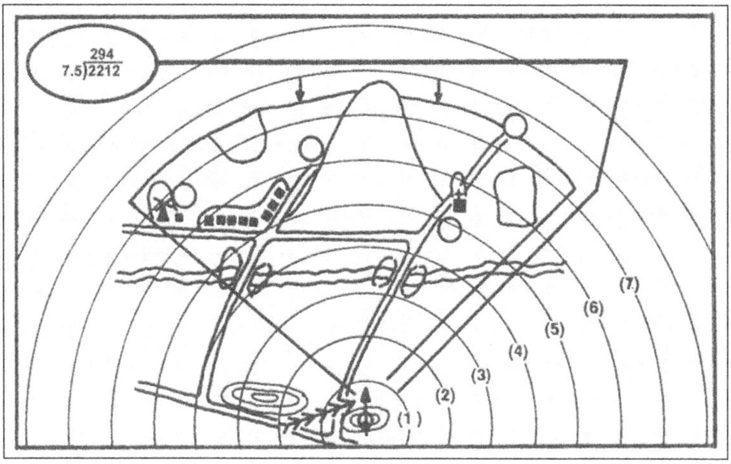

Figure 4-24. Determination of range the circles represent.

- **Number (No.).** Starting with number 1, write the numbers as listed for TRPs and target engagement locations.
- **Direction/deflection.** Write the mils from the traversing bar and handwheel.
- **Elevation.** Write the elevation from the elevating screw plate scale and the elevating handwheel.
- **Range.** Write distance in meters from the weapon to the TRP or target engagement area.
- **Ammunition.** Write the type of round issued for the mission.
- **Description.** List the name of the object; for example: road, windmill, or church. If the item is a TRP, list the TRP number also.

Section II. RANGE FIRE

Range firing gives hands-on MK 19 firing experience to the soldier who has completed the Intermediate Gunnery Test. Use the procedures in this section to set up and conduct range firing. The desired width for each lane is 100 meters at 800 meters from the firing line but can be modified due to local range or terrain limitations. Range firing includes zeroing procedures, observation and adjustment for fire, instructional firing, and qualification firing. Active fighting-force gunners qualify semiannually and participate in exercises and unit-level live fires quarterly. Active supporting-force gunners qualify and participate in instructional fire annually. All reserve component gunners qualify annually. See Appendix E, Annual Gunnery Training Program, for details.

4-5. DESCRIPTION

There are two types of ranges that may be used for MK 19 live-fire practice and qualification. Regardless of which range is used, the MK 19 can be fired from either a tripod or a vehicle.

a. **Multipurpose Gunnery Range for the 40-mm Grenade Machine Gun** (Figures 4-25 and 4-26). This range is designed for the conduct of individual firing exercises and qualifications against hull type targets. Details about this range can be found in TC 25-8.

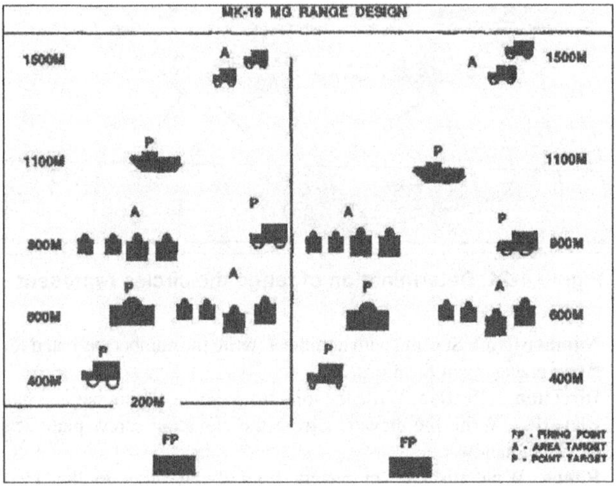

Figure 4-25. Multipurpose gunnery range for the 40-mm grenade machine gun.

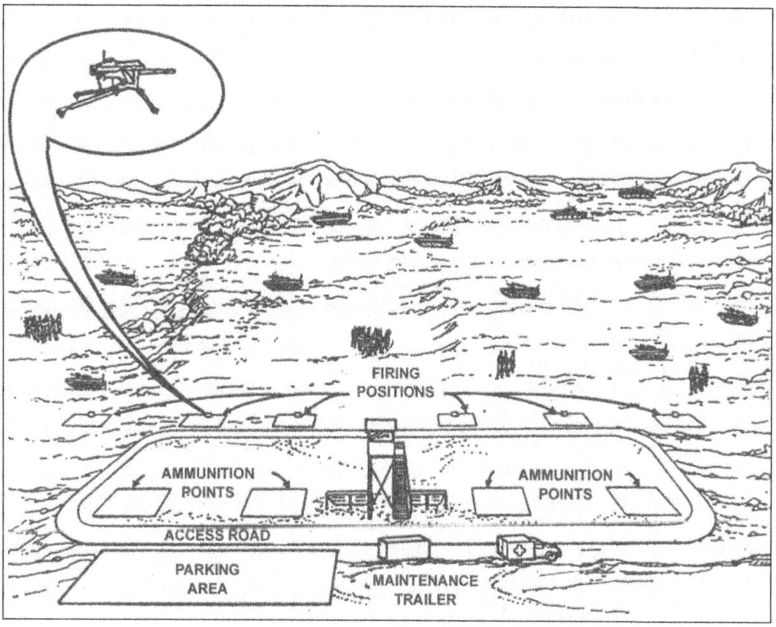

Figure 4-26. Multipurpose gunnery range for the 40-mm grenade machine gun (continued).

b. **Multipurpose Machine Gun Range.** The MPMG is the range suggested for MK 19 live-fire training. It is used to train mechanized, motorized, combat support, and combat service support units. The design and traits of this type of range are described in TC 25-8. Only a non dud-producing practice round is used on this range.

4-6. OPERATION

Use range operation procedures any time LFXs are conducted on the range. These procedures are the same for all range firing exercises.

a. Brief the gun crews on the range setup, fire commands, and safety procedures. Explain exercise needs as defined by the unit commander, or as they are to be done by the firing tables.

b. Divide the crews into groups and assign each group a firing point and an assistant instructor (AI). Ensure they are directed to their firing points and are shown their assigned sectors of fire. Have the AIs ensure that all crews have the equipment needed to perform the task.

c. Conduct the exercise. For example you can command GUNNER, WHEN I SAY BEGIN, PLACE THE MK 19 INTO OPERATION ON THE TRIPOD (ORGANIC CARRIER) WITHIN ____ MINUTES. ARE THERE ANY QUESTIONS? (pause) BEGIN.

At the end of the given time period, direct STOP. AIs, CRITIQUE YOUR CREWS AND GIVE ME YOUR "UP" SIGNAL WHEN YOU HAVE FINISHED.

4-7. COACHES

The instructor may arrange for a coach to be present at each gun during the following preparatory training, exercises, and instructional firing. Successful marksmanship training depends on the coaches. Well-trained and well-rehearsed gunners or crew members may be assigned as coaches. The coaches:

- Require the gunner to inspect his gun and equipment.
- Help the gunner place the gun on target.
- Require the gunner to explain the exercise that he is about to perform.
- Observe the gunner's position, grip, and manipulation during simulated firing.

4-8. FIRE PROCEDURES

Firing the MK 19 is not hard. However, to be effective, the gunner must be able to skillfully manipulate the trigger and T&E mechanism; the weapon must be held down and to one side while aiming and adjusting the T&E mechanism. The following steps are simple, but the gunner must remember to estimate the distance to the target, set the sights for distance, manipulate the T&E mechanism, press the trigger, and fire a single round; since the MK 19 is a fully automatic weapon, this can only be done by "popping" the trigger once. He must spot the impact of the first round and adjust to the target area. He should begin firing six- to nine-round bursts. This whole process can be done in a matter of seconds, the longest part being the flight time of the round. To fire the weapon:

(a) Place the safety on F (FIRE).
(b) Ensure the charger handles are in the forward and upright position.
(c) Place your hands on the control grips, and your thumb(s) on the trigger.
(d) Hold the weapon down and to one side, and check sight picture.
(e) Press the trigger to fire.

WARNING

Because of the gun's recoil, the first burst is the hardest to control. Ensure that the gun's recoil does not cause the barrel of the weapon to drop and rounds to strike short of the target. Keep it elevated.

4-9. ZEROING PROCEDURES

Zeroing procedures are crucial for hitting targets at ranges of 600 meters or more. It is strongly recommended that a target at 400 meters be used to zero. The following is the correct way to zero the MK 19 to the gunner.

a. Loosen the range plate screw. Move the range plate to the midpoint between the two studs. Tighten the range plate screw. Move the rear sight slide to the meter mark that represents the distance to the target. For example, move to the 400-meter mark to zero on a target known to be 400 meters away. Set the windage knob at the zero index line.

b. Align the sights on the base of the target using the T&E mechanism.

c. Fire a single round and spot the impact of the round. If it is on the target, fire another short burst to confirm the zero. If the round is not on target, go on to the next step.

d. To adjust for a round that is not on target, do the following:

(1) If the impact of the round is short or over, adjust the elevation knob. Estimate how short or over the round is. If the round is short, use this estimate to adjust the elevation knob clockwise, which moves the sights up onto the knob counterclockwise to bring the sights down to the target. For example, if the rounds impact 10 mils short, adjust the elevation knob 10 mils up by turning it clockwise. Realign the sights and adjust the gun back on target using the T&E mechanism before the next round is fired.

NOTE: 1. If the adjustment was correct, the second round should be on target. If so, fire the rest of the rounds to confirm the zero. If not, repeat the previous step.
2. If the impacts are not observed, bold adjustments may be necessary.

(2) If the round is to the right or left, adjust the windage knob. Estimate how far to the right or left the sight needs to move to bring the rounds on target. Turn the windage knob clockwise to adjust to the right; turn the windage knob counterclockwise to adjust to the left. For example, if the rounds impact 10 mils to the right, adjust the sight 10 mils to the left by turning the windage knob counterclockwise. Realign and adjust the sights back on target using the T&E mechanism before firing the next round.

(3) Once the zero is completed, align the range plate scale at the exact range of the target used to zero, and tighten it.

e. Point out errors and explain their effect.

f. When the gunner maintains the same sight picture, the type of firing position does not alter the zero.

4-10. OBSERVATION AND ADJUSTMENT OF FIRE

Observation and adjustment gives the soldier a chance to practice adjustment of fire by observing trajectory and impact or by re-laying often on the target using the sights. Live-fire training is held on the firing range after the methods are explained in the classroom.

a. When firing on targets, adjust by moving the burst into the target. Based on the strike of the rounds, adjust the number of clicks for elevation and direction needed to move the center of impact onto the target. This does not call for the use of sights. For example, fire at a target 500 meters away; the rounds impact 20 meters short and 50 meters right. Use the T&E handwheels, moving the muzzle left and up the proper number of clicks, to manipulate the gun onto the target.

b. Another procedure is the adjusted aiming point method. Use the sights and select an aiming point calculated to place the second burst on target. For example, fire at a 500-meter target; the rounds impact 20 meters short and 10 meters right. Rapidly select another aiming point about 20 meters beyond and 10 meters to the left of the target, lay on that aiming point, and fire.

4-11. INDIVIDUAL GUNNER EXERCISES AND QUALIFICATION

The individual gunnery exercises train and qualify MK 19 gunners. There are four scorecards available and they are used based on the type of target (hull or pop-up silhouettes) and whether the practice-qualification is during the day or during limited visibility. Each scorecard has two tables, one for practice and one for qualification. The tables have versions for hull or pop-up silhouette engagements and for the type of NVD used. Sample scorecards are shown in Figures 4-27 through 4-30 (and reproducible scorecards are provided in the back of this manual). MK 19s will be mounted or in the tripod configuration based on the range constraints and the commander's guidance. These tables are *recommended* tables for the infantry MK 19 gunner and crew. The first task in each table is a field zeroing evaluation, which allows the gunner to ensure his weapon is zeroed (even if he boresighted the MK 19). If the gunner fails to zero within four rounds, he is removed from the line and given additional training before attempting the table again. Refer to Section 4-5 for more detailed description of the ranges. The rest of the table consists of firing at individual and multiple targets.

a. **Day and Night Practice and Qualification.** Specific scorecards have been developed for different targets and NVDs. Gunners will only fire one day practice-qualification and one night practice-qualification. Units should select the practice and qualification based on the light conditions, type of targets available, and type of NVDs used. The following table shows which scorecards would be used:

Conditions	Target	Night Vision Devices	Scorecard
Day	Hull	NA	Scorecard I
Day	Pop-Up	NA	Scorecard III
Limited/Night	Hull	**AN/PEQ-2A** mounted on the TWS mounting bracket. **AN/PAS-13** mounted on the TWS mounting bracket. **AN/TVS-5** upgraded with the 3d generation tube mounted on the TWS mounting bracket.	Scorecard II
Limited/Night	Pop-Up or E-Type	All night vision devices.	Scorecard IV
Limited/Night	Any type	No night vision device.	Scorecard IV

Table 4-1. Scorecard matrix.

NOTE: 1. Both the MPMG range and the multipurpose gunnery range (MK 19) can be used for practice and qualification.

2. The MPMG range, modified with stationary armor targets at the required ranges, can be used as pop-up targets if a non dud-producing practice round is used. Scorecards III and IV are used on the MPMG range.

3. The multipurpose gunnery range (MK 19) can be used with any round authorized for the range. Scorecards I and II are used on the multipurpose gunnery range (MK 19).

Every target listed in each task is a point target; however one or two targets between the range of 600 and 900 meters should be changed to an area target. The number of rounds and engagement times should be the same for both point and area targets. The commander's guidance and the local range configuration should determine the location of the area targets.

(1) ***Day Practice and Qualification.*** Due to the types of targets available for practice and qualification, there are two scorecards for day practice and qualification. One (Table I) is used when engaging hull-type targets and the other (Table III) is used when engaging pop-up silhouette targets.

(a) *Hull-Type Targets.* Hull-type targets provide height, width, and depth, and give the MK 19 gunner a realistic target. The engagement ranges for practice and qualifications can therefore be set for the full range of the gun and are set at ranges up to 1,500 meters.

(b) *Pop-Up Silhouette Targets.* Pop-up silhouettes provide a target with width and height but very little depth. Due to the high angle of fall of 40-mm rounds at ranges greater than 800 meters, it is difficult to hit this type of target beyond that range. Therefore, the engagement ranges for practice and qualifications are set at 800 meters or less.

(c) *Engagement Times.* There is a 30 second difference for the completion of each task between the practice and qualification tables. Practice tables allow thirty additional seconds for each engagement.

b. **Night Practice and Qualification.** The MK 19 night practice and qualification tables are shown in Scorecards II and IV. Units with AN/PEQ-2A, AN/PAS-13, and AN/TVS-5 night aiming devices *and* engaging hull-type targets use Scorecard II. Units without a MK 19/sight combination or engaging pop-up silhouettes use Scorecard IV. Gunners do not fire both. Infantry gun crews are required to qualify at night. Other types of units may determine that day qualification is adequate due to their wartime missions.

c. **Scoring.** Scoring is done on a GO/NO GO basis for each task within the practice or qualification table.

(1) Zeroing the gun, the first task in each table, is scored as a GO/NO GO. Giving a score for the zero emphasizes the importance of a proper zero to effectively engage targets at 600 meters and beyond. However, if the gunner fails to zero within four rounds, he is removed from the line and given additional training before attempting the table again. This step reduces the waste of ammunition.

(2) On point target engagements (lightly armored vehicle targets such as BRDMs, threat scout cars, etc.), the gunner receives a GO if he meets or exceeds the engagement standard of one or two rounds hitting the target.

(3) If area targets are included (infantry squads, RPG teams, etc.), the gunner receives a GO when at least the number of rounds stated in the engagement standard for that task impact within ten meters of the area target and thus suppresses it.

(4) At the end of each table, the scorer adds up the number of GOs and NO GOs, places that number in the "Totals block, and checks the box to the left of the appropriate qualification (expert, sharpshooter, marksman, or unqualified).

d. **Range Setup.** Targets should be within the ranges provided on the table scorecard.

(1) Because the gunner has to be able to observe the impact of the round to make adjustments, there should be no dead space within 100 meters of the selected targets.

(2) Every target listed in each task is a point target; however, one or two targets between the range of 600 and 900 meters should be changed to an area target. The number of rounds and engagement time will be the same for both point and area targets. The commander's guidance and the local range configuration should determine the location of the area targets.

(3) For area targets, multiple E type personnel targets may be placed on line or in wedge formations. Multiple personnel targets, indicating area targets, should not be more than 5 meters apart, and not extend more than 30 meters in width or 20 meters in depth.

(4) During night firing using hulls as targets, no modification to the target is needed to assist the gunner in identifying the target. If pop-up silhouette targets are used however, a thermal source is needed on each target to enable the gunner to acquire it with the thermal weapon sight (TWS) and a light source is needed on each target if the AN/TVS-5 is being used. The thermal source can be two chemical lights on targets between 400 meters and 600 meters and three chemical lights on targets over 600 meters.

e. **Grading.** One grader will be required at each firing point.

(1) ***Grading Equipment***. During the day, the grader will need a set of binoculars. At night equipment will vary according to the type of range being used. With an impact range with hull targets, the grader will need a NVD (examples, the AN/PVS-14,7B with the 3X magnifier, the AN/TAS-4, or the AN/PAS-13 [heavy]) to observe the strike of the round. The same equipment is needed if the pop up targets do not provide feedback. The grader also needs the order in which targets are engaged and a means to provide the gunner with the range to the target for that particular firing point. The grader must be able to identify which target is to be engaged by using, for example, a range card including a diagram of the range with targets numbered and ranges listed.

(2) ***Start and End Time.*** Time will start when the target is exposed and the grader has provided the target range (the graders will provide all information before the target is exposed). If hull targets are used and exposed at all times, then the time will start once the grader has told the gunner which target to engage and provided the range to the target. Time ends when the time indicated for that task expires, the target has been successfully engaged, or the target is no longer exposed.

f. **Ammunition.** Ammunition is broken down by task. The assistant gunner places each belt in its order of use. The number of rounds authorized for each task will be the number of rounds per belt. For example, if there are ten engagements, there should be ten belts of ammunition placed within reach of the assistant gunner in the order they are to be fired.

(1) HE rounds cannot be fired at pop-up silhouette targets because the lift mechanism will be damaged.

(2) Training practice tracer (TPT) rounds can be fired at both types of targets.

(3) The impact of a HE round is much easier to see than that of the TPT round.

g. **Fire Control.** Controlling and observing a target engagement with the MK 19 is not a problem with a range set up with a firing lane for each firing point. Especially with hull targets however, each point will not have an individual firing lane. Some ranges must use the same target for more than one lane, which may be a potential problem for grading. The grader must be able to identify which round impact is from which firing point. This is especially true for the 400-meter target. To prevent this problem, ensure that only one gunner at a time engages each target. The order in which the targets are engaged can be changed to allow more than one gunner to fire at the same time. Engagement start times can also be staggered so that gunners can engage targets at different ranges at the same time. This requires a great deal of coordination and communication between the graders and the personnel controlling the range.

h. **Day Practice and Day Qualification Firing Exercise** (Figures 4-27 and 4-28). Table I contains the tables for day practice and day qualification for hull targets and Table III contains the tables for day practice and day qualification for pop-up targets. Other than a 30 second difference in the engagement times for each task, the practice and the qualification tables are the same. It is held twice a year, or as often as the commander feels is needed to maintain gunner proficiency

(1) The day practice firing exercise allows the gunner to fire on a range engaging hull or pop-up targets to test his skills before qualification firing.

(2) The qualification LFX tests skills practiced during day firing exercise. It is scored on time taken and target hits made based on the firing tables.

(3) During scorecard preparation the grader selects the correct scorecard (Table I for hull targets or Table III for pop-up targets) and enters the gunner's name, rank, and unit in blocks "1" through "3." He also fills in blocks "4" through "7" with the range name, the firing lane, his name, and the date.

(4) The grader positions himself so that he can observe both the gunner and the target. Once live fire commences, he:

- Observes and informs the gunner the strike of each round.
- Observes and records a GO or NO GO for each task.

(5) At the end of the practice, the grader sums the total of GO/NO GOs in the "Totals" block, checks the appropriate qualification in block "9," has the gunner sign the scorecard in block "10," and signs the card in block "11."

(6) During the qualification phase, the grader repeats the steps above by filling in the appropriate blocks, summing the scores, and assigning the correct qualification.

(7) The grader can use the comment section in either table to enter remarks such as the operation of the gun, condition of the targets, and weather conditions to name a few.

MK 19, 40-mm GRENADE MACHINE GUN, MOD 3 FIRING TABLE I
DAY PRACTICE AND QUALIFICATION WITH HULL TARGETS SCORECARD
For use of this form, see FM 3-22.27; the proponent agency is TRADOC.

PRIVACY ACT STATEMENT

AUTHORITY: 10 USC 3012(g)/Executive order 9397
PRINCIPAL PURPOSE: To aid individual training on targets at various ranges.
ROUTINE USES: To evaluate individual proficiency.
DISCLOSURE: Voluntary. However, mass rating and recording require some tracking method.

1a. LAST NAME	1b. FIRST NAME	1c. MI	2. RANK	3. UNIT
PEVOSKI	RICHARD	M	SSG	1/26th IN

TABLE I (A). DISMOUNTED AND MOUNTED DAY PRACTICE

4. RANGE	5. LANE	6. GRADER	7. DATE
DOBOL	5	JOULWAN	4 SEP 03

TASK	RANGE (Meters)	AMMO	TIME (Minutes)	ENGAGEMENT STANDARDS	GO	NO GO
ZERO	400	4	NA	2 ROUNDS HIT	✓	
2	1,100 (+/- 200)	8	2.5	2 ROUNDS HIT		✓
3	1,500 (+/- 200)	10	3.5	2 ROUNDS HIT		✓
4	600 (+/- 100)	6	2	2 ROUNDS HIT	✓	
5	800 (+/- 100)	6	2	2 ROUNDS HIT	✓	
6	400	4	1.5	2 ROUNDS HIT	✓	
MULTIPLE TARGETS						
7	1,100 (+/- 200)	10	4	1 ROUND HIT	✓	
8	600 (+/- 100)			1 ROUND HIT	✓	
9	800 (+/- 100)	14	4.5	1 ROUND HIT	✓	
10	1,500 (+/- 200)			1 ROUND HIT	✓	
				TOTALS	8	

8. COMMENTS: USED HE

9. NUMBER OF ENGAGEMENT MET *(Choose One)*
- [ ] 10 - EXPERT
- [ ] 9 - SHARPSHOOTER
- [x] 8-7 - MARKSMAN
- [ ] 6 AND BELOW - UNQUALIFIED

10. GUNNER'S SIGNATURE: Richard M. Pevoski
11. GRADER'S SIGNATURE: George Joulwan

TABLE I (B). DISMOUNTED AND MOUNTED DAY QUALIFICATION

12. RANGE	13. LANE	14. GRADER	15. DATE
DOBOL	8	JOULWAN	4 SEP 03

TASK	RANGE (Meters)	AMMO	TIME (Minutes)	ENGAGEMENT STANDARDS	GO	NO GO
ZERO	400	4	NA	2 ROUNDS HIT	✓	
2	1,100 (+/- 200)	8	2	2 ROUNDS HIT	✓	
3	1,500 (+/- 200)	10	3	2 ROUNDS HIT		✓
4	600 (+/- 100)	6	1.5	2 ROUNDS HIT	✓	
5	800 (+/- 100)	6	1.5	2 ROUNDS HIT	✓	
6	400	4	1	2 ROUNDS HIT	✓	
MULTIPLE TARGETS						
7	1,100 (+/- 200)	10	3.5	1 ROUND HIT	✓	
8	600 (+/- 100)			1 ROUND HIT	✓	
9	800 (+/- 100)	14	4	1 ROUND HIT	✓	
10	1,500 (+/- 200)			1 ROUND HIT	✓	
				TOTALS	9	

16. COMMENTS: USED HG

17. NUMBER OF ENGAGEMENT MET *(Choose One)*
- [ ] 10 - EXPERT
- [x] 9 - SHARPSHOOTER
- [ ] 8-7 - MARKSMAN
- [ ] 6 AND BELOW - UNQUALIFIED

18. GUNNER'S SIGNATURE: Richard M. Pevoski
19. GRADER'S SIGNATURE: George Joulwan

DA FORM 7518-R, AUG 2003 APD V1.00

Figure 4-27. Example of completed DA FORM 7518-R, MK 19 day practice and qualification (hull targets).

MK 19, 40-mm GRENADE MACHINE GUN, MOD 3 FIRING TABLE III
DAY PRACTICE AND QUALIFICATION WITH POP-UP TARGETS SCORECARD
For use of this form, see FM 3-22.27; the proponent agency is TRADOC.

PRIVACY ACT STATEMENT

AUTHORITY:	10 USC 3012(g)/Executive order 9397
PRINCIPAL PURPOSE:	To aid individual training on targets at various ranges.
ROUTINE USES:	To evaluate individual proficiency.
DISCLOSURE:	Voluntary. However, mass rating and recording require some tracking method.

1a. LAST NAME	1b. FIRST NAME	1c. MI	2. RANK	3. UNIT
GUNNING	ROBERT	T	SGT	1/26TH IN

TABLE III (A). DISMOUNTED AND MOUNTED DAY PRACTICE

4. RANGE	5. LANE	6. GRADER	7. DATE
SEITZ	2	MULLENIX	9 SEP 03

TASK	RANGE (Meters)	AMMO	TIME (Minutes)	ENGAGEMENT STANDARDS	GO	NO GO
ZERO	400	4	NA	2 ROUNDS HIT	✓	
2	600 (+/- 100)	6	2	1 ROUND HIT	✓	
3	800 (+/- 100)	8	2.5	1 ROUND HIT	✓	
4	400	4	1.5	1 ROUND HIT	✓	
MULTIPLE TARGETS						
5	800 (+/- 100)	12	4	1 ROUND HIT		✓
6	400			1 ROUND HIT	✓	
7	400	10	3	1 ROUND HIT	✓	
8	600 (+/- 100)			1 ROUND HIT	✓	
				TOTALS	7	

8. COMMENTS
CHECK TARGET @ BOOM

9. NUMBER OF ENGAGEMENT MET *(Choose One)*
☐ 8 - EXPERT ☐ 6 - MARKSMAN
☑ 7 - SHARPSHOOTER ☐ 5 AND BELOW - UNQUALIFIED

10. GUNNER'S SIGNATURE
Robert T. Gunning

11. GRADER'S SIGNATURE
[illegible]

TABLE III (B). DISMOUNTED AND MOUNTED DAY QUALIFICATION

12. RANGE	13. LANE	14. GRADER	15. DATE
SEITZ	5	MULLENIX	9 SEP 03

TASK	RANGE (Meters)	AMMO	TIME (Minutes)	ENGAGEMENT STANDARDS	GO	NO GO
ZERO	400	4	NA	2 ROUNDS HIT	✓	
2	600 (+/- 100)	6	1.5	1 ROUND HIT	✓	
3	800 (+/- 100)	8	2	1 ROUND HIT	✓	
4	400	4	1	1 ROUND HIT	✓	
MULTIPLE TARGETS						
5	800 (+/- 100)	12	3.5	1 ROUND HIT		✓
6	400			1 ROUND HIT		✓
7	400	10	2.5	1 ROUND HIT	✓	
8	600 (+/- 100)			1 ROUND HIT	✓	
				TOTALS	6	

16. COMMENTS
T:6 CAME LOOSE
T 5:6

17. NUMBER OF ENGAGEMENT MET *(Choose One)*
☐ 8 - EXPERT ☑ 6 - MARKSMAN
☐ 7 - SHARPSHOOTER ☐ 5 AND BELOW - UNQUALIFIED

18. GUNNER'S SIGNATURE
Robert T. Gunning

19. GRADER'S SIGNATURE
[illegible]

DA FORM 7520-R, AUG 2003 APD V1.00

Figure 4-28. Example of completed DA FORM 7520-R, MK 19 day practice and qualification (pop-up targets).

i. **Night Practice and Night Qualification Firing Exercises** (Figures 4-29 and 4-30). Table II contains the tables for night practice and night qualification for hull targets using the AN/PEQ-2A, AN/PAS-13, or the AN/TVS-5 NVDs. Table IV contains the tables for night practice and night qualification with pop-up targets or hull targets without using NVDs. Other than a 30 second difference in the engagement times for each task, the practice and the qualification tables are the same. It is held twice a year, or as often as the commander feels is needed to maintain gunner proficiency.

(1) The night practice firing exercise allows the gunner to fire on a range engaging hull or pop-up targets to test his skills before qualification firing.

(2) The qualification exercise tests skills practiced during night firing exercise. It is scored on time taken and target hits made based on the firing tables.

(3) The grader selects the correct scorecard (Table II for hull targets using the NVDs listed on the scorecard, or Table IV for pop-up targets or hull targets without a NVD) and enters the gunner's name, rank, and unit in blocks "1" through "3." He also fills in blocks "4" through "7" with the range name, the firing lane, his name, and the date. He also checks the appropriate NVD in block "8" (Table II only).

(4) The grader positions himself so that he can observe both the gunner and the target. Once live fire commences, he:

- Observes and informs the gunner the strike of each round.
- Observes and records a GO or NO GO for each task.

(5) At the end of the practice, the grader sums the total of GO/NO GOs in the "Totals" block, checks the appropriate qualification in block "9," has the gunner sign the scorecard in block "10," and signs the card in block "11."

(6) During the qualification phase, the grader repeats the steps above by filling in the appropriate blocks, summing the scores, and assigning the correct qualification.

(7) The grader can use the back of the form or the comment section to enter remarks such as the operation of the gun, condition of the targets, and weather conditions to name a few.

MK 19, 40-mm GRENADE MACHINE GUN, MOD 3 FIRING TABLE II
NIGHT PRACTICE AND QUALIFICATION WITH HULL TARGETS SCORECARD
For use of this form, see FM 3-22.27; the proponent agency is TRADOC.

PRIVACY ACT STATEMENT

AUTHORITY:	10 USC 3012(g)/Executive order 9397
PRINCIPAL PURPOSE:	To aid individual training on targets at various ranges.
ROUTINE USES:	To evaluate individual proficiency.
DISCLOSURE:	Voluntary. However, mass rating and recording require some tracking method.

NOTE: Use these tables if the targets being used are hulls, and any of the following applies:
AN/PEQ-2A mounted on the TWS mounting bracket.
AN/PAS-13 mounted on the TWS mounting bracket.
AN/TVS-5 with the 3d generation tube mounted on the TWS mounting bracket.

1a. LAST NAME	1b. FIRST NAME	1c. MI	2. RANK	3. UNIT
PEVOSKI	RICHARD	M	SSG	1/26 IN

TABLE II (A). DISMOUNTED AND MOUNTED NIGHT PRACTICE

4. RANGE	5. LANE	6. GRADER	7. DATE
DOBOL	2	CROSS	2 SEP 03

TASK	RANGE (Meters)	AMMO	TIME (Minutes)	ENGAGEMENT STANDARDS	GO	NO GO
ZERO	400	4	NA	2 ROUNDS HIT	✓	
2	1,100 (+/- 200)	8	2.5	2 ROUNDS HIT	✓	
3	1,500 (+/- 200)	10	3.5	2 ROUNDS HIT		✓
4	600 (+/- 100)	6	2	2 ROUNDS HIT	✓	
5	800 (+/- 100)	6	2	2 ROUNDS HIT		✓
6	400	4	1.5	2 ROUNDS HIT	✓	
MULTIPLE TARGETS						
7	1,100 (+/- 200)	10	4	1 ROUND HIT		✓
8	600 (+/- 100)			1 ROUND HIT	✓	
9	800 (+/- 100)	14	4.5	1 ROUND HIT	✓	
10	1,500 (+/- 200)			1 ROUND HIT		✓
				TOTALS	6	

8. TYPE DEVICE *(Choose One)*
☐ AN/PEQ-2A ☐ AN/PAS-13 ☑ AN/TVS-5

9. NUMBER OF ENGAGEMENT MET *(Choose One)*
☐ 10 - EXPERT ☐ 9 - SHARPSHOOTER ☐ 8-7 - MARKSMAN ☑ 6 AND BELOW - UNQUALIFIED

10. GUNNER'S SIGNATURE	11. GRADER'S SIGNATURE
Richard M. Pevoski	Arnie Cross

TABLE II (B). DISMOUNTED AND MOUNTED NIGHT QUALIFICATION

12. RANGE	13. LANE	14. GRADER	15. DATE
DOBOL	3	CROSS	2 SEP 03

TASK	RANGE (Meters)	AMMO	TIME (Minutes)	ENGAGEMENT STANDARDS	GO	NO GO
ZERO	400	4	NA	2 ROUNDS HIT	✓	
2	1,100 (+/- 200)	8	2	2 ROUNDS HIT	✓	
3	1,500 (+/- 200)	10	3	2 ROUNDS HIT		✓
4	600 (+/- 100)	6	1.5	2 ROUNDS HIT	✓	
5	800 (+/- 100)	6	1.5	2 ROUNDS HIT		✓
6	400	4	1	2 ROUNDS HIT	✓	
MULTIPLE TARGETS						
7	1,100 (+/- 200)	10	3.5	1 ROUND HIT		✓
8	600 (+/- 100)			1 ROUND HIT	✓	
9	800 (+/- 100)	14	4	1 ROUND HIT	✓	
10	1,500 (+/- 200)			1 ROUND HIT		✓
				TOTALS	6	

16. TYPE DEVICE *(Choose One)*
☐ AN/PEQ-2A ☐ AN/PAS-13 ☑ AN/TVS-5

17. NUMBER OF ENGAGEMENT MET *(Choose One)*
☐ 10 - EXPERT ☐ 9 - SHARPSHOOTER ☐ 8-7 - MARKSMAN ☑ 6 AND BELOW - UNQUALIFIED

18. GUNNER'S SIGNATURE	19. GRADER'S SIGNATURE
Richard M. Pevoski	Arnie Cross

DA FORM 7519-R, AUG 2003 APD V1.00

Figure 4-29. Example of completed DA FORM 7519-R, MK 19 night practice and qualification with hull targets and NVDs.

MK 19, 40-mm GRENADE MACHINE GUN, MOD 3 FIRING TABLE IV
NIGHT PRACTICE AND QUALIFICATION WITH POP-UP TARGETS SCORECARD
For use of this form, see FM 3-22.27; the proponent agency is TRADOC.

PRIVACY ACT STATEMENT

AUTHORITY: 10 USC 3012(g)/Executive order 9397
PRINCIPAL PURPOSE: To aid individual training on targets at various ranges.
ROUTINE USES: To evaluate individual proficiency.
DISCLOSURE: Voluntary. However, mass rating and recording require some tracking method.

NOTE: Use this table if you do not have a MK 19/sight combination that applies to Table II, and you are using hull targets OR if targets are pop-up/E type silhouettes.

1a. LAST NAME	1b. FIRST NAME	1c. MI	2. RANK	3. UNIT
GUNNING	ROBERT	T	SGT	1/26th IN

TABLE IV (A). DISMOUNTED AND MOUNTED NIGHT PRACTICE

4. RANGE	5. LANE	6. GRADER	7. DATE
SEITZ	2	MULLENIX	9 SEP 03

TASK	RANGE (Meters)	AMMO	TIME (Minutes)	ENGAGEMENT STANDARDS	GO	NO GO
ZERO	400	4	NA	2 ROUNDS HIT	✓	
2	600 (+/- 100)	6	2	2 ROUNDS HIT	✓	
3	800 (+/- 100)	6	2	2 ROUNDS HIT	✓	
4	400	4	1.5	2 ROUNDS HIT	✓	
MULTIPLE TARGETS						
5	800 (+/- 100)	10	3.5	1 ROUND HIT		✓
6	400			1 ROUND HIT		✓
7	400	14	2.5	1 ROUND HIT	✓	
8	600 (+/- 100)			1 ROUND HIT	✓	
				TOTALS	6	

8. COMMENTS: OVERCAST BUT TARGETS VISIBLE.

9. NUMBER OF ENGAGEMENT MET *(Choose One)*
- [ ] 8 - EXPERT
- [x] 6 - MARKSMAN
- [ ] 7 - SHARPSHOOTER
- [ ] 5 AND BELOW - UNQUALIFIED

10. GUNNER'S SIGNATURE	11. GRADER'S SIGNATURE
Robert T. Gunning	J. Mullenix

TABLE IV (B). DISMOUNTED AND MOUNTED NIGHT QUALIFICATION

12. RANGE	13. LANE	14. GRADER	15. DATE
SEITZ	4	CORLEY	9 SEP 03

TASK	RANGE (Meters)	AMMO	TIME (Minutes)	ENGAGEMENT STANDARDS	GO	NO GO
ZERO	400	4	NA	2 ROUNDS HIT	✓	
2	600 (+/- 100)	6	1.5	2 ROUNDS HIT	✓	
3	800 (+/- 100)	6	1.5	2 ROUNDS HIT	✓	
4	400	4	1	2 ROUNDS HIT		✓
MULTIPLE TARGETS						
5	800 (+/- 100)	10	3	1 ROUND HIT	✓	
6	400			1 ROUND HIT		✓
7	400	14	2	1 ROUND HIT	✓	
8	600 (+/- 100)			1 ROUND HIT	✓	
				TOTALS	6	

16. COMMENTS:

17. NUMBER OF ENGAGEMENT MET *(Choose One)*
- [ ] 8 - EXPERT
- [x] 6 - MARKSMAN
- [ ] 7 - SHARPSHOOTER
- [ ] 5 AND BELOW - UNQUALIFIED

18. GUNNER'S SIGNATURE	19. GRADER'S SIGNATURE
Robert T. Gunning	Thomas Corley

DA FORM 7521-R, AUG 2003 APD V1.00

Figure 4-30. Example of completed DA FORM 7521-R, MK 19 night practice and qualification with pop-up targets or without NVDs.

CHAPTER 5

TECHNIQUES OF FIRE

Firing techniques allow the gunner to deliver controlled, effective fire on target. This chapter covers the characteristics and classes of fire; range determination; application of fire; firing under degraded conditions; and predetermined fires.

Section I. CHARACTERISTICS AND CLASSES OF FIRE

To effectively employ the MK 19, gunners must understand the definitions of the terms used to describe the actions and effects of firing the MK 19. This section discusses the characteristics and classes of fire.

5-1. CHARACTERISTICS OF FIRE

The following are definitions of the characteristics of fire:

a. **Trajectory.** The path of the projectile in flight. The path is curved due to gravity and elevation of the barrel. The trajectory increases as the sights are elevated for far targets.

b. **Line of Sight.** An imaginary line from the gun to the target, as seen through the sights.

c. **Ordinate.** The vertical distance that occurs anywhere between the line of sight and the trajectory.

d. **Maximum Ordinate.** The highest point of trajectory, at which the vertical distance between the trajectory and line of sight is greatest.

e. **Danger Space.** The area in which the impact of the round or the dispersal of fragmentation causes injuries to personnel or destruction of the target.

f. **Dead Space.** The area(s) in which personnel or targets are safe from the gun's fire. Dead space can be, ditches, depressions, or ravines.

g. **Cone of Fire.** The dispersion of the rounds as they leave the gun.

h. **Beaten Zone.** The area in which the cone of fire strikes the ground or target. Terrain slope and range to the target affect the beaten zone. If the cone of fire falls on upward-sloping ground, the beaten zone is shortened, and vice versa. As range increases, the beaten zone is shortened and widened.

i. **Center of Impact.** The area where the heaviest concentration of fire impacts.

5-2. CLASSES OF FIRE

Fire is classified with respect to the ground, target, and the MK 19.

a. **Respect to the Ground.** Normally, this means either plunging or grazing fire. However, since grazing fire is not practical for use with the MK 19, only plunging fire will be considered. Plunging fire strikes the target from a high angle and confines the danger space to the beaten zone. For example, when fired from the top of a hill, projectiles follow an arcing trajectory and land in the valley (Figure 5-1).

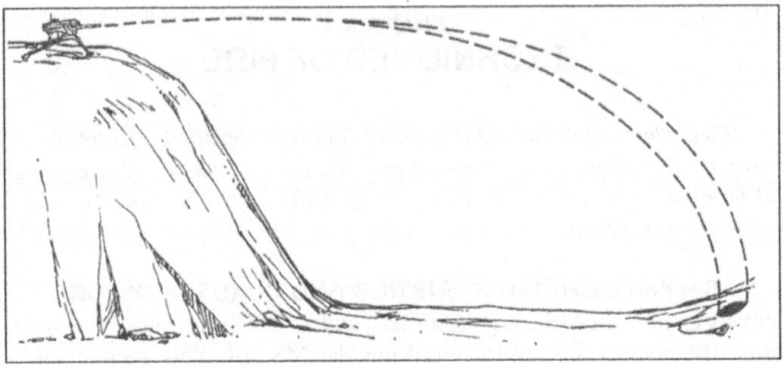

Figure 5-1. Plunging fire.

b. **Respect to the Target.** This class of fire is divided into four types of fire (Figure 5-2a and 5-2b):

(1) ***Frontal.*** The long axis of the beaten zone is at a right angle to the long axis of the target.

(2) ***Flanking.*** This type of fire is delivered against the flank of the target.

(3) ***Oblique.*** This type of fire is directed at a target moving at any angle other than directly toward or perpendicular to the gun.

(4) ***Enfilade.*** This type of fire occurs when the long axis of the beaten zone coincides with the long axis of the target. Enfilade fire may be frontal or flanking, depending on which way the target is facing. For example, frontal enfilade occurs if the MK 19 is in the middle of a road and the target is approaching on the same road. Flanking enfilade occurs if the target is moving either way, left or right, yet is still in a long axis configuration away from the MK 19.

Figure 5-2a. Classes of fire with respect to the target: frontal, flanking, and oblique fires.

Figure 5-2b. Classes of fire with respect to the target: flanking and frontal enfilade fires.

c. **Respect to the MK 19.** Fire from the MK 19 may be conducted in six ways (Figure 5-3).

(1) ***Fixed.*** Fire is delivered against a target with one aim point, which concentrates the beaten zone.

(2) ***Traverse*** Fire is moved from left to right or right to left, with no range change. It may be used against frontal or flanking targets.

(3) ***Search.*** Fire is directed against a deep target. Elevation changes are made, but direction changes are not. Searching fire is used against enfilade targets.

(4) ***Traverse and Search.*** This is a combination of traversing and searching fires used against a target with depth and width (most likely an oblique target).

(5) ***Swinging Traverse.*** This fire is slightly different from traversing and searching. Although it is delivered against a wide target, with major changes in direction, no elevation changes are made. To deliver fire using a swinging traverse, the gunner releases the traversing slide lock, allowing the gun to travel freely across the traversing bar.

(6) ***Free Gun.*** Fire is delivered with the T&E mechanism removed and is used when quick changes in direction and depth are needed to engage moving targets.

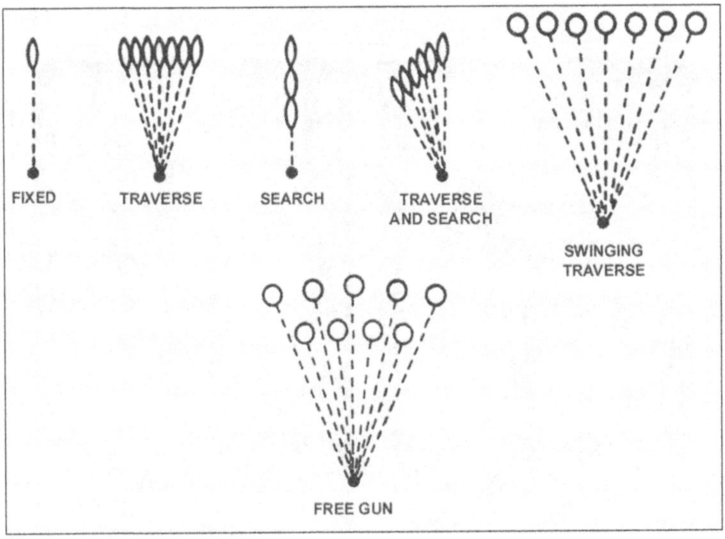

Figure 5-3. Classes of fire with respect to the MK 19.

Section II. RANGE DETERMINATION

Range determination is the process of finding the distance between the gunner's position and his target. The gunner's ability to engage a target effectively depends on determining the correct range to the target.

5-3. MEASUREMENT BY VISUAL ESTIMATION

Range is often determined by this method. There are two ways to estimate range visually.

a. Using the 100-meter unit-of-measure method, visualize 100 meters on the ground (this takes practice) and determine how many units of 100 meters there are to the target (Figure 5-4). For targets more than 500 meters (5 units) away, pick a point about halfway, estimate the range to that point, and double it (Figure 5-5).

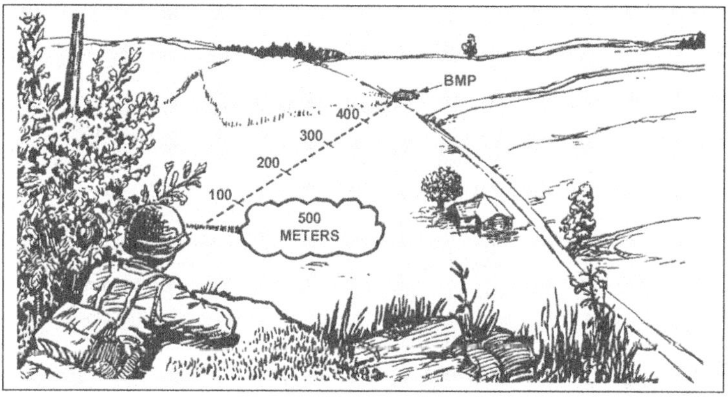

Figure 5-4. Application of the 100-meter unit-of-measure method for ranges up to 500 meters.

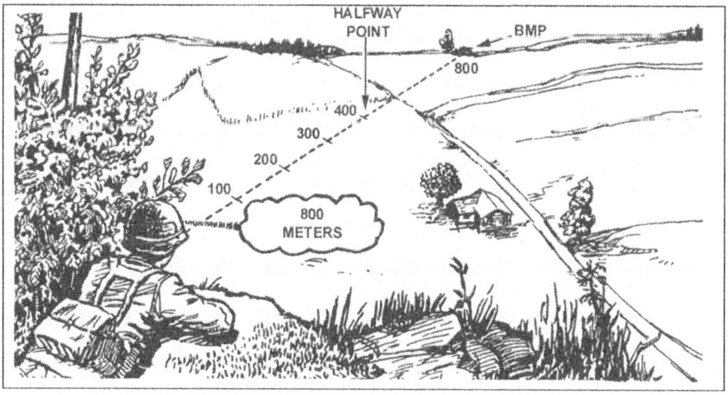

Figure 5-5. Application of the 100-meter unit-of-measure method for ranges greater than 500 meters.

b. Use the appearance-of-objects method, which is fairly reliable, to determine range (this also takes practice). Memorize the sizes and shapes of familiar objects at different ranges. Consider the factors in Table 5-1, which affect the appearance of objects.

FACTORS AFFECTING RANGE ESTIMATION	FACTORS CAUSING UNDERESTIMATION OF RANGE	FACTORS CAUSING OVERESTIMATION OF RANGE
The clarity of outline and details of the target. Nature of terrain or position of the observer. Light and atmosphere.	When most of the target is visible and a clear outline can be seen. When looking across a depression that is mostly hidden from view. When looking downward from high ground. When looking down a straight, open road or along a railroad. When looking over uniform surfaces like water, snow, desert, or grain fields. In bright light or when the sun is shining from behind the observer. When the target is in sharp contrast with the background or is silhouetted because of its size, shape, or color. When seen in the clear air or high altitudes.	When only a small part of the target can be seen or the target is small in relation to its surroundings. When looking across a depression that is totally visible. When looking from low ground toward high ground. When vision is narrowly confined as in streets, draws, or forest trails. In poor light such as dawn and dusk; in rain, snow, fog; or when the sun is in the observer's eyes. When the target blends into the background or terrain.

Table 5-1. Factors affecting visual range estimation.

5-4. MEASUREMENT FROM A MAP

Another way range may be determined is by using a military map. The gunner locates his position and the target's position on the map. He measures the distance and uses the legend scale at the bottom of the map to find the range.

5-5. MEASUREMENT BY PACING

When using this method, the leader ensures there is no immediate danger to the gun's crew. The crew sets up its position, and the gunner selects the target. The assistant gunner walks to the target in a straight line, counting the number of paces it takes to reach the target. As the distance is paced off, the gunner can determine dead space when the assistant gunner disappears from view.

5-6. MEASUREMENT USING BINOCULARS

Binoculars can be used to determine range. The recently adopted M19 binocular, unlike other models, is a lightweight, compact instrument intended for use in general field observation and fire direction. The binocular's left lens includes horizontal and vertical reticles graduated in 10-mil increments (Figure 5-6). When using binoculars to determine range, the soldier must understand the mil relationship (Figure 5-7).

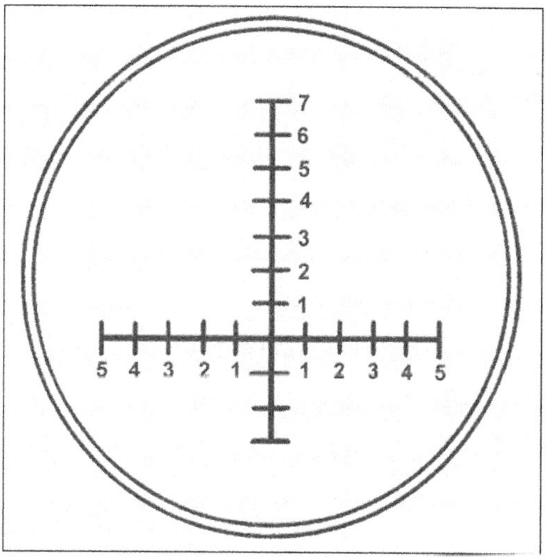

Figure 5-6. M19 binocular reticle.

a. The mil (₥) is the unit of angular measurement used in adjusting fire. A circle has 6,400 mils. At a distance of 1,000 meters, an object 1 meter wide measures 1 mil. Change mils to meters by multiplying the number of mils times the range (distance) in thousands of meters. Obtain the unknown width or range to an object using a similar method (Figure 5-7).

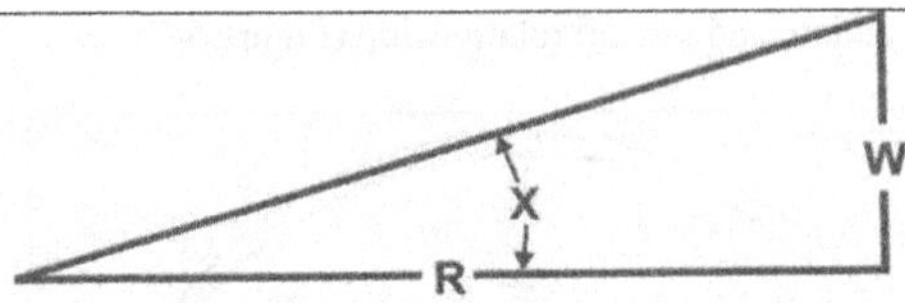

X - IS THE ANGULAR WIDTH OF THE OBJECT IN MILS (₥).

W - IS THE WIDTH OF THE OBJECT IN METERS.

R - IS THE RANGE OR DISTANCE IN THOUSANDS OF METERS.

FORMULA: $X = \frac{W}{R}$

Figure 5-7. Mil relationship.

b. The mil relationship, as shown in Figure 5-7, is X = W ÷ R; where X is the angular width of the object in mils (₥), W is the width of the object in meters, and R is the range or distance in thousands of meters.

(1) To find Width when the known Range is 4,000 meters and the object is 15 ₥ wide, multiply R times X ₥ or 4 (range in thousands) times 15 (mils). The answer is 60 (meters), or 4 x 15 = 60 meters.

$$W = \frac{\text{R times}}{\text{X₥}}$$

$$W = \underline{\text{4 times 15}}$$

$$W = 60$$

(2) To find Range when the known Width in meters (between two bursts or two objects) is 60 and the angular measurement for the same width, when measured with binoculars, is known to be 15 mils, divide (W: ₥) 60 (meters) by 15 (mils).

The answer is 4,000 meters:

$$R = \frac{W}{X \text{ ₥}}$$

$$R = \frac{60}{15}$$

R = 4 (in thousands of meters) = 4,000 meters

(3) To find the angle A in ₥ when the known Width in meters between a reference point and the target is 60 meters and the known Range to the target is 4,000 meters, divide W by: R or 60 (meters) by 4 (range in thousands) and the answer is 15 mils (mils):

$$A \text{ ₥} = \frac{W}{R}$$

$$A \text{ ₥} = \frac{60}{4}$$

$$A \text{ ₥} = 15$$

c. The observer uses the mil relation in computing direction shifts as indicated in Figure 5-8.

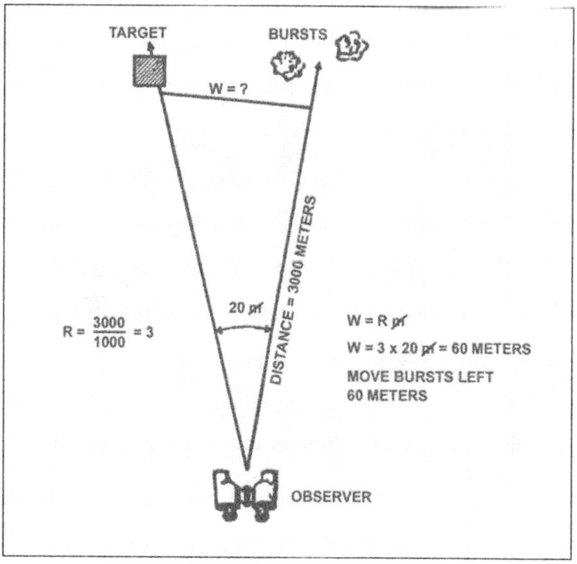

Figure 5-8. Computation of direction shift.

5-7. MEASUREMENT USING AN/GVS-5 LASER RANGE FINDER

The handheld AN/GVS-5 laser range finder provides observers and small-unit leaders the ability to more accurately determine range to targets and other known points. This range finder allows first-round fire for effect, which results in increased enemy casualties and saves ammunition. Friendly units can use the range finder to precisely locate themselves using intersection and modified resection techniques. The range finder aids in the determination of sector depth, range to suspect enemy locations, distance to targets along avenues of approach, and information for overlapping fires. When an AN/GVS-5 is available, it should be used to construct range cards. Information about placing the AN/GVS-5 into action can be found in TM 11-5860-201-10.

5-8. MEASUREMENT BY FIRING

The gunner can determine range by firing a zeroed MK 19. The gunner fires a burst and uses the T & E mechanism to "walk" the rounds on target. The gunner sets and reads the sights on the target, and notes the range. This measurement method is used frequently in combat situations.

5-9. MEASUREMENT OF LATERAL DISTANCE

The gunner can use the finger method to find distance. He extends his arm toward the target with his palm out, elbow locked, one eye closed, and index finger raised (Figure 5-9). He sights along the edge of the finger, adding extra fingers to fill in the space (the average finger is 30 mils wide). One finger equals 30 mils, two fingers equals 70 mils, three fingers equals 100 mils (Figure 5-10).

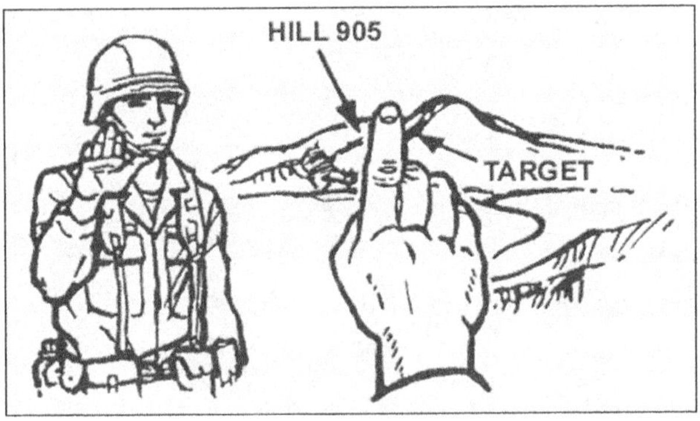

Figure 5-9. Use of fingers to measure deviation.

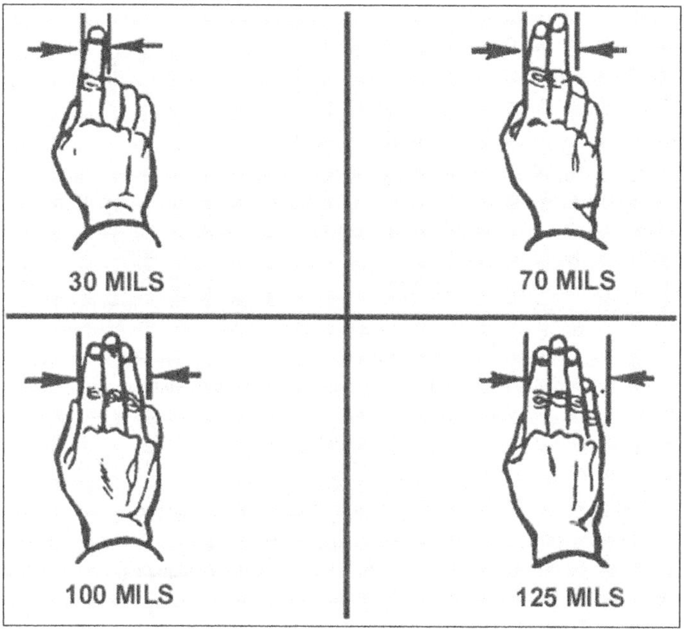

Figure 5-10. Hand-and-finger value estimates.

Section III. APPLICATION OF FIRE

This section includes information about MK 19 fire control and target engagement. Fire control includes all operations connected with the preparation and application of fire to a target.

5-10. FIRE CONTROL

To correctly exercise fire control, the gunner must be able to:

- Open fire at the instant desired.
- Adjust fire on the target.
- Regulate the rate of fire.
- Shift fire from one target to another.
- Cease fire.

Failure to exercise correct fire control results in danger to friendly troops, loss of surprise, premature disclosure of position, misuse of fire on unimportant targets, loss of time in securing adjustments, and waste of ammunition. The main factors to consider for effective fire control are as follows:

a. **Sector of Fire.** This is an area of terrain that an assigned unit covers. It has specific, designated boundaries. Sectors of fire vary in size but are generally limited areas that can be engaged without movement of the tripod.

b. **Rate of Fire.** Surprise and shock effect may be obtained by firing all MK 19s at the same time, using the rapid rate for at least the first few bursts. Engage fleeing targets as soon as possible and with all available fire. Use the rapid rate for the initial delivery of fire to make adjustment of fire easier. Also, whenever the rate of fire is not stated, use the rapid rate. In all cases, unless otherwise ordered, fire the first few bursts at the rapid rate; thereafter, use the prescribed rate.

c. **Adjustment of Fire.** Adjustments may be made before or after firing.

(1) ***Initial Adjustment.*** Set the sights on the range to the target, lay on the target, fire an aimed burst, and observe the impact. When the initial burst is correct, continue to fire (manipulating if necessary) until the target is covered. When the initial burst is not correct, determine from its impact the amount of traverse and search required to place the next burst on the target. Manipulate the gun with the T&E handwheels, making large range corrections by resetting the sights and re-laying on the target.

(2) ***Subsequent Correction and Adjustment.*** Observation and adjustment of fire is the most important element of fire control. It is continuous throughout the action. The gunner is trained to observe and adjust fire without command and to check the lay of the gun frequently. The gunner is also trained to foresee the action of the enemy after he opens fire, and to shift fire to cover any changes in the formation or location of the target. If the gunner fails to do these things, the leader promptly corrects him by announcing or signaling subsequent fire commands. This responsibility to adjust fire continues up through the chain of command. When subsequent fire commands are issued, the gunner makes the required adjustments. When the gun is fired from the tripod mount, subsequent commands are given as adjustments to the elevation or deflection at which the last burst was fired. The adjustments are given in mils and are announced or signaled as ADD, DROP, RIGHT, or LEFT so many mils. Changes in deflection and elevation such as RIGHT 2 or ADD 5 are made with the T&E handwheels.

5-11. FIRE COMMANDS

Fire commands are the means by which leaders control fires. A fire command is a set of instructions, given in a prescribed format that enables the gun crew to properly engage the desired targets. There are two types of fire commands: initial and subsequent. Initial fire commands are used for engagement of targets and the shifting of fire to new targets. Subsequent fire commands are used to adjust fire, change the rate of fire, and cease-fire. The explanation below is based on two gun crews. A good fire command is brief and clear, and in the proper sequence. It is given clearly at a rate that is easily understood by the gunner. Gunners repeat each element to ensure understanding. It is unlikely for a complete initial fire command to be issued during a fire fight. The leader determines which elements of a fire command are obvious to the gunners and which elements must be given to them. Some targets may be engaged with only the alert, range, and command of fire. For example: FIRE MISSION, SEVEN HUNDRED, FIRE. The procedures outlined below are used to accustom gunners to issuing and executing instructions in a logical sequence. The use of complete fire commands in training prepares the gunners to receive fragmentary fire commands in combat.

a. **Initial Fire Command.** This contains the following elements:

- ALERT.
- DIRECTION (as needed).
- RANGE.

- DESCRIPTION (as needed).
- ASSIGNMENT/METHOD (division, manipulation, rate; all only as needed).
- CONTROL.

(1) ***Alert.*** The alert is the first element of the initial fire command. Its purpose is to choose the gun crews and ready them to receive and execute the fire command. FIRE MISSION is announced for all targets. When two or more guns are to fire, the leader announces, FIRE MISSION. If only one gun is to fire, then NUMBER 1 (2) FIRE MISSION is given. When the leader wants to alert all guns, but only wants one gun to fire, the announcement is FIRE MISSION, NUMBER 1 (2).

(2) ***Direction.*** There are several ways to designate target direction:

(a) *Oral.* When the target is not obvious, the leader must tell the gunners where to look. The leader gives direction as: FRONT, RIGHT FRONT, LEFT FLANK, and so on. The leader may identify an indistinct target by the use of a reference point. The selected reference point, announced as REFERENCE, must be a distinctive terrain feature or object in or near the target area. The word TARGET precedes the target description when a reference point is used.

- When the reference point is within the target area, the leader may describe the target as extending so many mils, meters, or fingers from that point. He announces range to the reference point. For example:
  REFERENCE: BUNKER, FIVE ZERO ZERO.
  TARGET: TROOPS EXTENDING RIGHT TWO FIVE, LEFT TWO FIVE.
- If the selected reference point is outside the target area, the leader announces the range to the target. For example:
  REFERENCE: LONE TREE.
  TARGET: MACHINE GUN IN EDGE OF WOODS; FIVE ZERO ZERO.
- The leader may identify the direction to a target that is not obvious by selecting an obvious feature, and (by naming successive reference points) leading the gunner to the target step by step. For example:
  REFERENCE: RED-ROOFED HOUSE, RIGHT OF HOUSE, HEDGE, CENTER OF HEDGE, GATE, ABOVE GATE.
  TARGET: MACHINE GUN.
- For a tripod-mounted MK 19, the interval between the reference and the target is measured by laying the gun on the reference point and manipulating the gun a given number of mils to the target.

(b) *Firing a MK 19.* Pointing out an obscure target by firing a MK 19 is simple, fast, and accurate. However, it may cause loss of surprise or premature disclosure of the gun's position. The leader announces the general direction of fire, if it is not obvious. He then lays one gun on the target, commands: WATCH MY BURSTS, and fires one or more on the target. The leader completes the designation orally using the target's midpoint or extremes: RIGHT FLANK, MIDPOINT, or NEAR END. Firing rifle tracers at the target also makes it easier for the gunners to find.

(c) *Laying a MK 19.* Laying the gun on a target is a simple and accurate method that does not sacrifice surprise. The leader goes to each gun, lays it on the target, and has the gunners check the lay. The gunners open fire at the same time on command.

(3) ***Range.*** This element follows the target description. The leader announces it in even digits, hundreds or thousands. For example: FOUR FIVE ZERO; THREE HUNDRED; ONE ONE HUNDRED; ONE THOUSAND.

(4) ***Description.*** The leader gives a brief target description to inform the gunners of the nature of the target. Examples of target descriptions include TROOPS, which refers to any dismounted enemy personnel; MACHINE GUN, which refers to any automatic gun; and TANK, which refers to any armored vehicle. If several targets are in view, the leader may describe the target or part of a target to be engaged as LEADING TRUCK, RIGHT BUILDING, FAR END, HALTED COLUMN, and so on. If the target is obvious, no description is necessary.

(5) ***Assignment Method.*** This element is used only when specific assignments are required to divide or subdivide the target, assign class of fire, or announce a rate of fire.

(a) *Division (or subdivision).* The leader gives this element only when it is required (paragraph 5-12). Division may be announced as follows:

NUMBER 1 RIGHT HALF, NUMBER 2 LEFT HALF.
NUMBER 1 RIGHT TWO-THIRDS, NUMBER 2 LEFT TWO-THIRDS.
NUMBER 1 RIGHT TWO-THIRDS, NUMBER 2 LEFT ONE-THIRD.

(b) *Manipulation.* The leader gives this element to prescribe the class of fire with respect to the gun, and gives it only if the required manipulation is not obvious. For example:

FIXED.
TRAVERSE.
SEARCH.
TRAVERSE AND SEARCH.
SWINGING TRAVERSE.

(c) *Rate.* The greatest effect results from having two guns open fire at the rapid rate, at the same time. Regardless of the rate announced, gunners open and adjust fire at the rapid rate, and use the announced rate (sustained, rapid, or cyclic) thereafter. Size of target, type of target, and ammunition supply are the factors that influence selection of a rate of fire. Rates of fire are distinguished as follows:

- **Sustained.** The sustained rate of fire is measured in rounds for each minute and is the rate at which a MK 19 may be fired indefinitely without damage from overheating (40 rounds per minute). The leader announces SUSTAINED.
- **Rapid.** The rapid rate is measured in rounds for each minute and is the rate at which a MK 19 may be fired for a limited time without danger of overheating (60 rounds per minute). This rate permits a high volume of fire to be delivered for a short, set time. Gunners use the rapid rate unless another rate is announced.
- **Cyclic.** The cyclic rate is the rate at which a MK 19 can be fired for a limited time only, at the fastest rate the gun will fire (325 to 375 rounds per minute). This rate may be used when the target or area needs the greatest possible suppression. The leader announces CYCLIC.

(6) ***Control.*** The leader uses this element to give the command to open fire. For immediate engagement of the target, the leader gives the command, FIRE, or the arm-and-hand signal to fire without pause. Fire is often withheld for surprise and effect, and both guns open fire at the same time. To ensure this, the leader may preface the command or signal to fire with the words AT MY COMMAND or ON MY SIGNAL. When the gunners are ready

to engage the target, they report UP or announce: NUMBER 1 (2) UP, READY, and the leader gives the command or signal to fire.

b. **Subsequent Fire Commands.** The leader uses these to repeat or correct a fire command, to adjust fire, to cease or commence fire, or to terminate the alert. The deflection correction must always be given first.

(1) ***Repeat Fire Commands.*** If the gunner fails to understand any element of the fire command, a repetition of the element may be requested by announcing the misunderstood element with rising inflection to denote a question. When repeating any portion of the fire command, the leader will preface it with the words THE COMMAND WAS.

(2) ***Correct Fire Commands.*** The leader changes or corrects a portion of the fire command by announcing CORRECTION and giving a command. For example, to change an incorrect range command of 500 meters to 600 meters, the command is CORRECTION, SIX HUNDRED.

(3) ***Adjust Fire.*** The leader adjusts fire if the initial burst was not on target. The leader does this by announcing the direction and elevation needed to get on target ADD 100, RIGHT 20; or DROP 200, LEFT 50.

(4) ***Cease and Commence Fire.*** CEASE FIRE is announced if the leader wishes to interrupt for any reason. This type of subsequent fire command informs the MK 19 crew that it will remain on the alert and that more instructions will follow. Firing is renewed by announcing a subsequent fire command or by announcing a new fire command. Firing is resumed with the same data by using the command FIRE.

(5) ***Terminate the Alert.*** To allow the MK 19 crew to relax between fire missions, termination of the alert is announced as CEASE FIRING, END OF MISSION.

5-12. TARGET ENGAGEMENT

The method chosen depends on terrain, target presentation, type of target, and tactical situation.

a. **Distribution of Fire.** To be effective, fire must be distributed over an entire target. Improper distribution results in gaps between beaten zones and allows some of the enemy to escape or to use guns without effective opposition.

(1) ***Factors Affecting Distribution of Fire.*** No fixed rule can be given as to the widest target that a single MK 19 may effectively engage. Ideally, the target should be no more than 100 mils wide. The traversing screw on the T&E mechanism limits the gun's traverse unless the traversing slide lock lever is unlocked. Wider targets require more traversing time, which prevents the gunner from placing a continuous volume of fire on the whole target. The engagement of a wide target by a single MK 19 requires excessive ammunition.

(2) ***Fire Unit.*** This consists of a pair of MK 19s. If possible, at least two guns should be assigned to the same mission, although sometimes a single MK 19 may be effective. The assignment of a pair to a single mission ensures continuous fire in case either gun is put out of action. Two guns can provide a great volume of fire on the target, and can reduce the time required to cover it.

(3) ***Manipulation of the Tripod-mounted MK 19.*** Traversing fire is moved in 5-mil increments. The MK 19s are fired after each manipulation to ensure the beaten zones overlap. Searching fire is often used on level or evenly sloping ground. When the ground is irregular, however, the amount of search to apply between bursts in order to ensure overlap of the beaten zones is determined by observation.

b. **Engagement of Point Target.** Any target no larger than the beaten zone is a point target, and is engaged by fixed fire. The command is FIXED. MK 19 crews are trained to follow any movement or change in formation made by the enemy after the initial burst of fire. An example of a fire command for a point target is as follows:

FIRE MISSION.
FRONT.
MACHINE GUN.
SIX HUNDRED.
FIXED.
RAPID.
FIRE.

c. **Engagement of Linear Target.** There are several ways to engage a linear target. The method used is chosen based on the number of MK 19s available and whether or not the entire target is visible to the gunner.

(1) ***Linear Engagement with One MK 19***. A single MK 19 engages the target the same as either one of a pair. The MK 19 is laid just outside either flank (or on a reference point within the target area) and the gunner adjusts fire on the flank (or point). The gunner traverses back and forth across the entire area until told to cease firing. The leader may designate where he wants initial fire. For example, if he directs a single gunner to engage a target with width he may announce:

FIRE MISSION.
FRONT.
TROOPS EXTENDING FROM DEAD TREE RIGHT 20 MILS.
FIVE HUNDRED.
TRAVERSE.
RAPID.
AT MY COMMAND.
FIRE (given after the gunners announce "Up").

(2) ***Linear Engagement with Two MK 19s.*** A linear target may be engaged two ways when a pair of MK 19s is used, depending on the width of the target in mils.

(a) *Target 100 mils or Less in Width.* The normal traversing method is used in this case. Each MK 19 delivers the initial burst of fire on its corresponding flank of the target. Fire is adjusted on that point. Each MK 19 is traversed across the target to the other flank, covering the entire target, until the gunners are told to cease firing (Figure 5-11). The command for this type of fire is TRAVERSE. An example of a fire command used for this situation follows:

FIRE MISSION.
RIGHT FRONT.
TROOPS, EXTENDING FROM DEAD TREE RIGHT TO CLEARING.
SEVEN FIVE ZERO.
TRAVERSE.
RAPID.
AT MY SIGNAL.
FIRE (given after the gunner announces "Up").

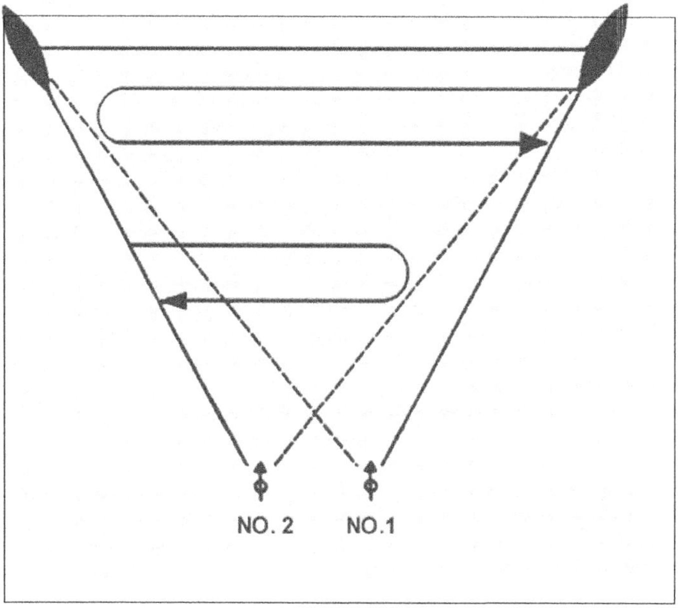

Figure 5-11. Traversing method, pair of MK 19s, target 100 mils or less in width, both flanks visible.

(b) *Target More than 100 mils in Width.* In this case, the leader assigns part of the target to each MK 19 (Figure 5-12). One part may be smaller than the other to ensure it receives a heavier concentration of fire. Each gun fires an initial burst on its respective flank, and covers its assigned part of the target as described in subparagraph (a). An example of a fire command follows:

FIRE MISSION.
RIGHT FRONT.
TROOPS, EXTENDING FROM DEAD TREE RIGHT TO CLEARING, ROW.
EIGHT HUNDRED.
NUMBER 1, RIGHT ONE-THIRD; NUMBER 2, LEFT TWO-THIRDS.
TRAVERSE.
RAPID.
AT MY COMMAND.
FIRE (given after the gunners announce "Up").

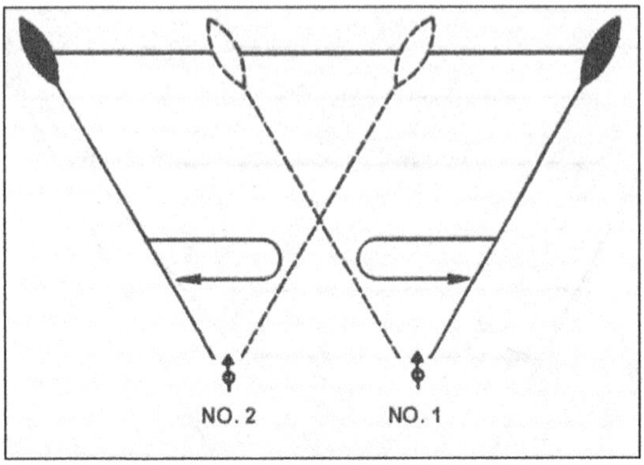

Figure 5-12. Traversing method, pair of MK 19s, target more than 100 mils in width.

(3) ***Linear Engagement with Four MK 19s.*** Four guns may be needed to concentrate a heavy volume of fire into a wide target area. Methods used are similar to the ones described for a pair of MK 19s. The width of the target in mils determines which type of linear engagement is used.

(a) *Target 100 to 200 mils in width*:

- Each pair of MK 19s engages the whole target (Figure 5-13).

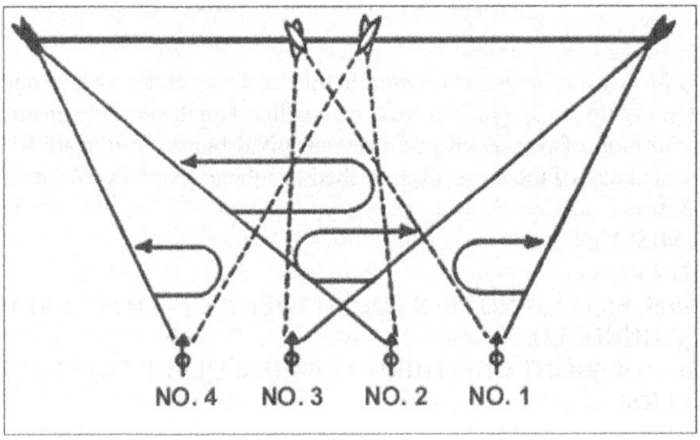

Figure 5-13. Four MK 19s, target 100 to 200 mils in width.

- Each pair of MK 19s engages half the target (Figure 5-14).

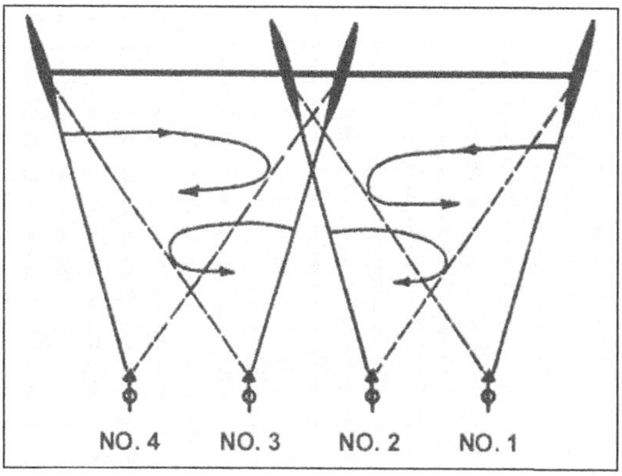

Figure 5-14. Two pairs of MK 19s engaging each half of a target 100 to 200 mils in width.

(b) *Target Greater than 200 mils in Width.* The leader may assign part of the target to each gun in this case. The target may be divided in half for each fire unit or in four parts, one for each MK 19 (Figure 5-15).

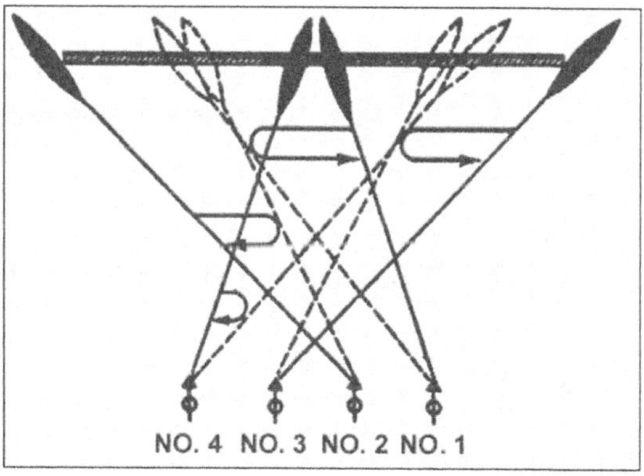

Figure 5-15. Four MK 19s, each pair engaging part of target with extreme width.

(4) ***Linear Engagement with the Flanks of Target Not Identifiable.*** The target may be located by firing tracers from an observer's gun, laying MK 19s, or using a reference point. The reference point may be in or near the target area.

(a) When a reference point is outside the target area, the leader may direct the gunner to it by announcing the interval (right or left, long or short) between the reference point and the target. With the MK 19 on a tripod mount, the gunner measures the interval right or left between the reference point and the flank of the target by laying on the reference point, with the sights set at the range to the target, and then shifting the MK 19 the designated number of mils. The interval long or short may be measured in mils using the computed search method, or it may be estimated in meters. When the gunners lay each MK 19 on its respective flank, they adjust fire and engage the target.

(b) When a reference point is in the target area, the leader may identify the flanks of an obscure target as extending so many mils from the reference point. In this case, gunners use the swinging traverse. With the swinging traverse each MK 19 is laid on the announced reference point (initial aiming point), adjusted for fire, and traversed the given distance to its corresponding flank and back, firing after each manipulation (Figure 5-16). Each gunner continues traversing back and forth across the entire target until told to cease-fire. Each gunner stops firing while traversing past the reference point. Examples of fire commands that may be used are as follows:

- The leader designates the target by firing one gun:
  FIRE MISSION.
  LEFT FLANK.
  WATCH MY BURSTS (OR TRACERS).
  (Lays and fires MK 19 at left flank) LEFT FLANK.
  (Lays and fires MK 19 at right flank) RIGHT FLANK.
  TROOPS.
  NINE HUNDRED.
  TRAVERSE.
  RAPID.
  AT MY COMMAND.
  FIRE (given after gunners announce "Up").
- The leader designates the target by using a reference point outside the target area (MK 19s on tripod mounts):
  FIRE MISSION.
  RIGHT FRONT.
  REFERENCE: DEAD TREE.
  RIGHT SIX ZERO MILS; DROP THREE FIVE ZERO METERS
  (or drop so many mils, if using computed search).
  TARGET: TROOPS EXTENDING RIGHT FIVE ZERO MILS.
  SIX HUNDRED.
  TRAVERSE.
  RAPID.
  AT MY COMMAND.
  FIRE (given after the gunners announce "Up").

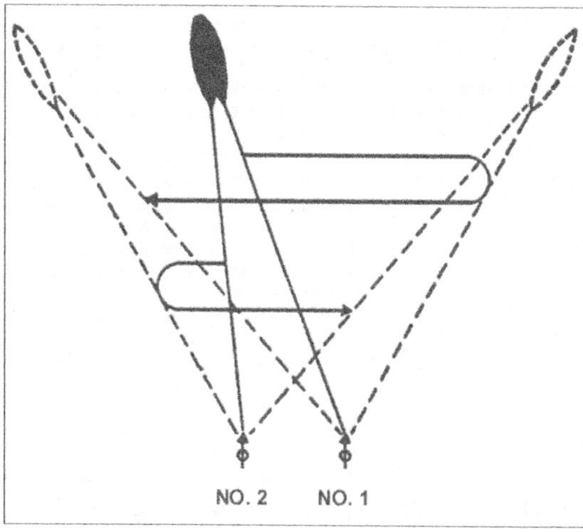

Figure 5-16. Two MK 19s, deep enfilade target, ends not visible.

- The leader designates the target by using a reference point within the target area (MK 19s on tripod mounts):

  FIRE MISSION.
  FRONT.
  REFERENCE: LONE TREE.
  TARGET: TROOPS EXTENDING RIGHT TWO ZERO MILS, LEFT, THREE ZERO MILS.
  SEVEN HUNDRED.
  TRAVERSE.
  RAPID.
  AT MY COMMAND.
  FIRE (given after the gunners announce "Up).

(5) ***Linear Engagement with Swinging Traverse.*** As previously described, a single gun uses this method against massed or rapidly moving targets at short ranges. The leader's fire command for a swinging traverse can be:

FIRE MISSION.
LEFT FRONT.
TRUCKS.
SWINGING TRAVERSE.
FIRE.

d. **Engagement of Deep Targets Using Searching Fire.** There are several ways for leaders to use two guns to engage a stationary deep target, depending on whether the ends of the target are visible to the gunner. If the target location is unknown, the computed search method may be used to engage the target. A deep target that is stationary or that has limited mobility may be engaged with searching fire.

(1) ***Target Ends Visible to Gunner.*** The Number 1 gun is adjusted first on the near end and the Number 2 gun on the far end. Each gunner searches to the opposite end and back again repeatedly, until told to cease firing (Figure 5-17). Target depth is considered in determining sight settings. The command is SEARCH.

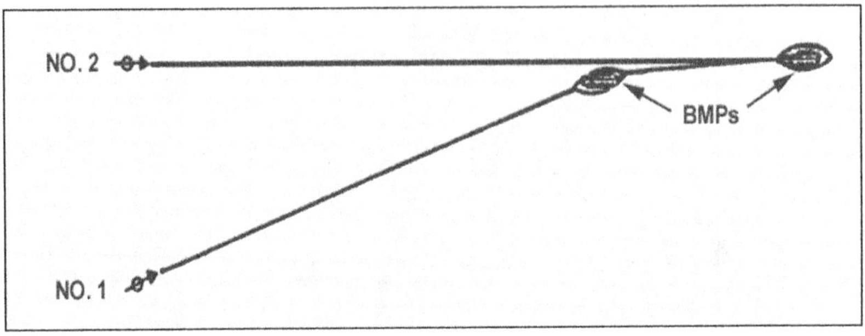

Figure 5-17. Two MK 19s, deep enfilade target, ends visible.

(a) *Target Less than 200 meters Deep.* The leader announces the range to the midpoint of the target for both MK 19s, using the length of the beaten zones to ensure the initial bursts impact on the target. An example of a fire command for this situation follows:

FIRE MISSION.
FRONT.
HALTED COLUMN.
SEVEN HUNDRED (midrange).
SEARCH.
RAPID.
AT MY COMMAND.
FIRE (given after gunners announce "Up").

(b) *Target Depth Greater than 200 meters.* The leader announces the range to the near end for the Number 1 gun, and to the far end for the Number 2 gun. An example of a fire command for this situation follows:

FIRE MISSION.
LEFT FRONT.
TROOPS ALONG RIGHT EDGE OF WOODS.
NUMBER 1, SIX HUNDRED; NUMBER 2. NINE HUNDRED.
SEARCH.
RAPID.
AT MY COMMAND.
FIRE (given after gunners announce "Up").

(2) ***Target Ends Not Visible to Gunner.*** The target may be located by having an observer fire tracers at the target, by laying the MK 19s, or by using a reference point. The reference point must be visible to the gunner and in the vicinity of the target. When a selected reference point is outside the target area, the leader may direct the gunner to the target and announce the interval to the right or left (long or short) between the reference

point and the target. If the MK 19 is on a tripod, the gunner lays on the reference point with the sights set at the range to the midpoint of the target; he measures the interval and shifts the appropriate number of mils. Using the computed search method, the gunner may measure the long or short interval in mils, or he may estimate the interval in meters. When the gunners lay their MK 19s on the right or left flank (near or far end), respectively, they adjust fire and engage the target. When the gunners lay their MK 19s on the announced release point (initial aiming point), they set their sights at the range to the reference point, and adjust fire. The gunner on the Number 1 gun searches down the designated number of meters (or mils), firing after each manipulation until the gun is set on the near end of the target. The gunner on the Number 2 gun searches up the designated number of meters (or mils), firing after each manipulation, until the gun is set on the far end of the target. When each gun has reached the near or far end, respectively, the gunners reverse the direction and both search up and down between the two limits, covering the entire target, until told to cease firing.

e. **Engagement of Deep Targets Using Computed Search Method.** When the depth of a target must be described in mils (search), the leader may compute the depth from the reference point using the computed search method. The leader determines the ranges to the near and far ends of the target and finds the required AE (angles of elevation) in mils for both ranges (Appendix F, Firing Table). The leader then computes the difference between the two. This is the amount of search required when the MK 19 and target are at the same elevation on level ground. If the fire is plunging, the amount of search should be increased. For example, when using M430 ammunition, a deep target has been sighted (the ends of which are not visible to the gunners). A reference point is also visible on the target at a range of 1,100 meters, the depth of the target is estimated at 200 meters, and the reference point appears to be midway between the ends of the target:

$\frac{200 \text{ meters}}{2}$ = 100 meters from the reference point to each end of target

1,100 + 100 = 1,200 meters to far end of target

1,100 - 100 = 1,000 meters to near end of target

AE 1,200 meters = 175 mils

-AE 1,000 meters = 132 mils

depth of target 43 mils

$\frac{43 \text{ mils}}{2}$ = 21.5 mils from release point to each end of target

To search in 5-mil increments, the gunners cover the target by searching 25 mils long and 25 mils short of the reference point, or a total of 50 mils. The leader describes the target to the gunner as TARGET; TROOPS EXTENDING LONG 25 MILS; SHORT 25 MILS. The following are examples of fire commands:

- Reference point in target area, ends of target not visible to the gunner.

  FIRE MISSION.
  RIGHT FRONT.
  REFERENCE: LONE BUSH.
  TARGET: TROOPS EXTENDING SHORT ONE HUNDRED YARDS, LONG TWO HUNDRED YARDS.
  SIX HUNDRED (range to reference point).
  SEARCH.
  RAPID.

AT MY COMMAND.
FIRE (given after gunners announce "Up").

- Reference point outside target area, gun on tripod.
  FIRE MISSION.
  LEFT FRONT.
  REFERENCE: DEAD TREE.
  RIGHT THREE ZERO MILS; ADD TWO HUNDRED METERS
  (or add so many mils, using computed search).
  TARGET: TROOPS EXTENDING OVER TWO HUNDRED METERS
  (or long so many mils, using computed search).
  NINE HUNDRED.
  SEARCH.
  RAPID.
  AT MY COMMAND.
  FIRE (given after gunners announce "Up").

f. **Engagement of Moving Targets.** Due to the slow flight of the MK 19 round, it is not practical to engage moving targets using the track-and-lead methods of engaging moving targets. In order to effectively engage moving targets with the MK 19, use the trapping method of engaging targets. The gunner chooses an aiming point forward of the target and on the target path, and presses the trigger before the target comes into the sights. After starting to fire, the gunner moves the MK 19 slowly toward the target. The target moves into the impact area, and is "trapped." The gunner must begin firing before the target is in the sights or in the impact area (Figure 5-18).

(1) ***Engagement of Deep Target Moving Away from the MK 19 Position.*** If the target is moving rapidly away, both gunners lay their MK 19s on the far end of the target with the range set to that point, and search up.

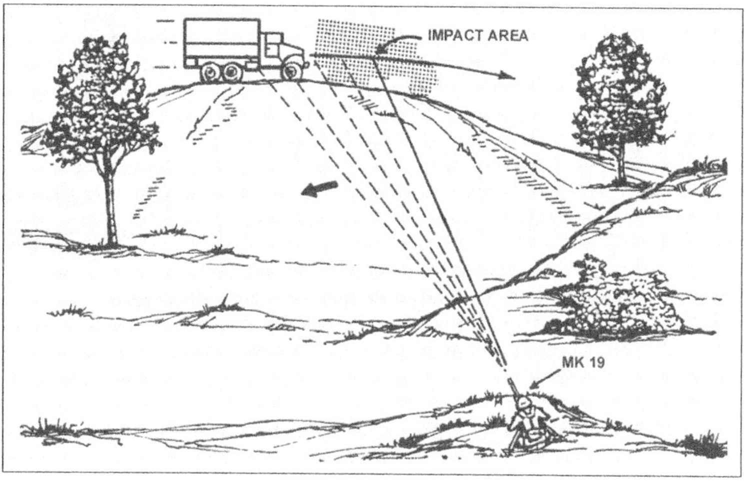

Figure 5-18. Target-trapping method.

(2) ***Engagement of Deep Target Moving toward the MK 19 Position.*** If the target is moving rapidly toward the MK 19 position:

(a) *One MK 19.* The gunner lays the MK 19, adjusts on the near end of the target with a range to the target's midpoint, and searches up and down the entire target. When a reference point within the target area is announced, the gunner lays on it with the range to that point. The gunner adjusts fire and searches up and down the entire target, beginning with the near end, until told to cease firing.

(b) *Two MK 19s.* The gunners lay both guns on the near end of the target, set the sights on the range to that point, and search down. The manipulation element of the fire command for a rapidly approaching or receding target is BOTH WEAPONS; FAR (NEAR) END; SEARCH.

(c) *Four MK 19s.* Each pair of gunners engages the target as if they were acting alone. If one pair must be switched to a different target, the other pair continues to cover the original target. A deep target should never be subdivided since the elevating mechanism on the M3 tripod allows enough search to cover any deep target within the maximum effective range for direct lay (1,500 meters).

g. **Engagement of Linear Target with Depth Using Traversing and Searching Fire.** The method used depends on the number of MK 19s available and whether the flanks of the target are visible to the gunner.

(1) ***One MK 19, Flanks Visible.*** The gunner lays initially on the near flank of the target with range set to the midpoint, adjusts fire on the near flank, and traverses and searches back and forth, covering the entire target, until told to cease-fire.

(2) ***Two MK 19s, Flanks Visible.*** Gunners use the same method as for engagement of a deep linear target: they must search to keep the center of impact on the target. The leader determines the range for the initial fire command the same as for a deep target (Figure 5-19).

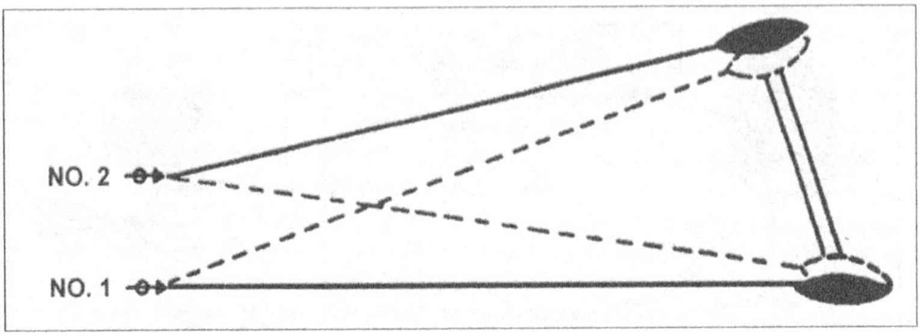

Figure 5-19. Two MK 19s, linear target with depth.

(3) ***Four MK 19s, Flanks Visible.*** Gunners use the same method for two pairs as is used for two guns. If one pair must be switched to another target, the other continues to cover the original target.

(4) ***Flanks Invisible.*** The leader may designate the target using a reference point. However, a reference point in the target area is not used as an initial aiming pointbecause of the difficulty in describing an oblique target.

(5) ***Fire Commands.*** The leader may use the following fire commands for linear targets with depth.

- When the target is not wide enough to require subdivision, is 200 meters or less in depth, and is to be engaged by either one or two MK 19s:

  FIRE MISSION.
  RIGHT FRONT.
  TROOPS EXTENDING FROM BLACK STUMP RIGHT TO LONE PINE.
  SEVEN HUNDRED (midrange).
  TRAVERSE AND SEARCH.
  RAPID.
  AT MY SIGNAL.
  FIRE (given after gunners announce "Up").

- When the target is 100 mils wide or less, is 200 meters deep or less, and is to be engaged by four MK 19s:

  FIRE UNIT, FIRE MISSION.
  FRONT.
  TROOPS TO THE RIGHT OF RED BANK.
  SEVEN HUNDRED (midrange).
  TRAVERSE AND SEARCH.
  RAPID.
  AT MY SIGNAL.
  FIRE (given after gunners announce, "Up").

- When the target is wide enough to require subdividing, is more than 200 meters in depth, and is to be engaged by either two or four MK 19s:

  FIRE MISSION.
  FRONT.

TROOPS EXTENDING FROM TRUCK, RIGHT TO BRIDGE.
NUMBER 1 (and 3 if four MK 19s are engaging the target),
SEVEN HUNDRED.
NUMBER 2 (and 4 if four MK 19s are engaging the target),
ONE THOUSAND.
NUMBER 1 (and 3), RIGHT HALF.
NUMBER 2 (and 4), LEFT HALF.
TRAVERSE AND SEARCH.
RAPID.
AT MY SIGNAL.
FIRE (given after gunners announce "Up").

h. **Engagement of Target Using Combined Sights.** Leaders may choose this expedient way to engage a linear target with depth. They may lengthen the beaten zone of a two- to four-gun fire unit by having gunners set the sights differently on each gun. The leader gives one gunner a setting 150 meters less than the estimated range, and the other a setting 50 meters more than the estimated range. When four guns are used, one pair of gunners sets sights 150 meters less than the range, the other 150 meters more than the range. This technique sacrifices precision for speed, but may be necessary for fast-moving or fleeing targets. A sample fire command for the use of combined sights follows:

FIRE MISSION.
RIGHT FRONT.
TROOPS EXTENDING FROM CROSSROADS TO HOUSE.
1,300 METERS (midrange).
COMBINED SIGHTS; NUMBER 1, 1,350 METERS; NUMBER 2,1,250 METERS.
TRAVERSE.
RAPID.
FIRE (given after gunners announce "Up").

i. **Engagement of Area Target.** This applies to a target that cannot be covered by either traversing fire, searching fire, or traversing and searching fire. Area target engagement requires that the leader mass fires from four to six guns. Examples of area targets include:

- Large troop or mechanized formations.
- Targets that must be suppressed (if the exact positions are unknown or are not visible).
- Large kill zones or engagement areas for which predetermined fires have been planned.

j. **Engagement of Target Using Massed Fires.** Leaders can mass fires in both offensive and defensive operations, depending on the time available to plan and conduct the engagement.

(1) ***Defensive Operations.*** There are two types of defensive operations.

- **Defense, Time Not a Factor.** Massing fires into an engagement area has advantages over other types of area fire engagements. Once TRPs are designated, gunners may fire upon them to register or rehearse the MK 19s. After the target has been subdivided and TRPs registered, the gunners traverse, search up, traverse back, and search down through their assigned parts of the target to assure full coverage of the target (Figure 5-20). The same method applies regardless of how many MK 19s are used. If four or six guns are used, the leader must subdivide the target and assign more TRPs than for a pair of MK 19s. An example of a fire command for this situation follows:

  FIRE MISSION.
  FRONT TRPs 1, 2, 3, 4.
  (MK 19s are laid on respective TRPs)
  AREA: ENEMY COMPANY ENTERING EA GOLD.
  (Range omitted MK 19s on TRPs.)
  TRAVERSE.
  SEARCH UP 50 MILS.
  RAPID.
  AT MY COMMAND.
  FIRE (given after the enemy is completely within the engagement area and the gunners announce "Up").

- **Defense, Time a Factor.** Leaders may not have the luxury of registering their MK 19s on TRPs before chance contacts due to time factors or the elements of surprise. However, the lethal effect of massed MK 19 fire on an area target should not be overlooked. Units should develop and practice SOPs that enable them to engage large-area targets on quick notice. For example, the leader may give each MK 19 a section or quadrant for all quick area engagements. He may assign the upper left side of the target to the Number 1 gun, the upper right to Number 2, and so on.

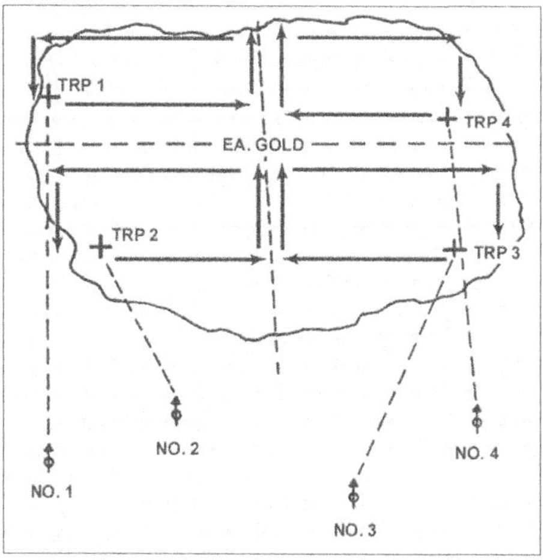

Figure 5-20. Use of massed fires, time not a factor.

(2) ***Offensive Operations.*** Massing fires into an area target can also be useful for offensive operations. During a movement to contact or an attack where time is a factor, massed fires may be used to suppress a suspected or actual enemy position, allowing the unit to flank or bypass the enemy. Although fast target acquisition and volume of fire are primary goals, the leader should subdivide the target for control, and adjust fires for greatest effect. Massed fires are especially useful in support of an attack where time is not a factor, which may require the use of overhead fire.

WARNING

Overhead fire should not be used unless the vertical interval between friendly soldiers and the target line is great enough to ensure safety.

5-13. DEFILADE POSITION

The MK 19 is most effective when laid directly on a target. Depending on the tactical situation, however, defilade could be the most desirable position.

a. **Definition.** A MK 19 is in defilade position when the gun and its crew are hidden by a landmass such as the crest of a hill from enemy ground (Figure 5-21). The position can be on the reverse slope of a hill, on the forward slope of the next high ground to the rear of a hill, or in a small fold in the ground. A defilade firing position does not necessarily reduce the effectiveness of fire against a stationary enemy target or preclude fire over the heads of friendly troops. Defilade may also be useful in providing predetermined fire into an EA.

b. **Advantages.** Some advantages of using the defilade position are that the MK 19 and a crew have cover and concealment from direct fire guns; the crew has some freedom of movement near the MK 19 position; control and supply are facilitated; and the characteristic smoke and flash of the MK 19 are partly concealed from observation.

c. **Disadvantages.** Disadvantages of using the defilade position are that targets close to the mask usually cannot be engaged, and (because adjustment of fire must be made through an observer) rapidly moving ground targets are not easily engaged.

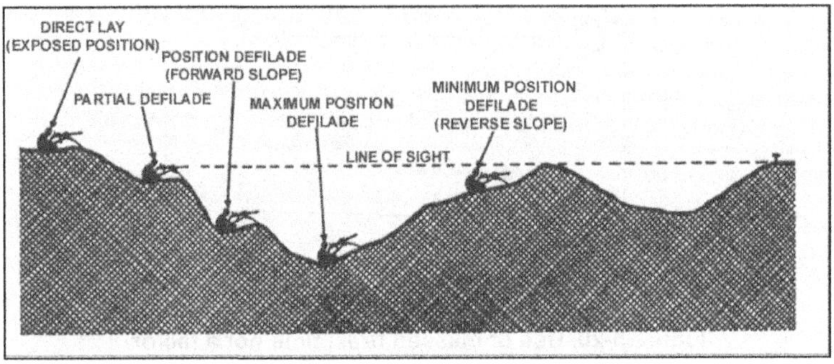

Figure 5-21. Defilade positions (and direct lay).

d. **Options.** A defilade position can either be on a forward or reverse slope.

(1) ***Reverse Slope Positions.*** These positions afford some protection from high-angle enemy fire and direct-fire cannon.

(a) *Partial Defilade Position.* The MK 19 is in partial defilade position when it is sited at the highest point on a slope on which it is defiladed. Partial defilade position provides great flexibility in engagement of new targets, although it does not provide the best cover. The MK 19 crew positioned in partial defilade can be grazed by enemy machine gun fire, but is concealed from enemy observation near the target area.

(b) *Maximum Defilade Position.* The MK 19 is in maximum defilade position when it is sited at the lowest point on a slope from which it can engage a target. The MK 19 crew positioned in maximum defilade has relatively good cover but lacks the flexibility to engage new targets.

(2) ***Forward Slope Defilade Position.*** A forward slope defilade position is not vulnerable to grazing fire from the target area, but it is open to attack or damage from mortar and artillery fire. The MK 19 is in partial defilade on a forward slope if the gun and gunner have some protection from direct fire, and if the gunner is able to engage the target using direct lay. Partial defilade is used only when a fire mission cannot be completed from defilade. The MK 19 is held in defilade and moved into partial defilade to fire.

e. **Considerations.** Direction, elevation, mask clearance, and adjustment of fire are four essential elements that must be considered when using the MK 19 in defilade position.

(1) ***Direction.*** There are four ways to lay the gun for direction.

(a) *Direct-alignment Method.* The observer selects a position on the gun-target line from which the target can be seen. The gunner loosens the traversing slide and, directed by the observer, moves the MK 19 right or left until it is aligned on the target.

(b) *Aiming-point Method.* For an aiming point, the gunner selects a prominent landmark visible through the sights (Figure 5-22). An aiming point on the gun-target line and at an equal or greater range than the target is desirable; however, the gunner may use an aiming point on the mask.

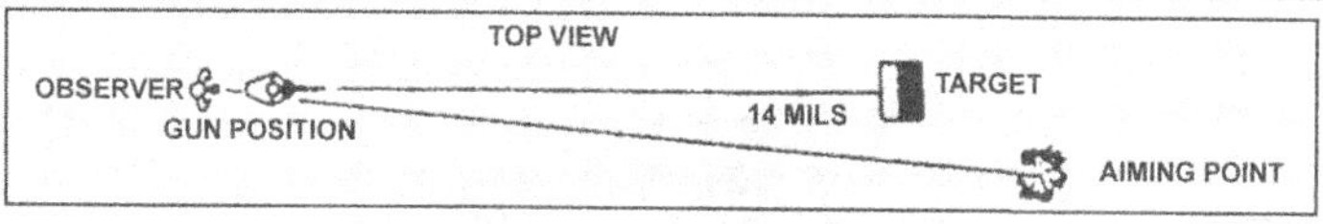

2 — AIMING POINT IS NOT ON LINE GUN-TARGET.
RANGE GUN-TARGET IS 1,000 METERS.
TARGET IS 14 MILS LEFT OF AIMING POINT.
DIRECTION: WITH REAR SIGHT SET AT 1,000 MILS,
LAY GUN ON AIMING POINT,
TRAVERSE GUN LEFT 14 MILS.
ELEVATION: DEPRESS GUN 12 MILS.

Figure 5-22. Aiming-point method.

- When the aiming point is on the gun-target line, the gunner lays the MK 19 on the aiming point, which aligns it on the target.
- When the aiming point is not on the gun-target line, the gunner measures the deflection with binoculars or compass, and adjusts the lay of the MK 19.

(c) *Aiming-stake Method.* When no natural aiming point is available, the assistant gunner sets out an aiming stake, and the gunner aligns the gun on the target.

(d) *Map-and-compass Method.* The leader locates the MK 19 position and target on a map; draws a line between the two points; orients the map to the terrain; and places the line of sight on the compass along the gun-target line drawn on the map. The leader then announces the magnetic azimuth at the compass index to the gunner as the direction of lay (Figure 5-23). Using this method with terrain-profiling techniques permits the MK 19 to be used in various defilade positions.

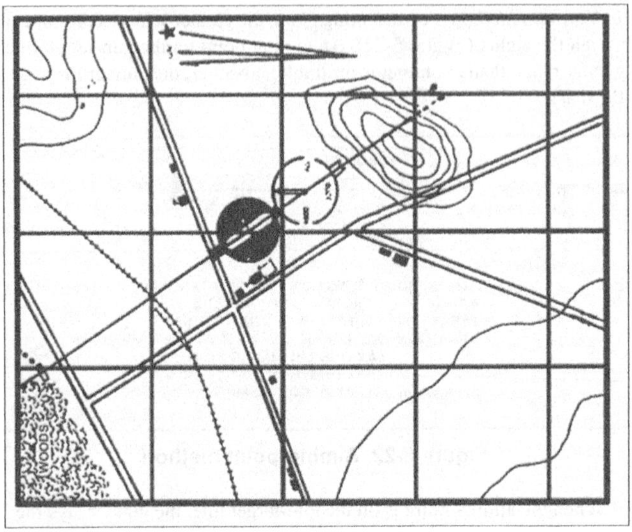

Figure 5-23. Map-and-compass method.

(2) ***Elevation.*** To lay the gun for elevation when engaging targets from defilade, knowledge of the trajectories of 40-mm rounds and of indirect lay is necessary. The MK 19 mounted on the M3 tripod with the T&E mechanism is not equipped to deliver indirect fire in the traditional sense like a mortar. Because the MK 19 has a high-arc trajectory when in the direct-fire mode, it can be fired effectively from defilade if the gunner positions it properly with the help of a well-trained observer. Discussion of laying the MK 19 for elevation includes those techniques and procedures that do not require a lot of data and calculations.

(a) The AE is the vertical angle between the bore line and the line of sight, when the gun and target are at the same elevation (Figure 5-24). The AE is always a positive (plus) and increases as the range increases. The AE for 40-mm ammunition, for each 100 meters of range up to 2,700 meters, is contained in Appendix F, Firing Table. For example, to hit a target at a range of 1,000 meters with M430 HEDP ammunition, the MK 19 must have an AE of +131.9 mils to the line of sight. Is a discussion on leveling the tripod necessary or can the gunner "eye-ball" it?

(b) When the MK 19 and target are not at the same elevation, an additional angle must be taken into consideration. The angle of sight (AS) is the vertical angle formed by the line of sight and a horizontal line from the base of the gun. When the target is at a higher elevation than the gun, the AS is positive (plus) (Figure 5-24). When the target is lower than the gun, the AS is negative (minus).

(c) The angle of quadrant elevation (QE) is formed between the bore line and the horizontal line through the base of the gun (Figure 5-24). The QE is positive (plus) whenever the gun is aimed above the horizontal, and negative (minus) whenever the gun is aimed below the horizontal. It is the algebraic sum of the AE and the AS; that is, if the angle of sight is positive, it is added to the AE; if the AS is negative, it is subtracted from the AE.

QE = AE + AS (target above horizontal line)
QE = AE - AS (target below horizontal line)

Other methods of elevation are as follows:

- **Computed quadrant elevation method.** The leader must determine the correct range to the target. Using the range, the leader finds the corresponding AE from Appendix F, Firing Table. The leader must find the AS using binoculars, by measuring in mils the vertical interval between the target and the estimated horizontal. The leader may assume the distant horizon to be at a zero AS, or at the same elevation as the MK 19 position. QE may be determined by algebraically adding this data as previously described.
- **Measured quadrant elevation method.** The gunner should locate the MK 19 in partial defilade and lay it on the target using direct-lay methods. The leader measures the QE with the M2 compass. The gunner moves the MK 19 into defilade position and places the measured QE on the gun. For each meter difference in elevation between the position in partial defilade and the firing position, the gunner adds 1 mil to the QE when firing at a range of 1,000 meters; 1/2 mil when firing at 2,000 meters, and so on.
- **Aiming-point method.** The gunner selects an aiming point visible from the MK 19 position; preferably a point at a greater range than the target and at a higher elevation than the target, and the leader finds the range to the target. Using binoculars, the leader measures the vertical angle in mils from the aiming point to the base of the target. The leader has the gunner lay the MK 19 on the aiming point, with the sight set to hit the aiming point, and then directs the gunner to manipulate the gun through the number of mils measured from the aiming point to the target. For example, the range to the target is 1,000 meters (Figure 5-25). The angle read with the binoculars from the aiming point down to the base of the target is 12 mils. The sight should be set at 1,000 meters, with the MK 19 laid on the aiming point, and the muzzle then depressed 12 mils.

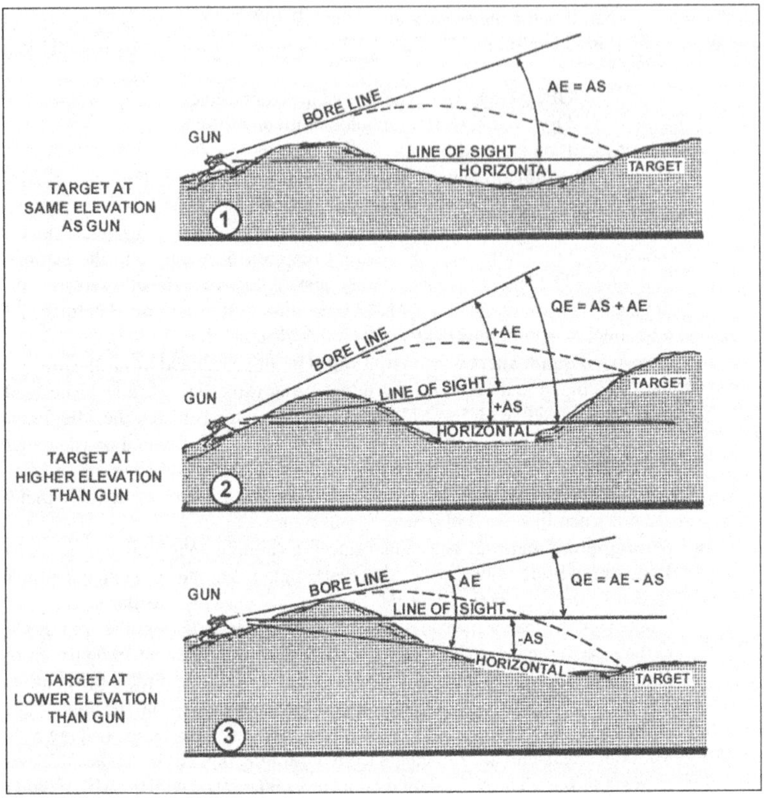

Figure 5-24. Angle of elevation, angle of sight, and quadrant elevation.

(3) ***Mask Clearance.*** After the gun(s) has been laid, the leader must determine whether or not the entire cone of fire will clear the mask, if mask clearance is not obvious. In order to ensure mask clearance, the leader has the gunner sight on the crest of the mask and re-lays the MK 19 for elevation on the target. The gunner ensures that the difference in elevation from the MK 19 to the target and from the MK 19 to the mask is at least +10 mils. The gunner checks this by sighting along the bottom of the receiver and the barrel.

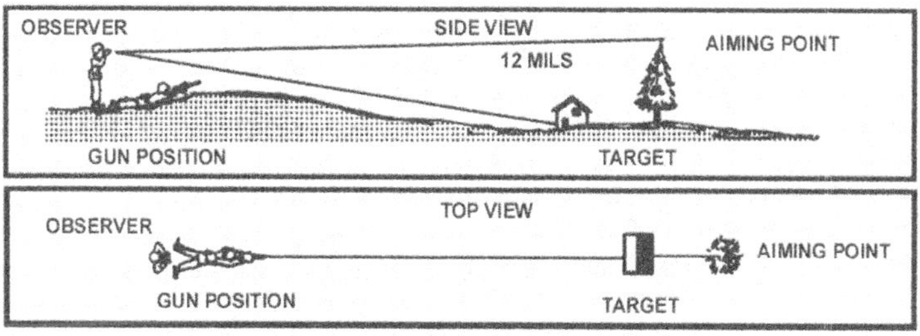

1 - AIMING POINT ON LINE GUN-TARGET.
RANGE GUN-TARGET IS 1,000 METERS.
DIRECTION: WITH REAR SIGHT SET AT 1,000 METERS.
LAY GUN ON AIMING POINT.
ELEVATION: DEPRESS GUN 12 MILS.

Figure 5-25. Aiming-point method for elevation.

NOTE: The procedures already discussed aid in the initial lay of the gun. The best tactical application of fire from the defilade position is in defense when time is not a factor, especially in a reverse slope defense. Leaders should try to register their MK 19s on selected TRPs, as previously discussed. Use trial and error to adjust for accurate fire from the defilade position. While the MK 19s are being registered, every precaution must be taken to guarantee troop safety.

(4) ***Adjustment of Fire.*** Under field conditions, the MK 19 that is quickly laid on the target seldom results in an initial impact directly on the target. Rapid adjustment of fire is essential. This is assured by thoroughly training the observer to estimate range and lateral distance. The observer should be in a position to see the initial impact. To ensure this, the burst (three to five rounds) should be long. Except for long ranges, or when for other reasons visual observation is hard, the observer does not ordinarily use binoculars for the initial burst. If the initial lay is incorrect, the binocular's limited field of view can cause the observer to miss the impact of the round. Adjustment of fire should be bold and aggressive. The observer should command large corrections and avoid creeping fire toward the target. When a burst is incorrect for deflection and elevation, the observer's next command should correct both.

f. **Construction of a Terrain Profile.** The leader may use a side view or cross section of the ground along a selected line or direction to determine where friendly and enemy forces can see each other. The leader can also use it to plan fires; that is, to locate dead space, to plan a defilade position, and so on. To construct a terrain profile, locate the gun position on the map and determine the direction of fire, or profile line (Figure 5-26).

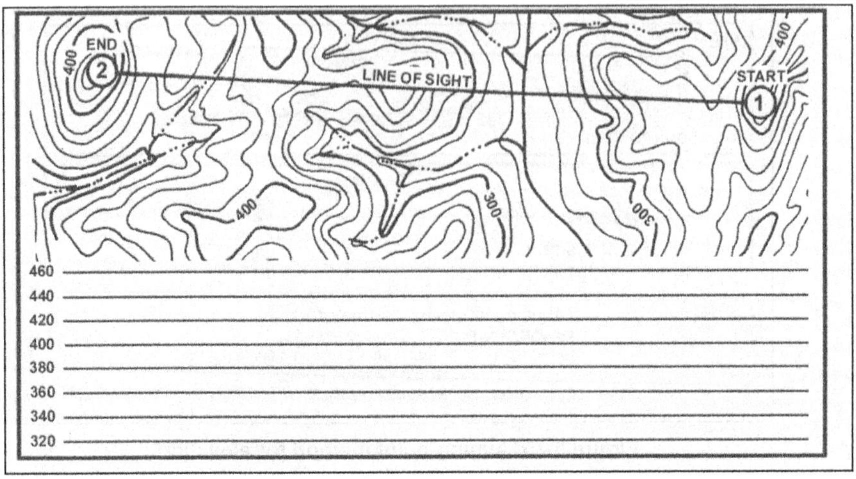

Figure 5-26. Profile line (map contour lines are 20 meters).

(1) Place the edge of a lined piece of paper along the profile line. Tick-mark all points of known elevation on the edge of the paper. These points are the contour lines. Any paper with evenly spaced horizontal lines, such as graph or notebook paper, may be used. The wider the spacing of the lines, the greater the vertical exaggeration in the profile; however this does not affect the information.

(2) Draw perpendicular lines, down across the horizontal lines for each marked point, and identify the high and low points along the profile (Figure 5-27).

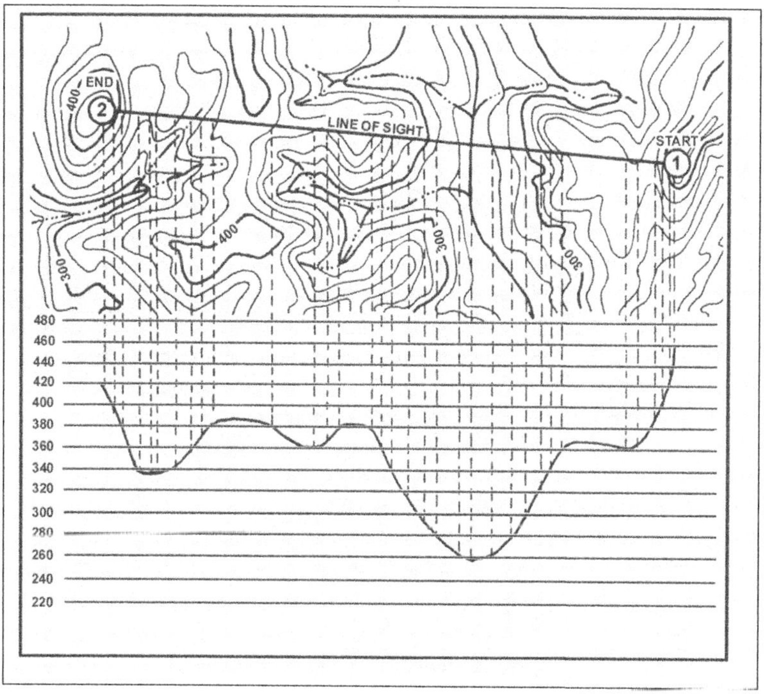

Figure 5-27. Marking of elevation.

(3) Connect all the points with a smooth curve (Figure 5-28).

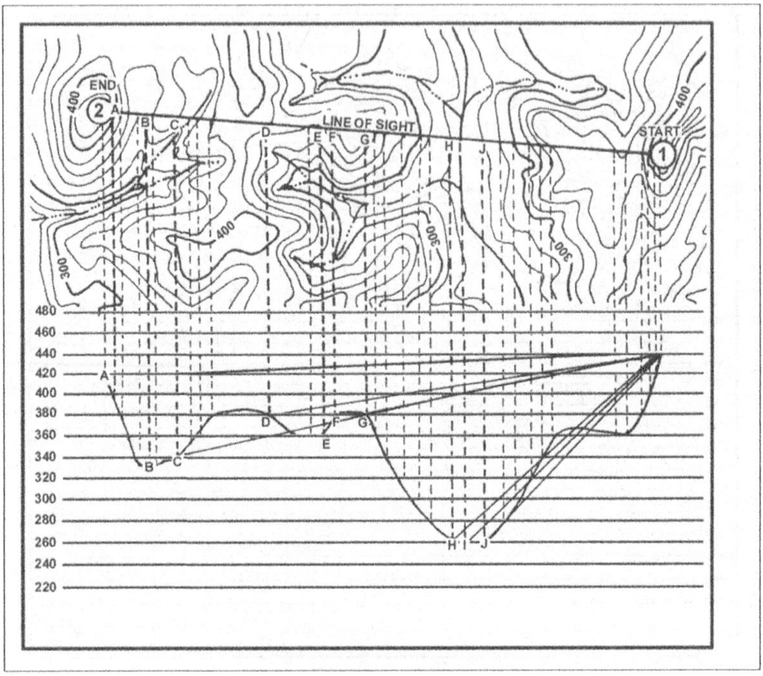

Figure 5-28. Terrain profile.

Section IV. DEGRADED CONDITIONS

The MK 19 is fired the same way under degraded conditions as it is fired under good conditions, except that degraded conditions limit the gunner's ability to observe the impact of rounds in relation to the target and to make adjustments. Refer to Appendix G, Aiming Devices, for additional information.

5-14. NIGHT FIRING

The use of range cards and predetermined fires are important at night or during degraded conditions because the targets or engagement areas are chosen and marked in advance, which enables the gunner to adjust onto targets quickly and easily. The gunner must know how to read his traversing slide bar and T&E mechanism. This task can be difficult at night or during degraded conditions. The gunner can mark targets on the slide bar with luminous tape, use mini-chemical lights, or use flashlights with red lens covers.

a. Firing with the use of illumination has proven to be much the same as firing during normal daylight conditions. Estimation of distances and adjustment of rounds are difficult at night or during degraded conditions. Gunners should be able, using illumination, to effectively engage targets out to 800 meters as well as they can in the daylight. Due to long shadows and dispersion of light past this distance however, visibility is poor even with illumination.

b. The AN/TVS-5 can be mounted on the MK 19 and used effectively to spot and adjust rounds to targets out to ranges of 800 meters. Past this range, the elevation of the scope on the gun is too high to see the impact of the rounds, so it can be handheld by an observer who calls out adjustments to fire.

5-15. FIRING IN A NUCLEAR, BIOLOGICAL, OR CHEMICAL ENVIRONMENT

Firing the MK 19 during NBC conditions is no different than firing in normal conditions. Wearing the protective mask during firing does not change the way the gunner sights and aims the MK 19 to engage targets. The assistant gunner should take care to avoid catching his clothes or gloves while feeding ammunition. The gunner should also avoid contact with the hot barrel.

5-16. FIRING IN FOG OR SMOKE

Neither the degraded conditions of fog or smoke change the way the gunner fires the MK 19. The only limitations are observation of rounds and adjustment onto targets. The use of predetermined fires, trip flares, PEWS, or forward observers can cut down on these limitations.

Section V. PREDETERMINED FIRES

Predetermined fires are planned to cover target areas such as enemy avenues of approach, likely sites for enemy guns, and probable enemy assault routes. Laying the MK 19 on predetermined targets by either of the following methods may be verified by firing the MK 19 and adjusting it on target.

5-17. T&E MECHANISM METHOD

When the MK 19 is laid on target in the primary sector, the direction and elevation is taken from the traversing bar and the T&E mechanism. Both direction and elevation, as well as the range to each target, are recorded on the range card. Each target in the sector of fire must be numbered and laid on in sequence.

5-18. FIELD-EXPEDIENT METHOD OF LAYING THE MK 19

Another method for laying the MK 19 on predetermined targets is to use field expedients (Figure 5-29). Field expedients must be used in the secondary sector, and consist of aiming and base stakes. They can be used in the primary sector to aid the gunner.

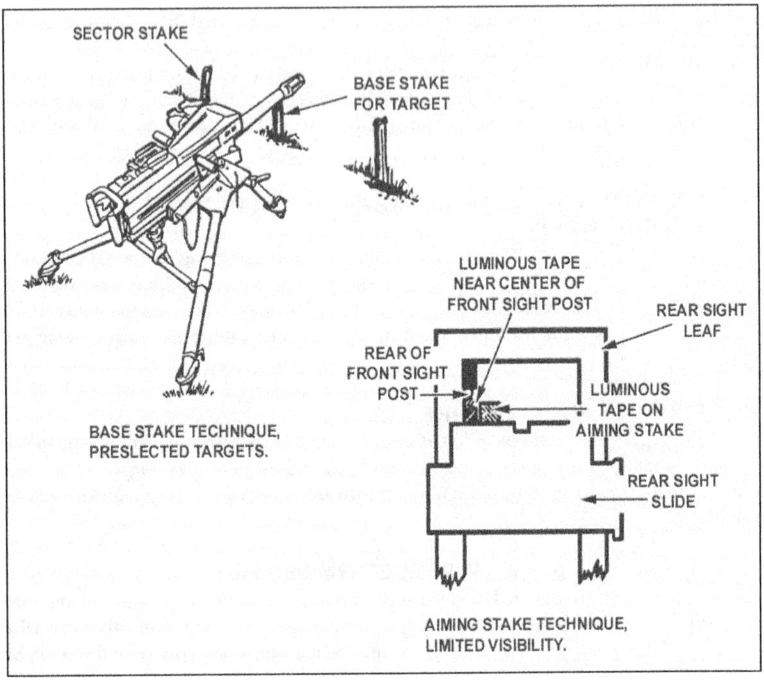

Figure 5-29. Techniques for laying the MK 19 using field-expedient method.

CHAPTER 6

INFORMATION FOR INSTRUCTORS

This chapter provides information to aid instructors in training MK 19 crews to become skilled teams. This chapter is organized to lead the trainer in a progressive manner through the material necessary to aid him in training units. Information is presented beginning with preliminary subjects; these areas include mechanical training, knowledge of the gun's capabilities, and the principles and fundamentals of MK 19 gunnery.

6-1. PRESENTATION

To present instruction in the time allotted, the instructor must be prepared and rehearsed. Assistant instructors and demonstrators must also be trained and rehearsed so they can conduct demonstrations using training aids that clearly illustrate the information. Use public address systems if the group is larger than a platoon.

a. **Organization.** Divide the soldiers into groups of appropriate size, while also considering the phase of instruction and the facilities available. If available, assign an assistant instructor to each crew; otherwise, position the assistant instructors so that they can closely supervise the crew.

b. **Presentation.** Explain and demonstrate each portion of the instruction, following with closely supervised practical work. Emphasize practice throughout training in order to develop skill and teamwork. Secure and critique all MK 19 firing during training.

c. **Examination.** Ensure the soldiers make good progress throughout each period of instruction. Those who do not progress well must receive more instruction under close supervision. Give a formal or informal participation-type examination following each phase of instruction to measure effectiveness of instruction.

6-2. MECHANICAL TRAINING

During mechanical training, the gun is mounted on the M3 mount, or installed on a table. Nomenclature charts (or mats) help the crewmembers learn the names of parts. If possible, assign one assistant instructor for each crew. Present a brief history of the gun. Stress the gun's combat role and assigned mission. Emphasize the purpose, scope, and importance of the instruction to be presented. Be sure to clearly define the term stoppage during the introduction, and emphasize the need for correct care and cleaning procedures.

a. **Organization.** Organize the class into groups of appropriate size. If enough assistant instructors are available, assign one to each crew; otherwise, position the assistant instructors so they can closely supervise the crew.

b. **Presentation.** Table 6-1 can be used as a guide in presenting instruction in stoppages and immediate action. Stress precision in detecting the cause and reducing stoppages. As the soldiers progress, concentrate on speed in applying immediate action and other methods of reducing a stoppage. Train the crew to function as a team.

STOPPAGES	INSTRUCTION PREPARATION
Misfire due to defective primer.	Place a dummy cartridge in the belt.
Short round.	Place a short round in the belt.
Bulged round.	Insert a bulged round in the belt.
Belt loaded improperly.	Pull a cartridge partly out of the belt.
Battered or thick-rimmed cartridge.	Place a battered or thick-rimmed cartridge in the belt.
Failure to remove a round from the chamber.	Place a dummy cartridge, with the rim filed off, in the chamber.
Separated case that is removed from the chamber by the new round when the bolt is pulled.	Drive the front portion of a cartridge securely on a dummy cartridge. Pull the bolt to the rear and place the cartridge properly on the face of the bolt. Ease the bolt forward.

Table 6-1. Preparation for instructing stoppages and immediate action.

6-3. CREW DRILLS

The demonstration crew must be well-trained and equipment must be complete. Use the following equipment for each crew: basic table of organization and equipment (TOE) and distinctive helmet markings that identify each crewmember according to his duties, and normal individual equipment. For more information on crew drills see Appendix H.

a. **Organization.** Present initial instruction and demonstration to one crew. For practical work, divide the class into crews of four men for each MK 19. To aid in control and supervision, mark off equipment with engineer tape or wire, and line up the crewmembers behind their equipment.

b. **Presentation.** Stress the fact that a crew drill is the first step in developing teamwork, and that precision is required of every member for the crew to function smoothly. During initial practice periods, work on precision. Talk each soldier through his duties as a crewmember. Require the soldiers to do each operation when directed with speed and precision.

c. **Advanced Training.** In advanced training, use a speed drill to create interest and stimulate competition. This tactic also furthers teamwork and tests individual crew performance. A well-trained crew should be able, without error, to place the gun into action, or take it out of action in 25 seconds.

6-4. MARKSMANSHIP

With gun and mount, the instructor makes arrangements for a demonstration crew to be available. He ensures that he has a sighting bar, an illustration of a correct sight picture, a sighting target (panel with holder), a stopwatch, control flags, cleaning rods, and patches.

a. **Organization.** Organize the crew and the training area. Organize each crew with four members, if possible. Use a centralized area, preferably with bleachers, for conferences and demonstrations. The area should be fairly level, and large enough for the guns to be placed 10 meters from the targets.

b. **Presentation.** Ensure marksmanship basics are learned and practiced. Make sure the soldiers know that marksmanship basics must be learned and practiced during the preparatory phase of the marksmanship training, in order for them to satisfactorily complete the marksmanship course. With the help of coaches and assistant instructors, ensure that steps are done correctly during the preparatory phase of marksmanship. Give a brief description of the operation, general data, and exterior nomenclature of the MK 19. Ensure that assistant instructors disassemble and assemble the MK 19 while the procedure is being explained.

NOTE: When the crews can skillfully disassemble and assemble the gun, demonstrate this phase of instruction to them blindfolded, using an assistant. Doing this encourages crews to practice during their free time, and develops individual skill and initiative.

Ensure the crew does not assemble or disassemble the MK 19 under stress against time because there is a danger of damaging the gun's parts.

c. **Operation.** Each crew requires a MK 19 mounted on an M3 mount and a belt of five dummy rounds. Emphasize the safety factors involved. Stress how the feed throat must be attached before loading the MK 19. Demonstrate the method of letting the bolt go forward.

d. **Functioning.** Training aids are useful if the class is platoon-sized; otherwise, training aids may be used for study and discussion during breaks. Use one gun for each crew, as in previous mechanical training instruction. Divide functioning into six phases (see Chapter 2, Operation and Function). Explain and demonstrate each phase of functioning. Assistant instructors should duplicate each demonstration on the guns within each crew. With these setups at the guns, each crewmember can see how the parts work. Setups may be used for each phase of functioning.

APPENDIX A
TRAINING STRATEGY

This appendix discusses the two primary components of training strategies: initial and sustainment training.

A-1. COMPONENTS

Training strategies have two primary components: *initial* and *sustainment* training. Both may include individual and collective skills, but *initial training is critical.* With initial training a soldier retains a correctly taught task that he has learned well. With such training, he can easily sustain or quickly regain his performance. Thus, leaders must conduct initial training correctly and well.

a. Trainers use standard, common, and current procedures. Regardless of how well skills are taught or learned, individual and unit skills decay over time. How fast and how much skills decay depends on many factors. Personnel turnover is a primary factor in the decay of collective skills. Other factors include the difficulty and complexity of the tasks.

b. When the unit loses critical crewmembers, it must quickly train new gunners to restore unit proficiency. If too much time passes between initial and sustainment training periods or if the unit alters the established training doctrine, skills erode and the unit might have to completely retrain its MK 19 gunners.

c. A critical step in the Army's overall training strategy is to train leaders and trainers first. Leader courses and unit publications help develop officer and noncommissioned officer (NCO) proficiencies needed to plan and conduct marksmanship training. These resources also help leaders learn how to evaluate the effectiveness of unit marksmanship programs. Proponent schools provide training support materials. These materials include field manuals, training aids, training devices, simulators, and programs of instruction that serve as the doctrinal foundation and guidance for training the force.

A-2. INITIAL TRAINING

The MK 19 training strategy starts with leader training and goes on to unit training. Because "MK 19 gunner" is a Skill Level 2 position, the MK 19 gunner receives most of his MK 19 training within the unit. The training institution provides the unit with a soldier who has received training in the basics of the MK 19. The unit plans minimal MK 19 training for advanced individual training (AIT) soldiers.

A-3. SUSTAINMENT TRAINING

Basic MK 19 gunnery skills and gunner qualification continues in units. Once soldiers learn individual and crew skills, the unit must help them sustain those skills. To do this, leaders follow the annual gunnery-training program provided in Appendix E, Annual Gunnery Training Program. In addition to sustaining skills, the unit incorporates the MK 19 into collective training exercises such as platoon and squad live-fire situational training exercises (STXs).

a. The strategy for sustaining basic MK 19 gunnery skills includes annual or semiannual instructional and qualification range firing. However, the unit must set up a year-round sustainment program. Key elements of the sustainment program include training the trainers, refreshing preliminary skills, and using devices to remedy the training of soldiers who have trouble qualifying on the MK 19.

b. The unit also trains the gunner in other skills such as night firing; mission-oriented protective posture (MOPP) firing; engaging moving targets, using camouflage, cover and concealment; firing and maneuvering; and preparing or selecting a fighting position. The unit also integrates these skills into tactical training. The references at the back of this manual contain additional guidance on the training for these skills.

c. The unit integrates individual and leader MK 19 tasks into collective training such as squad, section, and platoon drills and STXs as well as company LFXs and field training exercises (FTXs). Organizational mission training plans (MTP) and drill books include the tasks, conditions, and standards for the collective tasks in these exercises. These books also explain how to plan and conduct such collective tasks. Based on the type of organization, the unit evaluates collective exercises in leader and trainer after-action reviews (AARs). Objectively evaluating both individual and unit proficiency indicates the readiness and future training requirements.

d. Trainers schedule live fire after the soldiers demonstrate preliminary skills. For initial fire, the trainer conducts an instructional exercise. This improves the soldiers' ability to acquire targets with the MK 19. Once soldiers understand the gun and demonstrate the skills needed to acquire a target, the trainer conducts additional live-fire training and target-detection exercises at various ranges. To develop proficiency, soldiers have to master targets and scenarios of increasing difficulty. After initial individual training, the trainer conducts qualification fire on a range to evaluate each soldier's skills. When conducted in accordance with Chapter 4, Marksmanship, this evaluation also gives leaders some insight into the unit's skill and training effectiveness.

APPENDIX B

OPERATIONS CHECKLIST

This appendix consists of an operations checklist of procedures to be followed before, during, and after use of the MK 19.

WEAPON PART	BEFORE	DURING	AFTER
1. Bore	Ensure it is clear and clean.	________	Clean and oil lightly.
2. Moving parts	Oil lightly and test for worn or broken parts. The moving parts should function without excessive friction.	Lubricate working parts. Observe the functioning of the gun to anticipate failures.	Inspect, clean, and oil lightly.
3. Ammunition	a. Ensure the correct type of ammunition is used. b. Ensure the ammunition is clean and dent free. c. Ensure all ogives are tight. d. Have an adequate supply of ammunition on hand.	Keep correctly aligned in the feedtray. Check resupply. Protect from sun, moisture, and dirt. Watch for link stoppage.	Clean, store carefully, and replenish supply.
4. Top cover	Inspect the top cover for dents or damage.	Keep closed and locked down.	Lube properly after cleaning.
5. Line of fire	Ensure the line of fire is clear of all obstructions.	Cease fire if any obstruction appears in the line of fire.	________

Table B-1. Operations checklist.

APPENDIX C

STOPPAGES, MALFUNCTIONS, IMMEDIATE ACTIONS, AND REMEDIAL ACTIONS

A stoppage is any interruption in the cycle of operation caused by faulty action of the weapon or ammunition. A malfunction is a failure of the weapon to function properly. Neither defective ammunition nor improper operation of the gun by a crewmember is considered a malfunction of the MK 19. This appendix lists the actions to correct common stoppages and malfunctions. For more complete information on troubleshooting common malfunctions and stoppages, refer to TM 9-1010-230-10 Troubleshooting Index.

Section I. STOPPAGES, IMMEDIATE ACTIONS, AND REMEDIAL ACTIONS

A stoppage is cleared through immediate or remedial action. Immediate action is the action taken to reduce the stoppage without investigating the cause. The gunner takes remedial action if immediate action does not end the stoppage. Remedial action has three parts: the gunner first unloads and clears the MK 19, identifies the problem, and then applies corrective action to fix the problem. If the MK 19 still does not fire, or cannot be repaired at the user level, contact the unit armorer.

C-1. IMMEDIATE ACTION

Immediate action is different during peacetime and combat.

a. **Immediate Action During Peacetime and Training.** The gunner does the following immediate action to reduce a stoppage during peacetime.

> **CAUTION**
> Both charger handles must be forward and up for firing. If either handle is down, the gun will not fire.

(1) The gunner ensures the safety of other personnel and clears the weapon of ammunition by performing the following steps in sequence:

(a) Clear the area of personnel.

(b) Pull bolt to the rear. If the bolt does not go to the rear, go to BOLT JAMMED.

(c) Catch the live round as it is ejected.

(d) Push the charging handles forward and up.

(e) Put the gun on S (SAFE) and check for any bore obstructions.

(f) If bore is clear, move safety switch from the S (SAFE) to the F (FIRE) position, and attempt to fire.

(g) If nothing happens, put gun on S (SAFE) and wait 10 seconds.

(h) Pull the bolt to the rear.

(i) Catch the live round as it is ejected.

(j) Open the top cover and clear the ammunition.

(k) Check bore for any obstructions.

(2) Charge the MK 19 in the normal manner and check to see the primary pawls have clicked up behind the cartridge in front of the bolt face and that the secondary pawls have clicked up behind the next round before firing. The feed pawls should click up within the 1-inch of charging handle travel.

(3) If the primary and secondary pawls do not click up within the last 1-inch of charging travel, follow these procedures:

- Turn the MK 19 into unit level maintenance.
- Record ammo lot number, type of ammo, number of rounds fired, serial number of the MK 19 and indicate whether ammo is linked with one-piece or two-piece links.

WARNING

1. Do not use combat misfire procedures during peacetime or training. Serious injury can result if precautions are not observed.
2. Do not relink or fire ammunition that has been cycled through the weapon.

b. **Immediate Action During Combat.** To correct a misfire during combat, perform the following immediate actions:

(1) Press charger handle locks and rotate charger handles down.

(2) Pull charger handles to the rear until the bolt sears.

(3) Push charger handles forward and rotate charger handles up and lock.

(4) Relay and fire.

(5) Turn in live rounds that cycled through the weapon as instructed by current directives.

C-2. REMEDIAL ACTIONS TO CLEAR STOPPAGES

If the weapon still does not fire, the gunner clears the weapon, identifies the problem and then takes corrective action to fix or report the problem.

a. **Bad Ammunition.** Bad ammunition can cause a stoppage and is checked first.

(1) ***Primer on the Ejected Round is Indented.*** Clear the weapon and check to see if the primer is indented. If the primer is indented, it is a bad round and dispose of it as authorized. Reload the weapon and continue with the exercise.

(2) ***The Primer is Not Indented.*** Clear the weapon and check to see if the primer is indented. If the primer is not indented, then the firing pin can be bad. Turn the MK19 into unit level maintenance.

(3) ***Round on Bolt Face.*** Clear the weapon. If the round remains on the bolt face, remove the round and place the weapon on S (SAFE) with the bolt to the rear. Check the primer and see if it is indented. See the aforementioned steps for the corrective action.

b. **Ammunition Jammed in Feeder.** A misfire might occur due to ammunition jammed in feeder. Clear the weapon and put gun on S (SAFE).

(1) ***Feed Throat Improperly Attached.*** Check for proper attachment of the feed throat. If the feed throat is attached incorrectly, then re-attach it and reload. Try to recharge and fire the weapon. If charging is not possible, go to BOLT JAMMED.

(2) ***Rounds Crooked or Not Seated Firmly.*** Clear the weapon. If the rounds are crooked or not seated firmly, remove the linked ammunition from feeder. Ensure the link band is even and adjacent to copper band all around ammo. Reload and continue with the exercise.

(3) ***Broken Link.*** A broken link can cause the weapon to stop firing. Clear the weapon. If the link is inside the weapon, remove the link. If it is still on the round, then remove the round and dispose it as authorized. Reload and continue with the exercise.

(4) ***Link Off Rotating Band.*** Clear the weapon. Remove round from belt, dispose of round as authorized. Reload and continue with the exercise.

(5) ***Female Link Not First.*** A round loaded with the female link not first will cause a stoppage. Clear the weapon. Reload the weapon and ensure that the female link is first.

c. **Bad Firing Pin.** A stoppage might occur due to a bad firing pin or firing pin spring. A possible bad firing pin can be determined while checking for a bad round.

(1) ***Primer on Ejected Round Not Indented.*** Clear the weapon. If the primer on the ejected round is not indented, then the firing pin may be bad. Report the defect to the unit armorer.

(2) ***Firing Pin Tip Fails to Spring Forward.*** Clear the weapon. If the firing pin tip fails to spring forward then the firing pin spring can be bad. Report the defect to the unit armorer.

d. **Defective Feeder or Feed Slide Assembly.** A misfire might occur due to a defective feeder or feed slide assembly.

CAUTION
Clear feeder of ammunition. Make sure gun is on S (SAFE).

(1) ***Broken or Worn Feed Pawls.*** Clear the weapon and check to see if the primary and secondary pawls are broken, worn, without spring action, or the pin is missing or dislodged. If they are, turn the MK 19 into unit level maintenance.

(2) ***Badly Worn or Gouged Link Guides.*** Clear the weapon and check to see if the guides are worn or gouged. If they are, turn the MK 19 into unit level maintenance.

(3) ***Binding Feed Slide Assembly.*** Clear the weapon. Remove the feed slide assembly and tray. Clean, inspect and lubricate feed assembly and tray. If the feed slide assembly continues to bind, report the defects to armorer or support maintenance.

e. **Bolt Won't Pick Up the Round.** A misfire might occur because a bolt won't pick up the round.

(1) ***Charging Handles are Down.*** Clear the weapon. Raise the charging handles before firing. If the weapon still does not fire, report defects to the armorer or support maintenance.

(2) ***Dirt-Clogged, Weak, or Damaged Extractors.*** Clear the weapon and remove obstruction. Clean, lubricate, and reload weapon. Continue with the exercise.

(3) ***Feed Slide Out of Adjustment.*** Clear the weapon. The feed slide is out of adjustment if:

- Round fails to feed.
- Round drops.
- Extractors won't pick up round.
- Round stubs on face of chamber.

Report defects to armorer or support maintenance.

f. **Bolt Drops a Round Before Firing.** A stoppage may occur because a bolt drops a round before firing.

(1) ***Weak or Damaged Extractors or Bolt Fingers.*** Clear the weapon and check for weak or damaged extractors or bolt fingers. Report any defects to the armorer or support maintenance.

(2) ***Binding Receiver Rails.*** A misfire might occur because of binding receiver rails. Clear the weapon and pull the bolt to the rear without stopping or pausing. Ease it forward (holding onto one charging handle while you press the trigger) and check for binding. Place weapon on S (SAFE) and remove backplate pin. Lift up slightly on the backplate assembly and pull the bolt and backplate assembly to the rear. Remove the chargers. Check the charger rails and receiver rails for burrs. Report defects to armorer or support maintenance.

WARNING

Be sure bolt is forward before removing backplate pin assembly. If not, serious injury could result.

CAUTION

When installing bolt and backplate, ensure cocking lever is in the forward position. Damage to equipment could result if cocking lever is to the rear.

g. **Bad Cocking Lever.** Clear the weapon and remove the bolt and backplate assembly. Examine the cocking lever on the left side of the bolt for wear or damage. Report defects to armorer or support maintenance.

h. **Unknown Cause.** If the MK 19 stops firing and the procedures above do not identify and correct the problem, then turn it into the armorer or support maintenance.

Section II. MALFUNCTIONS, EMERGENCY ACTIONS AND REMEDIAL ACTIONS

A malfunction is a failure of the weapon to function properly. Neither defective ammunition nor improper operation of the gun by a crewmember is considered a malfunction of the MK 19. The two most common MK 19 malfunctions are sluggish action and runaway gun.

CAUTION

Do not try to break the ammunition belt; injury could result. Lower one charging handle to stop the gun.

WARNING

1. Before performing any non-firing procedure, ensure the weapon is clear of any ammunition.
2. Ensure all ammunition and non-essential personnel are at least 65 meters to the rear of the weapon.
3. If the bolt jams during firing, do not let the bolt slam forward as the top cover is being opened because a round could fire.
4. Be sure to put bolt in forward position before removing the backplate pin assembly. Serious injury can result if the pin assembly is removed with the bolt to the rear.
5. Be prepared to catch dropped/ejected live round from weapon.

C-3. SLUGGISH OR ERRATIC FIRING

Excessive friction from dirt, carbon buildup, lack of lubrication, or burred parts usually causes sluggish or erratic action. Once the gunner realizes that the gun is operating sluggishly or erratically, he should cease-fire and clear the weapon. The two primary corrective actions are cleaning or turning the weapon into the armorer. There are several reasons why a weapon may exhibit sluggish or erratic firing.

a. **Dirty Bore or Chamber.** A dirty bore or firing chamber can cause sluggish or erratic firing. Clear the weapon and clean its bore and chamber. Reload and continue with the exercise.

b. **Recoil Springs or Guide Rods.** Clear the weapon and remove the bolt and backplate assembly from gun. Push against springs to test for weakness. Note bent rods. Report defect to armorer or support maintenance.

c. **Bolt Sear Timing Adjustment.** Clear the weapon and turn it into direct support maintenance.

C-4. RUNAWAY GUN

A runaway weapon continues to fire after the trigger has been released. Worn parts or short recoil of the bolt assembly may cause a runaway gun. Consider the amount of ammunition left and the type of MK 19 mount used when determining the best way to stop the weapon.

a. **Emergency Action.** This paragraph addresses emergency actions that must be taken to stabilize uncontrolled automatic fire. The following procedures should be taken to control a runaway gun:

(1) If ammunition is not low and the MK 19 is used in the free gun mode, keep rounds on target until the all the rounds on the belt have been fired.

(2) If the MK 19 is mounted on either the M3 tripod or on a vehicle with the T&E mechanism attached, hold the grip with one hand. At the same time, press the charger handle lock and lower one charger handle. This action interrupts the cycle of operation, causing the MK 19 to cease firing. The gunner therefore:

(a) Keeps gun pointed downrange and slightly elevated.

(b) Presses charger handle locks.

(c) Lowers the charging handle(s) so the gun will stop firing.

(d) Places the gun on S (SAFE).

WARNING

Never try to break the ammo belt with your hands. Injury could result. Lower one charger handle to stop gun from firing.

b. Once the gun has stopped firing, clear the weapon and report the condition to the armorer or support maintenance.

C-5. BORE OBSTRUCTION

A bore obstruction is indicated by a muffled sound of round firing, excess smoke out of the chamber, and/or excess debris/gases below the gun.

a. **Emergency Action.** The emergency actions for an obstructed bore are the following:

(1) Place weapon on S (SAFE).

(2) Notify range safety officer (during training).

(3) Depress feed pawls, release ammunition belt, and clear feed.

(4) Move the ammunition belt and can to a safe area.

(5) Remove the empty case catch bag.

(6) Charge gun and hold bolt to rear.

WARNING

Do not relink or fire any ammunition that has been cycled through the weapon.

(7) Holding the bolt to the rear, insert a cleaning rod through the receiver rail to the top of the shell casing and as close to the face of the bolt as possible.

(8) Place left hand underneath as close to the round as possible. Raise cleaning rod upward forcing the round off the bolt face into the hand. Remove round to designated area for explosive ordinance disposal (EOD).

b. **Correcting an Obstructed Bore.** To correct an obstructed bore follow these procedures:

(1) Place selector lever on F (FIRE) and ease the bolt forward.

(2) Remove the backplate pin bolt, backplate assembly, vertical cam assembly, and primary drive lever.

(3) Check for any type of obstruction.

(4) Check and remove any case or round from bolt face.

(5) Insert bore obstruction detector into bore to check for a live round.

(6) Remove obstruction per round removal procedures (for exact bore obstruction instructions using the round removal tool, refer to WP 0014 00).

C-6. GUN FIRES TOO SOON

If the weapon fires too soon, or when the trigger is not engaged, the gunner applies emergency action and the weapon is taken to support maintenance. The gunner must not continue to fire the weapon. The following emergency actions must be taken if a weapon fires too soon:

> **WARNING**
> **Do not attempt to clear the weapon if the weapon fires too soon. Do not attempt to clear or fire the weapon until it is fixed.**

(1) Cease fire.

(2) Place weapon on S (SAFE).

(3) Clear area of personnel and ammunition.

(4) Notify the range safety officer.

(5) Check barrel for lodged round using the bore obstruction detector (BOD).

(6) After proper personnel have cleared weapon, evacuate weapon to support maintenance.

C-7. BOLT JAMMED

The following actions are taken if the gunner cannot pull the bolt to the rear.

> **WARNING**
> **The following procedures must be performed in sequence to open the top cover. The bolt could spring forward suddenly and fire a round causing severe injury. Be prepared to catch ejected round.**

a. **Emergency Action and the Bolt Can be Pulled to the Rear.** The following emergency actions must be taken if the weapon initially jams but the bolt can then be pulled to the rear.

(1) Put gun on S (SAFE).

(2) Press charger handle locks and rotate charger handles down.

(3) Pull charger handles to the rear as far as possible without stopping or pausing, until bolt locks. Ensure bolt will stay to the rear before releasing charger handles.

(4) Maintain rearward pressure on charging handles while assistant lifts top cover.

(5) Insert cleaning rod section through slot in side of receiver. Prepare to the catch ejected, live round.

(6) Raise cleaning rod to force live round down. Catch live round as it is ejected.

(7) Remove ammo belt from feeder.

(8) Reposition ammo belt in feeder.

(9) Put gun on F (FIRE).

(10) Ride the bolt forward by grasping one charging handle and depressing the trigger.

(11) Ensure feed slide assembly is to the left.

(12) Ensure secondary drive lever is engaged with the feed slide pin. If not, engage forked end with feed slide pin.

(13) Close top cover gently.

(14) Charge weapon and attempt to fire.

(15) If bolt still jams, repeat first seven steps. Put weapon on S (SAFE), and evacuate to support maintenance.

b. **Gunner's Actions.** The gunner performs the following action once emergency actions are completed:

(1) Clear the jam and make sure there is no bore obstruction.

(2) Charge the MK 19 in the normal manner, and check to see if the primary pawls have clicked up behind the cartridge in front of bolt face and the secondary pawls have clicked up behind the next round before firing. The feed pawls should click up within 1-inch of the charging handle travel.

(3) Check the ammo link to ensure it is even and touches the copper band all around the ammo. If the primary and secondary pawls do not click up within the last 1 inch of the charging handle then:

- Turn MK 19 into unit level maintenance.
- Record ammo lot number, type of ammo, number of rounds fired, serial number of the MK 19 and indicate whether ammo is linked with one-piece or two-piece links.

(4) If the MK 19 does not require feed slide adjustment and there appears to be no other deficiencies that would prevent the weapon from firing, it should be turned into the armorer or support maintenance.

C-8. SHORT RECOIL

A short recoil occurs when the bolt does not fully return to the rear.

WARNING

When firing HE or TP ammunition, the gunner should observe downrange and attempt to determine if the round left the barrel and also be alert to these three danger signals:

1. A muffled report from the gun.

2. Smoke and debris from the bottom of the receiver.

3. Failure of the projectile to leave the muzzle.

Any of these three symptoms mean a bore obstruction. Do not attempt to clear a bore obstruction.

a. **Emergency Actions after a Short Recoil.** The gunner does the following actions after a short recoil:

(1) Place weapon on S (SAFE).

(2) Clear area of personnel and ammunition.

(3) Notify range safety officer.

(4) Pull the charger handles to the rear, without stopping or pausing, until the bolt locks. Ensure that the bolt stays to the rear before releasing the charger handles.

(5) Insert cleaning rod section through slot in side of receiver. Prepare to catch the ejected live round.

(6) Raise cleaning rod to force live round down. Catch the live round as it is ejected.

(7) Raise top cover.

(8) Remove ammo belt from feeder.

(9) Check for bore obstruction using bore obstruction detector.

(10) If bore is obstructed, refer to the section above on clearing bore obstructions or refer to WP 0014 00 for round removal procedures.

(11) If there is no obstruction, reposition belt in feeder.

(12) Ensure feed slide assembly is to the left.

(13) Charge weapon and attempt to fire.

b. If a short recoil occurs again, repeat the first eight steps. Put weapon on S (SAFE) and turn it into support maintenance.

C-9. TOP COVER WILL NOT CLOSE

The gunner checks the following so that the top cover can be closed.

a. **Improper Position of the Feed Slide Assembly.** Move feed slide assembly all the way left. The spring should touch the cover.

b. **Bolt is Locked to the Rear.** Ride the bolt forward.

c. **Misaligned Ammunition.** Ensure rounds are straight and firmly seated in the feeder. Ensure links are evenly aligned in the link guide and on rounds. Clean dirt from the feeder.

APPENDIX D

GUNNERY TESTS

There are three gunnery tests: a diagnostic gunnery test, an intermediate gunnery test, and a gunnery skills test. The instructor is responsible for the conduct of all tests and must ensure that sufficient scorecards or answer sheets are available.

D-1. DIAGNOSTIC GUNNERY TEST

This written test, given before and during preparatory training, evaluates trainees' knowledge of the MK 19. It also serves as an evaluation tool for the instruction. Training units prescribe procedures for the test, and announce scores of those who satisfactorily complete the test. If this test is used, it should be given before range firing. To satisfactorily complete the test, the trainee must earn 80 out of 100 points.

		POINTS
1.	Name the five major assemblies of the MK 19.	5
2.	Explain how to clear and disassemble the MK 19.	15
3.	Explain how to assemble the MK 19.	15
4.	Explain how to perform a function check.	5
5.	Explain how to apply immediate action to a runaway gun.	5
6.	Explain how to apply immediate action when a malfunction occurs.	10
7.	Explain one phase of functioning (directed by the examining officer).	10
8.	Explain why the weapon should not be fired with the top cover assembly open.	10
9.	Explain how to inspect ammunition before loading (belt to have at least one each of the following: short round, bent round, loose bullet, broken link or round not pushed fully into its link).	10
10.	Explain how to load to half-load position.	5
11.	Explain and demonstrate sight alignment.	10
	TOTAL	100

D-2. INTERMEDIATE GUNNERY TEST

This written test, given at the end of preparatory marksmanship training, is used to evaluate the instruction as well as the soldier's gunnery skills. This test has 21 questions: 20 multiple-choice and 1 true or false. Each soldier must correctly answer 17 of the 21 in order to move on to the next phase of training. The instructor is responsible for the conduct of the test. The following is a sample of the standard Intermediate Gunnery Test.

INTERMEDIATE GUNNERY TEST

1. What are the MK 19's five major groups (field strip groups)?
   a. Receiver assembly, top cover assembly, sear assembly, bolt assembly, and feed slide assembly.
   b. Top cover assembly, receiver assembly, barrel, sear assembly and feed tray, and bolt and backplate assembly.
   c. Receiver assembly, feed slide assembly and feed tray, top cover assembly, sear assembly, and bolt and backplate assembly.
   d. Sear assembly with chargers, receiver assembly, feed tray assembly, top cover assembly, bolt and backplate assembly.

2. Where on the weapon is the safety located?
   a. On the right side of the receiver assembly.
   b. On the left side of the receiver assembly.
   c. On top of the receiver assembly.
   d. On the sear assembly at the bottom rear of the receiver assembly.

3. In a nonfiring situation to clear the weapon, in which order are the steps below taken?
   a. 2, 4, 3, 1.
   b. 1, 3, 4, 2.
   c. 3, 1, 2, 4.
   d. 1, 2, 4, 3.

   1. Ensure that the safety switch is in the S (SAFE) position, and the bolt is forward. Open the top cover by rotating the top cover latch counterclockwise. Lift the cover.
   2. Return the bolt to the full forward position. Rotate the charger handles upward.
   3. Inspect the feed tray and extractors on the top of the bolt to make sure no round is in the "pickup" position.
   4. Unlock and pull the charging handles to the rear two to three inches. Inspect the face of the bolt and the chamber.

4. In which substance(s) should you never immerse the bolt and backplate assembly?
   a. Cleaning solvent.
   b. Lubricant.
   c. Bore cleaner.
   d. All of the above.

5. Under which conditions should you apply a generous second coat of lubricant when cleaning the weapon?
   a. Hot, wet, salt air, sea spray.
   b. Hot, dry, sandy, dusty.
   c. Cold, icy, snow.
   d. All of the above.

6. Under which conditions should you use a second coat of lubricant?
   a. Hot, wet, salt air, sea spray.
   b. Hot, dry, sandy, dusty.
   c. Cold, icy, snow.
   d. All of the above.

7. During reassembly, what is the first major group you should place back onto the weapon?
   a. Sear assembly.
   b. Bolt and backplate assembly.
   c. Top cover assembly.
   d. Receiver assembly.

8. To ensure the primary and secondary drive levers align properly when you close the top cover of the weapon, in what positions should you place the feed slide assembly and bolt?
   a. Left, to the rear.
   b. Right, to the rear.
   c. Left, forward.
   d. Right, forward.

9. What is the maximum range of the weapon, in meters?
   a. 1,500.
   b. 2,212.
   c. 2,000.
   d. 1,800.

10. What is the maximum effective range of the weapon, in meters, for point targets?

a. 1,500.
b. 2,212.
c. 2,000.
d. 1,800.

11. What is the cyclic rate of fire for the weapon, in rounds per minute?
   a. 225 to 275.
   b. 275 to 325.
   c. 325 to 375.
   d. 375 to 425.

12. What is the main characteristic(s) of the MK 19?
   a. Fully automatic.
   b. Blowback operated.
   c. Air-cooled.
   d. All of the above.

13. Which mount(s) should you avoid attaching the weapon to?
   a. M3 tripod without the T&E mechanism.
   b. M3 tripod with the T&E mechanism.
   c. M4 pedestal.
   d. All of the above.

14. How much armor plating, in inches, can a M430 HEDP round penetrate when fired at zero degrees obliquity?
   a. 1.
   b. 2.
   c. 3.
   d. 4.

15. Within how many meters must the 40-mm HE round impact to cause casualties to exposed enemy personnel?
   a. 5.
   b. 10.
   c. 15.
   d. 20.

16. What steps must you take to stop a runaway weapon?
   a. Release the trigger and step away from the weapon.
   b. Grab and twist the ammunition belt to break it off.

c. Unlock and lower one charging handle while keeping control of the weapon.
d. None of the above.

17. When loading the weapon, place the female link end of the ammunition belt through the feed throat and draw the ammunition into the weapon feed area.
    a. True.
    b. False.

18. What term describes an interruption in the functioning cycle caused either by faulty action of the gun or by faulty ammunition?
    a. Blockage.
    b. Breakage.
    c. Malfunction.
    d. Stoppage.

19. What steps must you take to perform corrective action?
    a. Unload and clear the weapon.
    b. Inspect the weapon and ammunition to find the cause of the problem.
    c. Correct the problem, reload, and try to fire the weapon.
    d. All of the above.

20. When you engage a target, how many rounds should you fire in each burst?
    a. 3 to 5.
    b. 5 to 8.
    c. 7 to 10.
    d. 9 to 12.

21. In which position(s) must you place both charger handles before the weapon will fire?
    a. Forward and down.
    b. Forward and up.
    c. To the rear and down.
    d. None of the above.

D-3. GUNNERY SKILLS TEST

This is a guide for unit commanders to standardize and maintain MK 19 gunnery skills. It is a nonfiring, hands-on test given at the end of MK 19 training. The commander may authorize indoor testing if the facilities are available. Seven stations are used for the skill tests. The test is graded on a GO (pass) or NO-GO (fail) basis.

a. **Instructions for the Gunner's Examination.** The following is a guide for giving gunners the skill test.

(1) To organize for testing, assemble the entire unit in one area and orient them as follows:

> "During the next _____ hours, you will take part in a performance-type examination to test your knowledge of the MK 19 machine gun. Stay with your assigned group during the entire examination. Initially, we will assign each group to a specific station. At each station, an assistant instructor will explain exactly what task or tasks you will perform. After beginning the task, if you do not know a step or steps, ask the assistant instructor for help. He will tell you how to perform the step. He will deduct the necessary points from your grade, and then you may continue to work. You must achieve a score of seventy percent in order to pass the test."

(2) Organize the unit into seven equal groups that rotate from station to station until all individuals have been tested. Gunners must wear their helmet and load-bearing equipment (LBE) to be tested. Assign enough assistant instructors to each station to grade and critique each gunner's performance. For ease in marking, give score sheets to each grader and collect the sheets immediately after the test.

(3) At the end of testing for each group, have each grader assemble the soldiers and give them a thorough five-minute critique.

b. **Testing Stations.** There are seven stations for gunnery skills testing. All stations except one, six, and seven have intermediate training objectives.

c. **Scoring.** For tasks at stations one, six, and seven, score the soldier GO if all steps are passed (P). Score the soldier NO-GO if any steps are failed (F). If the soldier fails any steps, show him what was done wrong and how to do it correctly. For stations two through five, score the soldier GO on a task if he receives no more than one intermediate training objective NO-GO.

STATION ONE

TASK: Identify the characteristics of the MK 19.

CONDITIONS: In any environment, the gunner is given a complete tripod-mounted MK 19.

STANDARDS: The gunner will know characteristics of the MK 19.

PERFORMANCE MEASURES	GO	NO-GO
1. General description (nomenclature).		
2. Weapons sight.		
3. Weight of gun.		
4. Maximum effective range for a point target.		
5. Maximum effective range for an area target.		
6. Cyclic rate of fire.		
7. Muzzle velocity.		
8. Types of ammunition.		
9. Weight of the MK 64, MOD 7, gun mount.		
10. Types of mounts.		

STATION TWO

Station Two has seven intermediate training objectives.

TASK: Perform operator's maintenance on the MK 19.

CONDITIONS: In any environment, the gunner is given a complete MK 19, cleaning equipment, tools, workspace, and TM 9-1010-230-10.

STANDARDS: Disassemble the MK 19 within four minutes. Use the cleaning materials as specified in the operator's TM to clean the MK 19. Inspect, lubricate, and assemble the MK 19 within four minutes, and perform a function check.

Intermediate Training Objective 1: The gunner must disassemble the MK 19.

CONDITIONS: The gunner is given a location in which to work, an assembled MK 19, weapon-cleaning equipment, lubricant, and TM 9-1010-230-10.

STANDARDS: The gunner must disassemble the MK 19 within four minutes.

PERFORMANCE MEASURES	GO	NO-GO
1. Clear the gun.		
2. Remove the bolt and backplate assembly.		
3. Remove the secondary drive lever.		
4. Remove the feed slide assembly.		
5. Remove the top cover assembly.		
6. Remove the feed tray.		
7. Remove the primary drive lever and vertical cam.		
8. Remove the alignment guide assembly.		
9. Remove the ogive plunger.		
10. Remove the round-positioning block.		
11. Remove the charger assemblies (left hand [LH], right hand [RH]).		
12. Remove the sear assembly.		
13. Complete within four minutes.		

Intermediate Training Objective 2: Clean the MK 19.

CONDITIONS: The gunner is given a disassembled MK 19, with the proper cleaning equipment.

STANDARDS: The gunner must clean the MK 19.

PERFORMANCE MEASURES	GO	NO-GO
1. Receiver assembly:		
a. Wipe or brush away dirt from all parts, especially the interior of the receiver housing, receiver rails, and feeder area.		
b. Use a rag to apply the cleaning solvent.		
c. Swab out the bore and chamber using a bore brush and RBC.		
d. Wipe dry.		
2. Sear assembly:		
a. Wipe or brush away dirt.		
b. Use a rag to apply the cleaning solvent.		
c. Wipe dry.		
3. Alignment guide assembly:		
a. Wipe or brush away dirt.		
b. Soak assembly in cleaning solvent.		
c. Wipe dry.		
4. Ogive plunger assembly:		
a. Wipe or brush away dirt.		
b. Use a rag to apply the cleaning solvent.		
c. Wipe dry.		
5. Round-positioning block:		
a. Wipe or brush away dirt.		
b. Soak in cleaning solvent.		
6. Charger assemblies:		
a. Wipe or brush away dirt.		
b. Use a rag to apply cleaning solvent.		
c. Wipe dry.		
7. Vertical cam assembly:		
a. Wipe or brush away dirt.		

PERFORMANCE MEASURES	GO	NO-GO
b. Soak in cleaning solvent.		
c. Wipe dry.		
8. Primary drive lever:		
a. Wipe or brush away dirt.		
b. Soak in cleaning solvent.		
c. Wipe dry.		
9. Secondary drive lever:		
a. Wipe or brush away dirt.		
b. Soak in cleaning solvent.		
c. Wipe dry.		
10. Feed slide assembly:		
a. Wipe or brush away dirt.		
b. Use a rag to apply the cleaning solvent.		
c. Wipe dry.		
11. Feed tray:		
a. Wipe or brush away dirt.		
b. Soak in cleaning solvent.		
c. Wipe dry.		
12. Top cover assembly:		
a. Wipe or brush away dirt.		
b. Use a rag to apply the cleaning solvent.		
c. Wipe dry.		
13. Bolt and backplate assembly:		
a. Wipe or brush away dirt.		
b. Use a rag to apply the cleaning solvent.		
c. Wipe dry.		
14. Use only authorized cleaning materials.		

Intermediate Training Objective 3: Inspect the MK 19.

CONDITIONS: The gunner is given a disassembled and cleaned MK 19.

STANDARDS: The gunner must inspect the MK 19.

PERFORMANCE MEASURES	GO	NO-GO
1. Receiver assembly: a. Inspect the receiver housing for cracks (even hairline), rust, burrs. b. Inspect the receiver rails for burrs, deformity, or unusual wear. c. Inspect the feeder pawls for spring action, burrs, or broken pawls. d. Inspect the barrel for carbon buildup or pitting in the bore or chamber. e. Inspect the flash suppressor for dents, cracks, or loose pin. f. Inspect the rear sight for rust, and the sight scale for readability.		
2. Sear assembly: Inspect for burrs, rust, or wear on sear face, or binding in the safety switch movement.		
3. Alignment guide assembly: a. Inspect for deformed or cracked spring. b. Inspect for cracks around the pin. c. Inspect for loose pin. d. Inspect for burrs, dents, and cracks.		
4. Ogive plunger assembly: Inspect for cracks in the housing.		
5. Round-positioning block: a. Inspect for burrs, dents, or cracks. b. Inspect for broken, deformed, or missing springs. c. Inspect for bent, cracked, or broken parts.		
6. Charger assemblies (LH, RH): a. Inspect for burrs on grooved edges.		

PERFORMANCE MEASURES	GO	NO-GO
b. Inspect for chipped, cracked, broken, or missing handle assemblies.		
c. Inspect for spring tension in the charger handle lock.		
d. Inspect for spring tension in the lock plunger.		
e. Inspect for bent charger housing.		
7. Vertical cam assembly:		
a. Inspect for burrs or buckling.		
b. Inspect for scratches, chips, or aluminum buildup on the chromed edge.		
c. Inspect for lock binding.		
8. Primary drive lever: Inspect for burrs around the pivot post.		
9. Secondary drive lever:		
a. Inspect for burrs, cracks, chips, or buckling.		
b. Inspect for missing retaining pin.		
10. Feed slide assembly:		
a. Inspect for burrs or cracks.		
b. Inspect for broken feed pawls, binding, or no spring action.		
c. Inspect for deformed shuttle-spring tip.		
d. Inspect for loose mounting screws.		
11. Feed tray:		
a. Inspect for burrs or cracks.		
b. Inspect for lack of spring action.		
12. Top cover assembly:		
a. Inspect for burrs, cracks, or rust.		
b. Inspect for loose or missing latch.		
c. Inspect for bent pins, and for missing or broken cross pins.		
13. Bolt and backplate assembly:		
a. Inspect overall for burrs, rust, or cracks.		

PERFORMANCE MEASURES	GO	NO-GO
b. Inspect handles for cracks. c. Inspect for broken or bent recoil springs. d. Inspect for bent or broken guide rods. e. Inspect for missing safety wire. f. Inspect for broken or worn cocking lever. g. Inspect for cracked or chipped bolt sear. h. Inspect for cracked or binding cam followers. i. Inspect for cracked or chipped bolt fingers. j. Inspect for chipped or broken firing pin tip. k. Inspect for chipped, cracked, or missing extractors. l. Inspect for bent or cracked backplate or missing retaining ring.		

Intermediate Training Objective 4: Lubricate the MK 19.

CONDITIONS: The gunner is given a disassembled, cleaned, and inspected MK 19, and lubricant and brush.

STANDARDS: The gunner must lubricate the weapon.

PERFORMANCE MEASURES	GO	NO-GO
1. Receiver assembly: Apply a light coat of lubricant on the receiver rails, feeder, and feeder pawls.		
2. Sear assembly: Apply a light coat to all parts.		
3. Alignment guide assembly: Apply a light coat to all parts.		
4. Round-positioning block: Apply a light coat to the springs.		
5. Charger assemblies (LH, RH): Apply a light coat to all parts, especially the grooved edges of the rails.		
6. Vertical cam: Apply a light coat to all parts, especially the chromed edge.		
7. Primary drive lever: Apply a light coat to all parts, especially the pivot posts.		
8. Secondary drive lever: Apply a light coat to all parts, especially the pivot post and forked end.		
9. Feed slide assembly: Apply a light coat to all parts, especially the feed pawls.		
10. Feed tray: Apply a light coat to all parts, especially the pawl and guide rails.		
11. Top cover assembly: Apply a light coat to all parts, especially the latch, the pivot post hole, and the pin holes.		
12. Bolt and backplate assembly: Apply a light coat to all parts, especially the bolt face, bolt sear, guide rods, recoil springs, bolt rails, cam followers, and cocking lever.		

Intermediate Training Objective 5: Assemble the MK 19.

CONDITIONS: The gunner is given a disassembled, cleaned, inspected, and lubricated MK 19.

STANDARDS: The gunner must assemble the weapon in the proper sequence within four minutes.

PERFORMANCE MEASURES	GO	NO-GO
1. Assemble the sear assembly.		
2. Assemble the charger assemblies (LH, RH).		
3. Assemble the round-positioning block.		
4. Insert the ogive plunger assembly.		
5. Insert the alignment guide assembly.		
6. Attach the vertical cam.		
7. Attach the primary drive lever.		
8. Attach the feed tray.		
9. Attach the feed slide assembly.		
10. Attach the top cover assembly.		
11. Attach and engage the secondary drive lever.		
12. Insert the bolt and backplate assembly.		
13. Complete within four minutes.		

Intermediate Training Objective 6: Perform a function check.

CONDITIONS: The gunner is given an assembled MK 19.

STANDARDS: The gunner must perform a function check on the weapon.

PERFORMANCE MEASURES	GO	NO-GO
1. Close the top cover.		
2. Charge the weapon.		
3. Lock the charging handles forward.		
4. Place the gun on S (SAFE).		
5. Press the trigger.		
6. If the weapon fires, notify the supervisor.		
7. If the weapon does not fire, continue.		
8. Move the safety to F (FIRE).		
9. Press the trigger. Bolt should spring forward.		
10. Open the top cover.		
11. Check the tip of the firing pin. It should be exposed.		
12. Move the feed slide to the left.		
13. Close the top cover.		
14. Move the safety to S (SAFE).		

Intermediate Training Objective 7: Clean, inspect, and lubricate the MK 64, MOD 7, gun cradle.

CONDITIONS: The gunner is given a gun cradle, TM 9-1010-231-13&P, and lubricant.

STANDARDS: The gunner must clean, inspect, and lubricate the gun cradle.

PERFORMANCE MEASURES	GO	NO-GO
1. Clean the gun cradle.		
a. Brush or wipe away dirt.		
b. Use a rag to apply cleaning solvent.		
c. Wipe dry.		
2. Inspect the gun cradle for:		
a. Bent or cracked retaining pins.		
b. Damaged threads on depression stop.		
c. Damage to mounting surface on the gun cradle.		
d. Loss of mounting hardware (nuts, bolts, pins, or chains).		
e. Depression stop in place (training use only).		
f. Frozen pintle lock.		
g. Missing cotter pins.		
3. Lubricate the gun cradle. Apply a light coat to all pins and pivot points.		

STATION THREE

Station Three has four intermediate training objectives.

TASK: Load, unload, correct malfunctions, and clear the MK 19.

CONDITIONS: The gunner is given a location in which to work, a complete M3 tripod-mounted MK 19, and 10 rounds of linked 40-mm ammunition (dummy).

STANDARDS: The gunner must load, unload, and clear the MK 19, and take immediate and remedial action according to TM 9-1010-230-10 and this training program.

Intermediate Training Objective 1: Clear the MK 19.

CONDITIONS: The gunner is given an M3 tripod-mounted weapon in a designated location and 10 rounds of linked 40-mm ammunition (dummy).

STANDARDS: The gunner must clear the MK 19.

PERFORMANCE MEASURES	GO	NO-GO
1. Place the gun on S (SAFE).		
2. Lock the bolt to the rear.		
3. Remove the live round or spent case from the bolt face.		
4. Catch the live round as it falls out.		
5. Open the top cover.		
6. Remove the rounds from the feeder.		
7. Move the safety to F (FIRE).		
8. Ride the bolt forward.		
9. Move the safety to S (SAFE).		
10. Move the feed slide assembly to the left.		
11. Close the top cover.		

Intermediate Training Objective 2: Load the MK 19.

CONDITIONS: The gunner is given an M3 tripod-mounted weapon in a designated location and 10 rounds of linked 40-mm ammunition (dummy).

STANDARDS: The gunner must load the weapon.

PERFORMANCE MEASURES	GO	NO-GO
1. Place the weapon on S (SAFE).		
2. Ride the bolt forward.		
3. Return the charger assemblies to the original upright position.		
4. Open the top cover.		
5. Slide the rounds, female link first, through the feed throat.		
6. Insert the first round into the feeder and across the first pawl.		
7. Move the feed slide assembly to the left.		
8. Close the top cover.		
9. Charge the weapon.		
10. Lock the charger handles forward.		
11. Move the safety to F (FIRE).		
12. Press the trigger (bolt goes forward).		
13. Move the safety to S (SAFE).		
14. Charge the weapon (live round on bolt face).		
15. Lock the charger handles forward.		

Intermediate Training Objective 3: Correct malfunctions.

CONDITIONS: The gunner is given an M3 tripod-mounted weapon in a designated location, 10 rounds of linked 40-mm ammunition (dummy), and a simulated stoppage.

STANDARDS: The gunner must correct the malfunction.

PERFORMANCE MEASURES	GO	NO-GO
1. In a training situation, wait ten seconds after any stoppage.		
2. Charge the weapon and try to fire.		
3. If the weapon fails to fire, recharge and try to fire. If the weapon fails to fire (or when told that the gun did not fire) after the second attempt, take remedial action:		
4. Unload and clear the weapon.		
5. Inspect the weapon and ammunition and try to find the cause of the stoppage.		
6. If possible, correct the cause of the stoppage, reload, and try to fire the weapon.		

Intermediate Training Objective 4: Stop a runaway gun.

CONDITIONS: The gunner is given an M3 tripod-mounted weapon in a designated location, ten rounds of linked 40-mm ammunition (dummy), and a simulated runaway gun.

STANDARDS: The gunner must stop the runaway gun.

PERFORMANCE MEASURES	GO	NO-GO
Lower one of the charging handles.		

STATION FOUR

Station Four has four intermediate training objectives.

TASK: Mount and dismount the MK 19 onto one of the following: M3 tripod, M4 pedestal on a 1/4-ton vehicle, HMMWV with weapon platform, or commander's cupola on a M113.

CONDITIONS: The gunner is given a complete MK 19, a complete MK 64, MOD 7, gun cradle, a location in which to work, TM 9-1010-231-13&P, and one or more of the following: M3 tripod, 1/4-ton vehicle with an M4 pedestal, HMMWV with weapon platform, or M113.

STANDARDS: The gunner must mount and dismount the MK 19 on the gun cradle, mount the gun on other mounts as directed. The time limit for moving the MK 19 from one mount to another is four minutes.

TRAINING NOTE: When the MK 19 is mounted on a unit's organic carrier or the M3 tripod, it is called a weapon system. The weapon system is mounted using the three-man concept. The gunner is responsible for ensuring that it is mounted correctly, with the aid of the other two-team members. The steps for securing the MK 19 to the gun cradle are the same. The MK 19 is moved from one mount to another with the MK 64, MOD 7, gun cradle attached.

Intermediate Training Objective 1: Mount the MK 19 on the M3 tripod.

CONDITIONS: The gunner is given a location in which to work, an M3 tripod with T&E mechanism, the gun cradle, and TM 9-1010-231-13&P.

STANDARDS: The gunner must mount the gun cradle on the M3 tripod.

PERFORMANCE MEASURES	GO	NO-GO
MOUNTING:		
1. Set up the M3 tripod.		
2. Secure the MK 64, MOD 7, gun cradle onto the M3 tripod.		
3. Lift the gun into the gun cradle.		
4. Slide the front portion of the receiver into the gun cradle.		
5. Secure the rear of the gun to the gun cradle.		
6. Attach the T&E mechanism to the M3 tripod.		
7. Attach the T&E mechanism to the gun cradle.		
8. Attach the feed throat.		

PERFORMANCE MEASURES	GO	NO-GO
DISMOUNTING: 9. Remove the T&E mechanism retaining pin. 10. Insert the gun cradle stow pin (left side). 11. Unlock the M3 tripod pintle lock release cam. 12. Lift the MK 19, with the MK 64, MOD 7, gun cradle attached, from the M3 tripod.		

Intermediate Training Objective 2: Mount and dismount the MK 19 from the M4 pedestal.

CONDITIONS: The gunner is given a location in which to work, a vehicle with an M4 pedestal mount, an M3 tripod-mounted MK 19, the T&E mechanism, pintle adapter, ammunition container bracket, and tools.

STANDARDS: The gunner must remove the MK 19 from the M3 tripod, mount it on the M4 pedestal within four minutes, and dismount it from the M4 pedestal within four minutes.

PERFORMANCE MEASURES	GO	NO-GO
MOUNTING:		
1. Remove the T&E mechanism-retaining pin.		
2. Insert the cradle stow pin (left side).		
3. Unlock the M3 tripod pintle lock lever.		
4. Move the MK 19 to the M4 pedestal.		
5. Loosen the locking lever on the M4 pedestal.		
6. Insert the pintle adapter into the top part of the M4 pedestal.		
7. Remove the quick-release pin from the pintle adapter.		
8. Tighten the locking lever on the M4 pedestal.		
9. Lift up the MK 19 and insert the gun cradle pintle into the top of the pintle adapter.		
10. Insert the quick-release pin into the pintle adapter from either side.		
11. Attach the T&E mechanism to the gun mount.		
12. Center the elevation handwheel.		
13. Remove the stow pin.		
14. Attach the T&E mechanism clamp to the M4 pedestal.		
15. Attach the two support clamps to the M4 pedestal.		
16. Attach the ammunition container bracket (right side).		
17. Attach the empty case catch bag.		
DISMOUNTING:		
18. Remove the ammunition container bracket (right side).		
19. Remove the pin securing the T&E mechanism to the gun cradle.		
20. Insert the gun cradle stow pin.		
21. Remove the quick-release pin from the pintle adapter.		

PERFORMANCE MEASURES	GO	NO-GO
22. Lift the MK 19 from the M4 pedestal.		
23. Loosen the locking lever on the M4 pedestal.		
24. Lift the pintle adapter from the M4 pedestal.		

Intermediate Training Objective 3: Mount and dismount the MK 19 from the HMMWV weapon platform.

CONDITIONS: The gunner is given a mounted MK 19, a location in which to work, a HMMWV with weapon platform, pintle adapter assembly, T&E mechanism, and TM 9-1010-231-13&P.

STANDARDS: The gunner must remove the MK 19 from its mount and install it on the HMMWV weapon platform within four minutes.

PERFORMANCE MEASURES	GO	NO-GO
MOUNTING:		
1. Remove the MK 19 from its mount.		
2. Loosen the locking screws on the HMMWV pedestal.		
3. Insert the pintle adapter into the HMMWV pedestal.		
4. Tighten the locking screws.		
5. Remove the quick-release pin from the pintle adapter.		
6. Lift the MK 19, with MK 64, MOD 7, gun cradle attached to the pintle adapter and insert the gun cradle pintle into the pintle adapter.		
7. Insert the pintle adapter quick-release pin.		
8. Attach the train and elevating assembly to the gun cradle.		
9. Center the elevating handwheel.		
10. Remove the gun cradle stow pin.		
11. Attach the train and elevating mechanism to the HMMWV pedestal.		
12. Attach the ammunition container bracket.		
DISMOUNTING:		
13. Remove the ammunition container bracket.		
14. Remove the train and elevating mechanism from the HMMWV pedestal.		
15. Remove the T&E mechanism from the gun cradle.		
16. Insert the gun cradle stow pin.		

PERFORMANCE MEASURES	GO	NO-GO
17. Remove the quick-release pin from the pintle adapter assembly.		
18. Lift the MK 19 from the HMMWV pedestal.		
19. Remove the pintle adapter from the HMMWV pedestal.		

Intermediate Training Objective 4: Mount and dismount the MK 19 from the commander's cupola of an M113.

CONDITIONS: The gunner is given a mounted MK 19, a location in which to work, an M113 with the commander's cupola, pintle adapter, ammunition bracket, and TM 9-1010-231-13&P.

STANDARDS: The gunner must remove the MK 19 from its mount and install it on the commander's cupola of the M113 within four minutes.

PERFORMANCE MEASURES	GO	NO-GO
MOUNTING:		
1. Insert the pintle adapter into the mount receptacle.		
2. Remove the pintle adapter quick-release pin.		
3. Lift the MK 19 to the pintle adapter and insert the cradle pintle into the pintle adapter.		
4. Insert the pintle adapter quick-release pin.		
5. Attach the ammunition bracket.		
DISMOUNTING:		
6. Insert the gun cradle stow pin.		
7. Remove the ammunition bracket.		
8. Remove the pintle adapter quick-release pin.		
9. Lift the MK 19 from the commander's cupola.		

STATION FIVE

TASK: Identify characteristics of a defensive position for the MK 19.

CONDITIONS: The gunner is given an M3 tripod, a complete MK 19, a location in which to work, a direction of fire, and a partially dug fighting position.

STANDARDS: The gunner must identify the actions needed to complete the fighting position.

PERFORMANCE MEASURES	GO	NO-GO
1. Check assigned direction of fire.		
2. Check gun traverse.		
3. Check overhead cover.		
4. Check fighting position camouflage.		
5. Check for protection against small-arms fire.		
6. Check for protection against indirect fire fragments.		
7. Check for protection from aerial observation.		
8. Check for mutual support with other fighting positions.		
9. Check for a good field of fire.		

STATION SIX

TASK: Prepare a range card for the MK 19.

CONDITIONS: During daylight, in a defensive position, the gunner is given a tripod-mounted MK 19, a principal direction of fire and secondary sector of fire (both with recognizable targets), paper, pencil, and compass.

STANDARDS: The gunner must prepare range cards that include a data and sketch section for each sector of fire, with appropriate sketches and military symbols.

PERFORMANCE MEASURES	GO	NO-GO
1. Orient the range card.		
2. Sketch the terrain that is to the front of the position (prominent and man-made, including terrain features).		
3. Fill in or construct the marginal data section.		
a. Gun number.		
b. Unit designation.		
c. Date.		
d. Magnetic north arrow.		
4. Sketch machine gun symbol.		
5. Specify gun location in relation to prominent terrain feature or eight-digit grid coordinate.		
a. Determine azimuth in mils from the terrain feature to the gun.		
b. Determine distance between the terrain feature and the gun.		
c. Sketch and identify the terrain feature on the card.		
d. Connect the feature and gun with a barbed line.		
e. Write the distance in meters above the line.		
f. Write the azimuth in mils below the line from the feature to the gun.		
6. Sketch in the TRPs, if any.		
7. Label TRPs, if any.		

STATION SEVEN

TASK: Mount, place into operation, and dismount the AN/TVS-5 NVD from the MK 19.

CONDITIONS: The gunner is given a mounted MK 19 and an AN/TVS-5.

STANDARDS: The gunner must mount the AN/TVS-5 on the MK 19, place the sight into operation, and dismount the AN/TVS-5 from the MK 19.

PERFORMANCE MEASURES	GO	NO-GO
1. Mount the M2 mounting bracket onto the MK 19.		
2. Mount the AN/TVS-5 onto the M2 bracket.		
3. Install the batteries in the device.		
4. Press your eye against the eye guard.		
5. Turn the ON-OFF/TUBE BRIGHTNESS control to ON.		
6. Adjust the ON-OFF/TUBE BRIGHTNESS control, as needed.		
7. Turn the ON-OFF/RETICLE BRIGHTNESS control to ON.		
8. Turn the diopter focus ring until the clearest image of the reticle pattern is obtained.		
9. Turn the objective focus ring until the target in the field of view is sharply defined.		
10. Turn the ON-OFF/RETICLE BRIGHTNESS control to OFF.		
11. Turn the ON-OFF/TUBE BRIGHTNESS control to OFF.		
12. Remove the batteries from the AN/TVS-5.		
13. Remove the AN/TVS-5 from the M2 bracket.		
14. Remove the M2 bracket from the MK 19.		

APPENDIX E

ANNUAL GUNNERY TRAINING PROGRAM

This appendix contains four suggested annual training programs for the MK 19, MOD 3; one for each of the fighting and supporting active and reserve forces. Because of the high cost of M918TP and M430 HEDP rounds, live fire is kept to a minimum. Gunnery training programs have been prepared to adequately train gunners based on the mission profile of the unit. Refer to Chapter 4, Marksmanship, and Appendix I, Gunnery Exercises, for the firing tables that support these programs. A definition of each force, along with a summary of the annual gunnery training program designed for it, is included in this appendix.

E-1. LIVE-FIRE EXERCISES

LFXs allow the gunner to qualify on the MK 19 and to participate in collective gunnery exercises. Gunners practice and qualify using Tables I through IV in Chapter 4. The gun crews practice individual and collective skills using Tables I-1 through I-7 in Appendix I.

a. **Practice and Qualification Table Matrix** (Table E-1). The tables used for gunner practice and qualification are in Chapter 4. There are four scorecards available and they are used based on the type of target (hull or pop-up silhouettes) and whether the practice-qualification is during the day or during limited visibility. Each scorecard has two tables, one for practice and one for qualification. The tables have versions for hull or pop-up silhouette engagements and for the type of NVD used. Sample scorecards are shown in Figures 4-27 through 4-30 (and reproducible forms are provided in the back of this manual). Gunners will only fire one day practice-qualification and one night practice-qualification. Units should select the practice and qualification based on the light conditions, type of targets available, and type of NVDs used. The following table shows which scorecards should be used.

Conditions	Target	Night Vision Devices	Scorecard
Day	Hull	NA	Scorecard I
Day	Pop-Up	NA	Scorecard III
Limited/Night	Hull	**AN/PEQ-2A** mounted on the TWS mounting bracket. **AN/PAS-13** mounted on the TWS mounting bracket. **AN/TVS-5** upgraded with the 3d generation tube mounted on the TWS mounting bracket.	Scorecard II
Limited/Night	Pop-Up or E-Type	All night vision devices.	Scorecard IV
Limited/Night	Any type	No night vision device.	Scorecard IV

Table E-1. Practice and qualification tables.

b. **Gunnery Exercises.** Gunnery exercises allow individual gun crews to practice firing the MK 19 while performing other tasks such as mounting the gun and preparing range cards. They also allow multiple gun crews to coordinate their fires while engaging targets. Table E-1 outlines each of the seven exercises. Based on their unit's training needs and the resources available, commanders must decide on the number and type of gunnery exercises to include in their annual training program.

Table	Exercise	Number of Guns	Number of Rounds
I-1	Instructional Fire Exercise	1	26 rounds
I-2	Dismounted Range Card Exercise	1	42 rounds
I-3	Mounted Combat Exercise	1	26 rounds
I-4	Night Fire Exercise	1	22 rounds
I-5	Linear and Deep Targets	2 to 4	80 rounds per gun
I-6	Linear Targets with Depth	2 to 4	40 rounds per gun
I-7	Area Targets	4 to 6	80 rounds per gun

Table E-2. Gunnery exercises.

E-2. FIGHTING FORCES ACTIVE COMPONENT

These forces consist of infantry, military police, scout, and engineer units of the active Army. Gunners assigned to these units are required to qualify semiannually. They are also required to participate in squad/platoon-level live fires quarterly and combined arms live fire exercises (CALFEX) annually. Assistant gunners are also required to qualify or fire familiarization exercises annually.

a. *Every quarter* the unit conducts preliminary marksmanship instruction and crew drills. The Gunnery Skills Test is also conducted.

b. During the *first quarter*, semiannual gunner practice qualification and qualification exercises are performed (see Table E-1 for the correct tables to use). Assistant gunner familiarization (with a zeroed gun) using Tables I and III are also conducted (see table E-1). Unit LFXs are performed (see Tables I-5 through I-7).

c. During the *second quarter*, unit LFXs are performed (see Tables I-1 through I-4, and I-5 through I-7).

d. During the *third quarter,* semiannual gunner practice qualification and qualification exercises are performed (see Table E-1 for the correct tables to use). Unit LFXs are also performed (see Tables I-5 through I-7).

e. During the *fourth quarter,* conduct LFXs or CALFEX using exercises from Tables I-5 through I-7.

E-3. SUPPORTING FORCES ACTIVE COMPONENT

These forces consist of all other units in the active component not identified as fighting force units. These guns are used primarily in defense of battalion, brigade, and division rear operations. Gunners from these forces participate in instruction fire and qualification annually.

a. *Every quarter* the unit conducts preliminary marksmanship instruction and crew drills. The Gunnery Skills Test is also conducted.

b. During the *first quarter,* there is no additional training.

c. During the *second quarter,* perform LFXs (see Tables I-5 through I-7).

d. There is no additional training during the *third quarter.*

e. During the *fourth quarter,* annual day gunner practice qualification and qualification exercises are performed. Night practice qualification and qualification is conducted once every two years (see Table E-1 for the correct tables to use). Once every two years, assistant gunner familiarization (with a zeroed gun) using Tables I and III is also conducted (see table E-1).

E-4. FIGHTING FORCES RESERVE COMPONENTS

These forces consist of military police, engineer, and infantry units of the reserve components. Gunners for these forces must maintain a level of proficiency higher than that of the active support forces. At the same time, limited time and range resources must be considered. Gunners of these forces conduct sustainment on the trainer at least quarterly. Live fires consist of annual qualification and a unit level LFX. These forces are allocated 192 or 118 rounds each year for live fire and qualification.

a. *Every quarter* the unit conducts preliminary marksmanship instruction and crew drills. The Gunnery Skills Test is also conducted.

b. There is no additional training during the *first quarter.*

c. There is no additional training during the *second quarter.*

d. During the *third quarter,* annual gunner practice qualification and qualification exercises are performed. See Table E-1 for the correct tables to use. Once every two years, the assistant gunner fires familiarization (with a zeroed gun) using Tables I and III (see table E-1). Unit LFXs are performed (see Tables I-5 through I-7).

e. There is no additional training during the *fourth quarter only.*

E-5. SUPPORTING FORCES RESERVE COMPONENT

These forces consist of all reserve units not specifically outlined under the paragraph on fighting forces. They are allocated 78 rounds for each year to qualify.

a. *Every quarter* the unit conducts preliminary marksmanship instruction and crew drills. The Gunnery Skills Test is also conducted.

b. There is no additional training during the *first quarter.*

c. There is no additional training during the *second quarter.*

d. During the *third quarter,* gunner practice qualification and qualification exercises are performed once every two years. See Table E-1 for the correct tables to use. Once every two years, the assistant gunner fires familiarization (with a zeroed gun) using Tables I and III (see table E-1). Unit LFXs are performed (see Table I-7).

e. There is no additional training during the *fourth quarter.*

APPENDIX F

FIRING TABLE

The table in this appendix provides information about the weapon elevation, time of flight, and other factors of MK 19 projectiles.

NOTE: 40-mm TP, XM918 is reduced time-of-flight experimental ammunition.

ELEVATION (MILS)					FLIGHT TIME (SECONDS)				REMAINING VELOCITY			
									M430		M385	
RANGE (M)	M430	M385	M918	XM918*	M430	M385	M918	XM918*	FPS	MPS	FPS	MPS
0	0.0	0.0	0.0	0.0	0.00	0.00	0.00	0.00	790	240.8	790	240.8
100	9.0	9.0	8.9	4.6	0.43	0.43	0.43	0.31	747	227.7	744	226.8
200	18.6	18.7	18.5	9.5	0.88	0.88	0.88	0.63	706	215.2	701	213.7
300	29.1	29.3	28.8	14.9	1.36	1.37	1.35	0.97	667	203.3	660	201.2
400	40.4	40.8	39.9	20.6	1.87	1.88	1.85	1.33	630	192.0	622	189.6
500	52.6	53.4	51.8	26.7	2.40	2.43	2.38	1.70	596	181.7	586	178.6
600	65.9	66.9	64.7	33.3	2.97	3.07	2.93	2.10	564	171.9	552	168.2
700	80.3	81.8	78.5	40.4	3.58	3.63	3.52	2.52	533	162.5	520	158.5
800	96.0	98.2	93.5	48.1	4.22	4.29	4.14	2.97	505	153.9	491	149.7
900	113.1	116.1	109.8	56.3	4.90	4.99	4.80	3.43	478	145.7	463	141.1
1,000	131.9	135.8	127.5	65.2	5.63	5.74	5.50	3.93	453	138.1	438	133.5
1,100	152.5	157.6	146.8	74.9	6.41	6.55	6.24	4.45	430	131.1	414	126.2
1,200	175.2	181.9	167.8	85.3	7.24	7.42	7.03	5.00	409	124.7	393	119.8
1,300	200.4	209.0	191.0	96.6	8.14	8.37	7.88	5.58	389	118.6	373	113.7
1,400	228.6	239.6	216.6	108.9	9.11	9.40	8.79	6.19	372	113.4	356	108.5
1,500	260.3	274.6	245.0	122.2	10.1	10.53	9.77	6.84	356	108.5	341	103.9
1,600	296.7	315.3	277.1	136.7	7	10.53	10.84	7.53	343	104.5	329	100.3
1,700	339.2	364.2	313.7	152.6	11.3	11.80	12.01	8.27	333	101.5	320	97.5
1,800	390.8	426.4	356.5	169.9	4	13.25	13.34	9.05	325	99.1	314	95.7
1,900	458.3	518.4	408.5	189.0	12.6	15.01	14.88	9.88	322	98.1	314	95.7
2,000	570.1	X	477.2	210.0	5	17.44	16.82	10.77	324	98.8	X	X
2,100	x	X	757.3	233.4	14.1	X	23.66	11.73	X	X	X	X
2,200	X	X	X	259.6	7	X	X	12.77	X	X	X	X
2,300	X	X	X	289.3	16.0	X	X	13.91	X	X	X	X
2,400	X	X	X	323.4	6	X	X	15.18	X	X	X	X
2,500	X	X	X	363.8	18.9	X	X	16.62	X	X	X	X
2,600	X	X	X	413.8	7	X	X	18.32	X	X	X	X
2,700	X	X	X	482.4	X	X	X	20.54	X	X	X	X
2,800	X	X	X	X	X	X	X	X	X	X	X	X
					X							
*REDUCED TIME-OF-FLIGHT EXPERIMENTAL AMMUNITION.												

Table F-1. Comparison of direct-fire aiming data.

	REMAINING VELOCITY				ANGLE OF FALL (DEGREES)				MAXIMUM ORDINATE (FEET)			
	M918		XM918*									
RANGE (M)	FPS	MPS	FPS	MPS	M430	M385	M918	XM918*	M430	M385	M918	XM918*
0	790	240.8	1100	335.3	0.0	0.0	0.0	0.0	0.0	0.0	0.0	0.0
100	750	228.6	1044	318.2	0.5	0.5	0.5	0.3	0.7	0.7	0.7	0.4
200	712	217.0	991	302.1	1.1	1.2	1.1	0.6	3.1	3.1	3.1	1.6
300	675	205.7	940	286.5	1.8	1.9	1.8	0.9	7.4	7.5	7.3	3.8
400	641	195.4	892	271.9	2.6	2.7	2.6	1.3	14.0	14.2	13.8	7.1
500	608	185.3	847	258.2	3.6	3.7	3.5	1.8	23.3	23.7	22.8	11.7
600	578	176.2	804	245.1	4.7	4.8	4.5	2.3	35.7	36.5	34.7	17.9
700	549	167.3	763	232.6	5.9	6.1	5.6	2.9	51.7	53.1	50.1	25.7
800	521	158.8	724	220.7	7.3	7.6	6.9	3.6	72.0	74.4	69.4	35.6
900	496	151.2	688	209.7	8.9	9.3	8.4	4.3	97.4	101.0	93.4	47.7
1,000	471	143.6	653	199.0	10.7	11.3	10.1	5.2	128.7	134.1	122.7	62.5
1,100	449	136.9	620	189.9	12.8	13.6	12.0	6.1	167.0	174.8	158.2	80.3
1,200	428	130.5	590	179.8	15.2	16.2	14.1	7.2	213.6	224.9	201.2	101.6
1,300	409	124.7	561	171.0	17.9	19.1	16.5	8.4	270.3	286.3	252.9	126.8
1,400	391	119.2	534	162.8	21.0	22.5	19.2	9.8	339.3	361.8	315.0	156.6
1,500	376	114.6	508	154.8	24.5	26.4	22.3	11.4	423.6	455.3	390.0	191.6
1,600	362	110.3	484	147.5	28.4	30.9	25.8	13.1	527.6	527.9	480.8	232.7
1,700	350	106.7	462	140.8	32.9	36.1	29.7	15.0	658.4	724.6	592.2	280.8
1,800	341	103.9	442	134.7	38.2	42.2	34.1	17.2	828.4	931.6	731.6	337.0
1,900	334	101.8	423	128.9	44.4	50.0	39.2	19.6	1,065.6	1,259.3	912.3	402.8
2,000	332	101.2	406	123.7	53.2	X	45.3	22.4	1,486.5	X	1,166.1	479.9
2,100	350	106.7	392	119.5	X	X	63.6	25.3	X	X	2,298.7	570.7
2,200	X	X	379	115.5	X	X	X	28.6	X	X	X	678.0
2,300	X	X	368	112.2	X	X	X	32.3	X	X	X	806.2
2,400	X	X	359	109.4	X	X	X	36.3	X	X	X	961.6
2,500	X	X	354	107.9	X	X	X	40.8	X	X	X	1,154.6
2,600	X	X	351	107.0	X	X	X	45.8	X	X	X	1,405.6
2,700	X	X	352	107.3	X	X	X	52.0	X	X	X	1,787.3
2,800	X	X	X	X	X	X	X	X	X	X	X	X

*REDUCED TIME-OF-FLIGHT EXPERIMENTAL AMMUNITION.

Table F-1. Comparison of direct-fire aiming data (continued).

APPENDIX G

AIMING DEVICES

This appendix provides information on the employment of the AN/PEQ-2A target pointer illuminator aiming light (TPIAL), AN/TVS-5, and the AN/PAS-13B (V) 3 heavy weapon thermal sight (HWTS). These devices have the advantage of greater magnification and resolution for the employment of the MK 19 during periods of degraded visibility. These devices, when properly boresighted and employed under proper conditions, allow infantrymen to engage targets at the weapon's maximum effective ranges during periods of degraded visibility.

G-1. AN/PEQ-2A DESCRIPTION AND OPERATION

The TPIAL projects an infrared laser beam that cannot be seen with the eye but can be seen with NVD. It is also capable of projecting a much wider infrared illuminating beam from an integral illuminator. The TPIAL works with night vision goggles and mounts on various weapons with mounting brackets and adapters (Figure G-1 and Table G-1). Leaders can also use the AN/PEQ-2A in the hand-held mode to illuminate and designate targets.

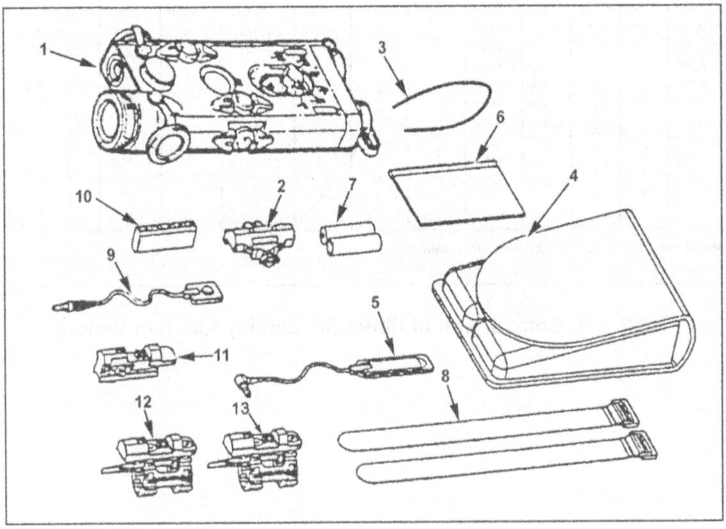

Figure G-1. AN/PEQ-2A major components.

ITEM	DESCRIPTION
1	TPIAL Assembly with Safety Block
2	Rail Grabber Mounting Bracket
3	Neck Cord
4	Carrying Bag
5	Cable Switch, 20 inch Remote, Button
6	Operator's Manual
7	Batteries, 2 AA
8	Straps, Retention
9	Cable Switch, 12 inch, Membrane
10	Bracket Adapter
11	Training Extender (Army only)
12	M4/M16A2 Bracket Assembly (Army only)
13	M16A2 Bracket Assembly (Marine only)

Table G-1. Names of the AN/PEQ-2A major components.

a. **Controls and Indicators.** The AN/PEQ-2A has controls and indicators that allow the user to operate the device and to select its different modes.

(1) ***Battery Installation.*** To install batteries in this device, first unscrew the battery caps and install two AA batteries. Orient the batteries as indicated by the markings on the AN/PEQ-2A body (Figure G-2).

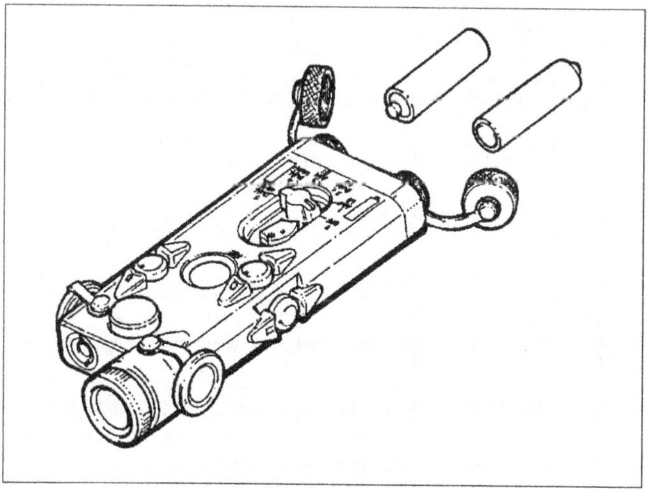

Figure G-2. AN/PEQ-2A battery installation.

(2) ***Safety Block Installation.*** The safety block installed in the training mode (blue side up) prevents the operator from accessing the non-eye safe modes (AIM HI, DUAL

LO/HI, DUAL HI/HI) (Figure G-3). A .050 hex-head Allen Wrench is needed to unscrew the block from the body and re-install it in the tactical mode (black side up).

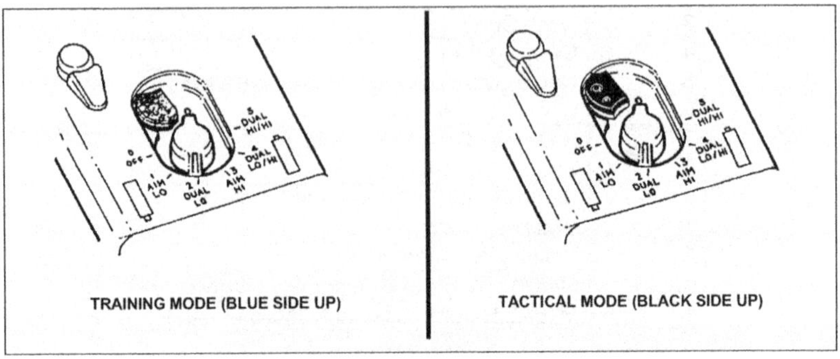

Figure G-3. Safety block installation.

(3) ***Mode Selector.*** The mode selector is used to set the mode in which the AN/PEQ-2A will operate when the cable switch button or push button is depressed. The mode selector has six positions (Table G-2).

KNOB POSITION		OPERATION
0	OFF	The AN/PEQ-2A will not operate.
1	AIM LO	The aiming beam operates at low power.
2	DUAL LO	The aiming beam and the illuminating beam operate at low power.
3	AIM HI	The aiming beam operates at high power.
4	DUAL LO/HI	The aiming beam operates at low power and the illuminating beam operates at full power.
5	DUAL HI/HI	The aiming beam operates at high power and the illuminating beam operates at full power.

Table G-2. Mode selector position.

(4) ***Button Switch.*** The button switch is used when the AN/PEQ-2A is hand held. Pressing the button switch operates the AN/PEQ-2A in the operational mode set by the

mode selector switch (Figure G-4). When the button is released, the AN/PEQ-2A turns off.

(a) A green LED is incorporated into the body of the AN/PEQ-2A to indicate that the unit is ON. Whenever the AN/PEQ-2A is activated, the green LED will light and stay lit until the unit is turned OFF.

(b) If continuous operation of the AN/PEQ-2A is desired, pressing the button switch twice in rapid succession will latch the unit ON. The unit will remain ON until the button switch is pressed the third time.

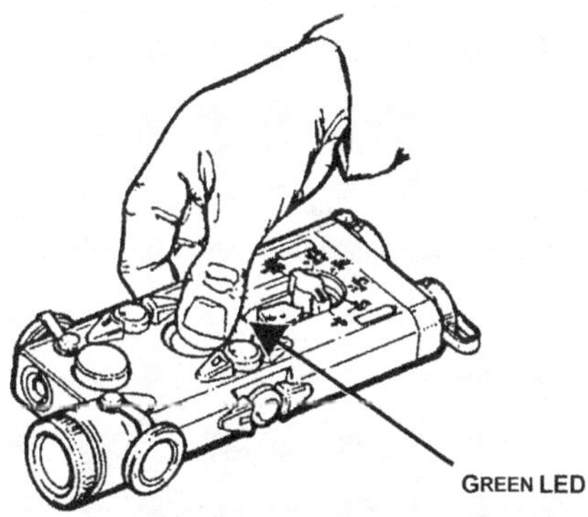

Figure G-4. Operation of the button switch.

(5) ***Cable Switch.*** The cable switch is used when the AN/PEQ-2A is mounted on a weapon (Figure G-5). The cable switch plugs into the back of the AN/PEQ-2A assembly. Pressing the button or pad at the end of the cable switch causes the unit to turn on in the operational mode selected by the mode select switch. When released, the AN/PEQ-2A turns off.

(a) If continuous operation of the AN/PEQ-2A is desired, pressing the cable switch twice in rapid succession will latch the unit ON. The unit will remain on until the push button is pressed a third time.

(b) When the cable switch plug is installed in the AN/PEQ-2A, it automatically locks into place. To remove the switch, pull back on the plug sleeve and pull the plug out.

CAUTION

Do not try to remove the plug by pulling on the cable.

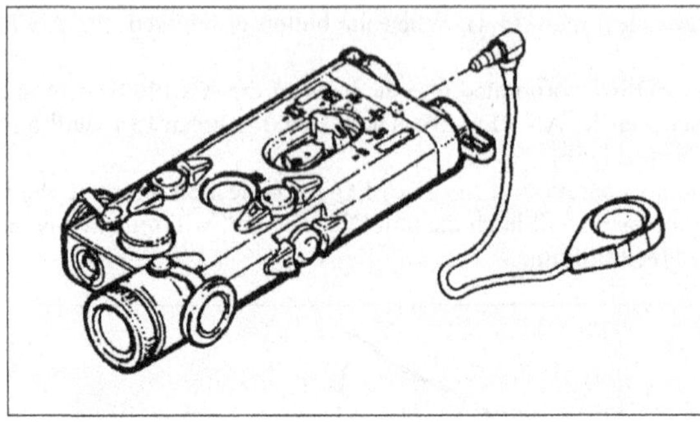

Figure G-5. Installation of the cable switch.

(6) ***Focus Knob.*** The focus knob is used to vary the spread of the illumination beam based on the range and size of the area to be illuminated (Figure G-6).

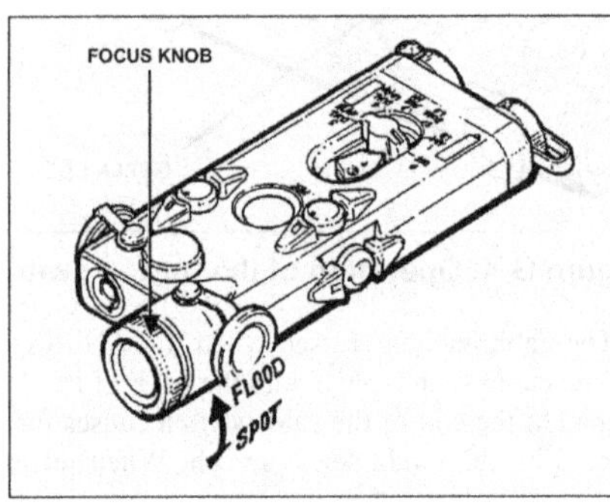

Figure G-6. Using the focus knob.

(7) ***Lens Caps.*** Several lens caps can be used with the AN/PEQ-2A for different purposes (Figure G-7).

(a) The *black lens cap* blocks the AN/PEQ-2A illuminator or aiming laser beam should the device be activated. To use the black lens cap, pull it from its stored location on the side of the unit and stretch it over the front of the focus knob or aiming beam so that it fits snugly in place.

(b) The *diffuser lens cap* enables the illuminator or aiming laser to emit in a 45-degree cone (10 feet at 10 feet). To use the diffuser lens cap, pull it over the front of the focus knob or aiming beam so that it fits snugly in place.

(c) The *neutral density lens cap* enables the AN/PEQ-2A illuminator or aiming laser to be operated in low power. To use the neutral density lens cap, pull it from its stored location on the side of the unit and stretch it over the front of the focus knob or aiming beam so that it fits snugly in place.

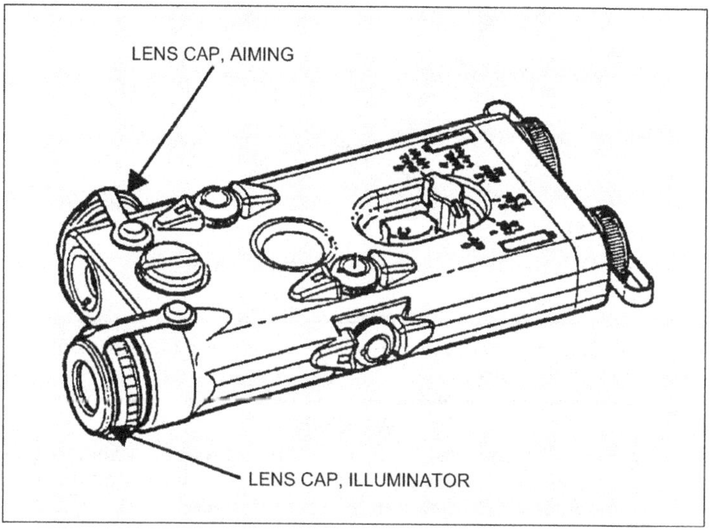

Figure G-7. Installing the lens caps.

(8) ***Boresight Adjusters.*** The AN/PEQ-2A is equipped with boresight adjusters for zeroing the aiming beam and illumination beam to the weapon (Figure G-8). The AN/PEQ-2A adjusters move the beam in true horizontal and vertical directions with the top adjusters used for elevation and the side adjusters for windage. When zeroing the AN/PEQ-2A, it is best to zero the aiming beam to the weapon and then align the illumination beam to the aiming beam.

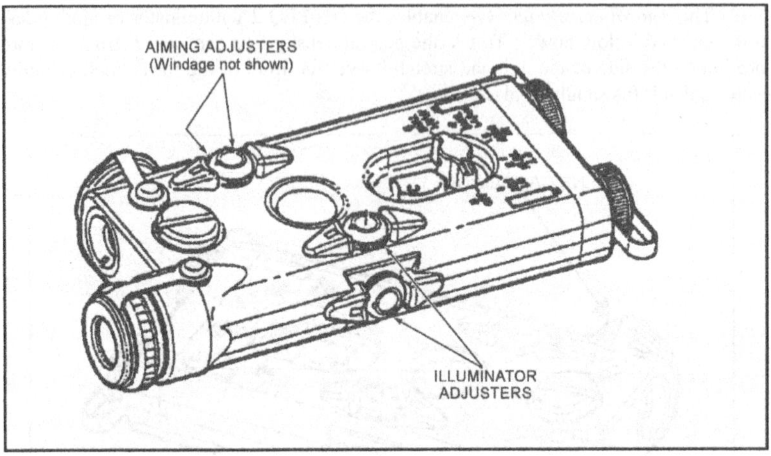

Figure G-8. Boresight adjusters for both aiming and illumination beams.

WARNING

Eye damage can occur if the laser is handled carelessly. The danger area is 15 meters in the training mode and 220 meters in the tactical mode.

b. **Mounting and Dismounting Procedures.** The AN/PEQ-2A can be mounted on various weapons or used in the hand-held mode to illuminate and designate targets.

(1) ***Mounting Procedures.*** Ensure that the mounting bracket is installed before proceeding.

(a) Secure the bracket adapter to the underside of the AN/PEQ-2A.

(b) Place the bracket adapter into the MK 19 bracket mounting groove, located in the center of the rail grabber.

(c) Tighten the bracket's lever screw into the bracket adapter hole.

(d) Plug the cable into the AN/PEQ-2A and run the cable to the front of the MK 19 bracket. The "cable switch, 12-inch, membrane" or the "cable switch, 20-inch, button" may be used for remote access.

(e) Once the cable is secure at the front of the MK 19 bracket (ensure the cable will not become damaged by the movement of the quadrant portion of the bracket) run it between the MK 19 and the bracket to the handles on the backplate assembly. If using the button cable, attach Velcro to the top of the night handle and secure excess cable to the handle. If using the membrane pad, run it the same way as with the button cable and attach the membrane to the inside of the right handle.

(2) ***Dismounting Procedures.*** Dismount the AN/PEQ-2A when the mission is complete, when it is needed on another weapon, or when it will be used in its hand-held mode.

(a) Untie the cable and disconnect the remote button from the backplate handles.

(b) Unplug the cable from the AN/PEQ-2A and place the cable back in the carrying case.

(c) Turn the AN/PEQ-2A mounting screw counterclockwise and remove the AN/PEQ-2A from the bracket adapter.

(d) Unscrew the bracket adapter from the MK 19 bracket rail and place the bracket adapter and the AN/PEQ-2A in its carrying case.

c. **Boresighting Procedures**. Boresighting aligns the sighting system to the bore of the weapon. The AN/PEM-1 Borelight is used for this procedure.

(1) Place the MK 19 in the ready to fire position 10 meters from the bore light offset zero target. Use the 10 meter distance gauge in the borelight kit to measure the distance. Ensure the target and weapon are level and stable (the bore sight zero will be off if not) before making any adjustments to the laser.

(2) Set the range on the MK 19 bracket to 500 meters.

(3) Mount the bore light and turn it on. While holding the MK 19 steady, zero the bore light. Follow the instructions in the AN/PEM-1 borelight operations manual for further details. The 40-mm MK 19 mandrel interface (NSN 3460-01-502-0575) is used with the 5.56-mm mandrel interface and has to be ordered separately.

(4) Make adjustments with the T&E mechanism until the bore light is on the bore light aiming point.

(5) Adjust the windage and elevation of the aiming laser until the laser is on the laser aiming point.

(6) Re-check the bore light aiming point and then the laser aiming point again.

(7) To zero the IR light source narrow the beam as much as possible and adjust its windage and elevation until the beam is 4 centimeters horizontally to the left of the laser aiming point while the aiming laser is still on the laser aiming point. The AN/PEQ-2A is now boresighted.

d. **Zeroing Procedures.** Zeroing the weapon aligns the line of sight of the AN/PEQ-2A with the trajectory of the round so that they intersect at 500 meters.

(1) Set the MK 19 mounting bracket to 500 meters by loosening the friction knob on the left side of the mount. Press and hold the friction knob in (towards the mount), then slide the mount assembly to the 500-meter increment. Release and tighten the friction knob.

(2) Identify a 500-meter target downrange and fire one round; note the impact of the round in relation to the target.

(3) Adjust the aiming light on the AN/PEQ-2A to the impact of the round by adjusting the elevation and windage adjusters.

(4) Manipulate the T&E mechanism so that the aiming light is center mass on the target. Repeat until the point of impact is the same as the point of aim.

(5) Adjust the illuminating beam's elevation and windage adjusters so that the aiming light is in the center of the illuminating beam.

G-2. THE AN/PAS-13B (V) 3 HEAVY WEAPON THERMAL SIGHT

The AN/PAS-13B (V) 3 HWTS is a silent, lightweight, compact, and durable battery-powered infrared imaging sensor that operates with low battery consumption (Figure G-9).

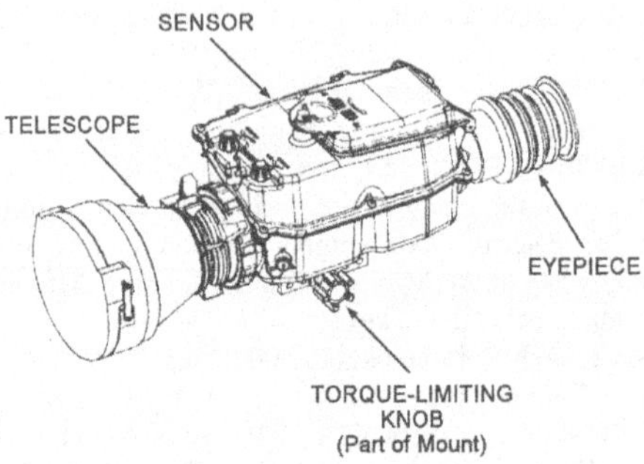

Figure G-9. AN/PAS-13B (V) 3 heavy weapon thermal sight.

a. **Components.** The HWTS is capable of target acquisition under conditions of limited visibility such as darkness, smoke, fog, dust, and haze. The HWTS operates effectively at night and can also be used in the daytime. Infrared light is received through the telescope, detected by an IR sensor, converted to digital data, processed, and then displayed for the user. Besides the carrying cases, the HWTS is composed of four major components: the telescope, the basic sensor, the eyepiece, and the mount.

(1) ***Telescope.*** The telescope receives IR light emitting from an intended target and its surroundings. The telescope magnifies and projects the IR light onto the basic sensor's scanner.

(2) ***Basic Sensor.*** The scanner reflects the IR light received from the telescope onto the detective assembly. The detective assembly senses the IR light and converts it to video. The sensor's electronics condition the video for display on the LED array. The LED array illuminates the IR image along with the reticle. The light from the LED array is reflected off the scanner to form an image at the eyepiece.

(3) ***Eyepiece.*** The eyepiece displays the thermal image and all system indicators on the cathode ray tube.

(4) ***Mount.*** A universal attachment interface between the HWTS and the MK 19 bracket (figure only shows the torque limiting knob).

b. **Modes of Operation.** The HWTS has three modes of operation: STANDBY, ON, and EMERGENCY.

(1) ***STANDBY Mode.*** When the system is first turned on, the HWTS begins a cool down period of approximately two minutes. After the cool down period, the HWTS

enters the STANDBY mode. During the STANDBY mode, power is not applied to the scanner or display in order to extend the life of the battery.

(2) ***ON Mode.*** When the HWTS is in the STANDBY mode and pressure is applied to the eyecup, the HWTS switches to the ON mode, and a switch engages to provide power to the scanner and display. After a three-second delay, the system is fully operational.

(3) ***EMERGENCY Mode.*** When switched to the EMERGENCY mode, the HWTS continuously applies power to the entire system. This allows the operator to bypass the three-second delay experienced when switching from the STANDBY to the ON mode. Since power is applied to the entire system while in the emergency mode, battery life is greatly reduced.

c. **Controls and Indicators.** Controls allow the gunner to configure the sight to the situation and individual preferences while the indicators display the current configuration (Figure G-10 and Table G-3).

(1) The CONTRAST CONTROL adjusts the contrast of the thermal image displayed on the raster. It has an automatic and a manual mode.

CAUTION

Ensure the brightness control switch is depressed before turning it on or off.

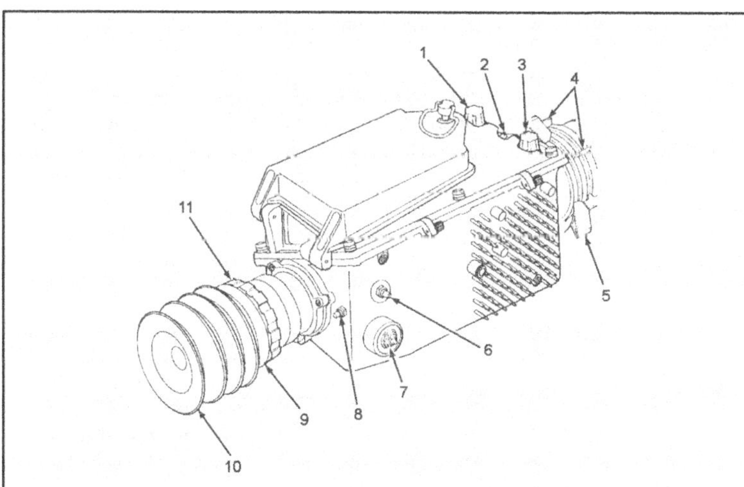

Figure G-10. Location of the HWTS controls and indicators.

ITEM	DESCRIPTION
1	Contrast Control
2	Emergency Switch Control
3	Brightness Control
4	Focus Ring
5	Field of View Ring
6	Zoom/Reticle Select Switch
7	Reticle Adjust Switch
8	Black/White Polarity Switch
9	Diopter Holding Device
10	Eyecup
11	Diopter Focus Ring

Table G-3. HWTS controls and Indicators.

(2) The EMERGENCY SWITCH CONTROL places the HTWS in emergency mode. The eyecup must be depressed in order to for the CRT to illuminate.

(3) The BRIGHTNESS CONTROL is a rotary switch with an off detent position (turned fully counterclockwise). The purpose is to turn the system on or off and adjust the brightness of the eyepiece display.

(4) The FOCUS RING adjusts the telescope focus from 20 meters to infinity. It requires a manual adjustment and affects both the wide and narrow fields of view.

(5) The FIELD OF VIEW RING (FOV) is located on the telescope. It has a wide and a narrow field of view. The wide FOV is for using low magnification during target detection, and the narrow FOV is for using high magnification during recognition and engagement.

(6) The RETICLE SELECT SWITCH selects one of the available reticles depending on the type of thermal sight (medium or heavy) and the weapon. It must be held for two seconds to enable reticle changes. After two seconds, release the switch to cycle to the next reticle. This control is disabled after ten seconds of inactivity.

(7) The RETICLE ADJUST SWITCH adjusts the reticle aiming features in azimuth and elevation. It is used during zeroing, and it must be held for two seconds to allow changes to be made. After two seconds, each press moves the reticle aiming features one increment. This control is also disabled after ten seconds of inactivity.

(8) The BLACK/WHITE POLARITY SWITCH selects the polarity of the thermal image displayed on the raster. The initial setting is "white hot." The polarity switch affects the appearance of the target.

(9) The DIOPTER HOLDING DEVICE allows the diopter to be adjusted when pulled back.

(10) The EYECUP controls the STANDBY/ON operating mode. When forward pressure is applied to the eyecup, the system is in the ON mode. When forward pressure is removed for more than 30 seconds, the system returns to the STANDBY mode.

(11) The DIOPTER FOCUS RING adjusts the focus of the raster and indicators to the operator's eye. It ranges from +2 to -6 diopters.

d. **Status Indicators and Raster** (Figure G-11). The status indicators and raster are visible when the user presses his eye against the eyecup. The indicators display the system status and configuration. The raster displays the thermal image, the selected reticle, and azimuth/elevation zeroing adjustment indicators.

(1) The status indicators are:

- NOT COOL (1): Displayed when the system is not cool enough for proper operation.
- WHT HOT/BLK HOT (2): Indicates the target polarity.
- EMER (3): Displayed when system is in emergency mode.
- LOW (4): Displayed when battery power is low (approximately fifteen minutes of useful power left).
- FIELD OF VIEW (7): Indicates when the wide field of view (WFOV) or the narrow field of view (NFOV) is in use
- ZOOM (8): Displayed when the zoom mode is selected
- RET ADJ (9): Indicates the reticle adjustment mode is selected
- RET SEL (10): Indicates the reticle select mode is selected

(2) The DISPLAY RASTER displays the thermal image with a superimposed reticle and indicators;

- RETICLE SELECT (5): Identifies the selected reticle/weapon.
- ELEVATION INDICATOR (6): Indicates the elevation zeroing adjustment of the reticle.
- AZIMUTH INDICATOR (11): Indicates the elevation zeroing adjustment of the reticle.
- RANGE SCALE (12): The range scale allows the operator to estimate the range to a target based on a known height or width. The numbers on the scale are the range in hundreds of meters (example: = 400 meters). The vertical lines on the scale reflect the height of a five-foot man at that range. The distance between two consecutive lines reflects the width of a 10-foot tank at the range specified by the vertical line on the left.
- CROSSHAIR/AIMING POINT (13): The vertical line at the aiming point reflects the height of a five-foot man at 800 meters. The horizontal line of the cross-hair reflects the width of a 10-foot tank at 800 meters.

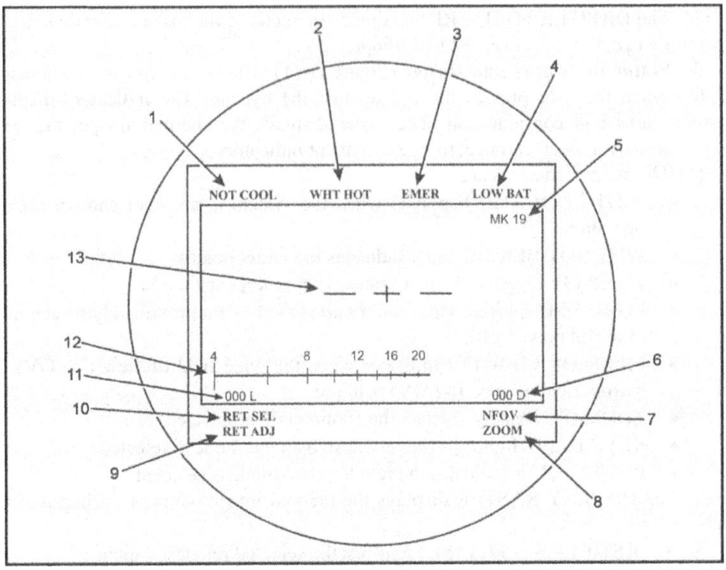

Figure G-11. Status indicators and raster.

e. **Mounting and Dismounting Procedures.** The HWTS is mounted on the MK 19 with a mounting bracket (Figure G-12). See TM 11-5855-312-10 for details.

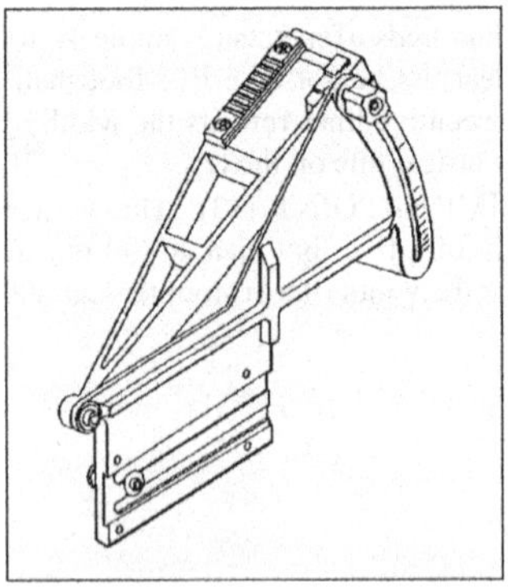

Figure G-12. MK 19 bracket.

(1) ***Mounting Procedures.*** Mounting the HWTS to the MK 19 is a two-step process (Figure G-13):

(a) Mount the bracket to the MK 19 by sliding the bracket into the dovetail slot until the locking pin engages the mount.

(b) Install the HWTS onto the bracket mount.

(c) Shake the HWTS to make sure the sight is secured properly.

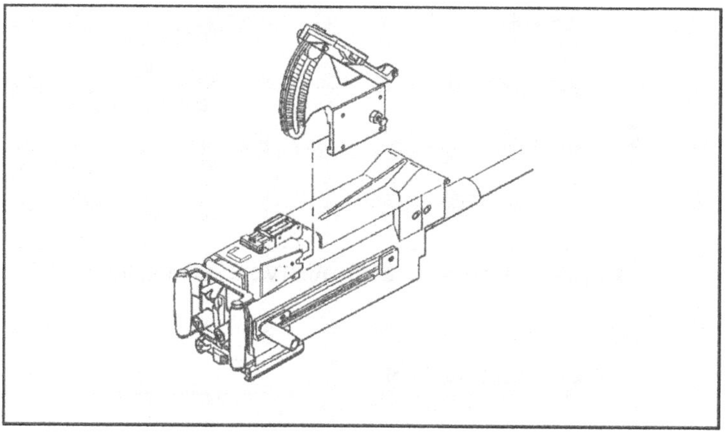

Figure G-13. Mounting the HWTS on MK 19.

(2) ***Dismounting Procedures.*** To dismount the HWTS:

(a) Take the HWTS off the mounting bracket rail and place the HWTS back in its carrying case.

(b) Pull out the locking pin and slide the bracket off the mount.

G-3. AN/TVS-5

The AN/TVS-5 is a portable, battery-operated electro-optical instrument used for observation and aimed fire of weapons at night (Figure G-14 and Table G-4). It amplifies reflected light such as moonlight, starlight, and sky glow so that the viewed scene becomes clearly visible to the operator. The sight does not emit visible or infrared light (except from the eyepiece) that can be detected by the enemy. By using this device, the gunner can observe the area and detect and engage any suitable target.

WARNING

Ensure the weapon is not loaded and is on S (SAFE) before installing the HWTS on the weapon. A loaded weapon may accidentally discharge causing severe injury or death.

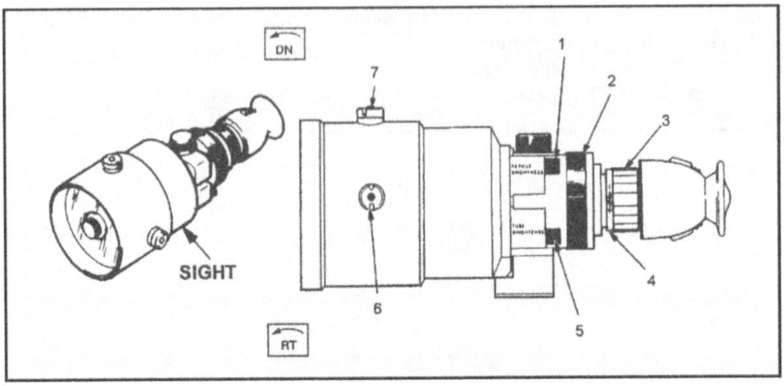

Figure G-14. Location of the AN/TVS-5 components.

ITEM	DESCRIPTION
1	ON-OFF/ Reticle Brightness
2	Objective Focus Ring
3	Diopter Focus Ring
4	Diopter Indicator
5	ON-OFF/ Tube Brightness
6	Reticle Azimuth Adjustment Acutator
7	Reticle Elevation Adjustment Acutator

Table G-4. AN/TVS-5 component names.

a. **Controls and Indicators.**Controls allow the gunner to configure the sight to the situation and individual preferences while indicators display the current configuration.

(1) The ON-OFF/TUBE BRIGHTNESS control applies power to the sight and controls the brightness of the image intensifier tube. It also enables the ON-OFF RETICLE BRIGHTNESS control to function.

(2) The ON-OFF RETICLE BRIGHTNESS control applies power to the reticle and controls the brightness of the reticle.

(3) The OBJECTIVE FOCUS RING adjusts the range focus from 25 meters to infinity.

(4) The DIOPTER FOCUS RING adjusts the focus of the eyepiece.

(5) The DIOPTER INDICATOR indicates the direction of rotation of the DIOPTER FOCUS RING for + and diopters.

(6) The RETICLE ELEVATION ADJUSTMENT ACTUATOR controls the reticle adjustment in the up and down direction. Each click of adjustment moves the strike of the round 1.0 inch at 100 meters.

(7) The RETICLE AZIMUTH ADJUSTMENT ACTUATOR controls reticle adjustment right or left. Each click or adjustment moves the strike of the round 1.0 inch at 100 meters.

b. **AN/TVS-5 Sight Reticle.** The MK 19 and the M2 machine guns share the same AN/TVS-5 sight reticle (Figure G-15).

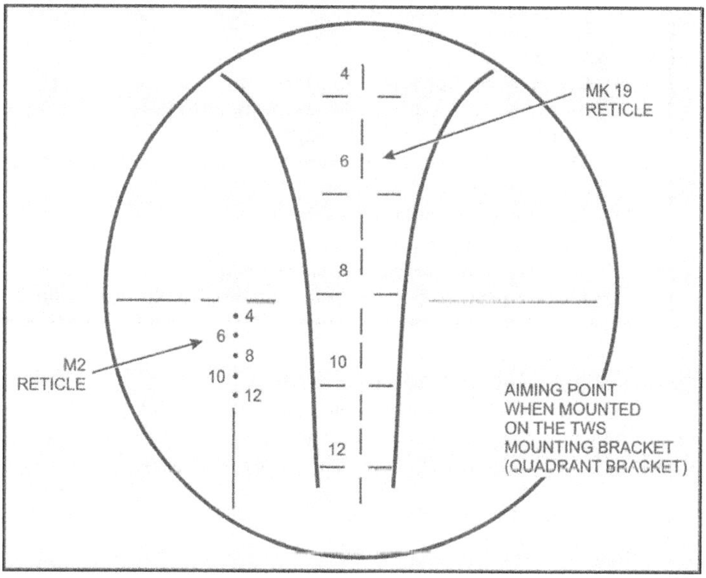

Figure G-15. AN/TVS-5 reticle for the MK 19 and M2 machine guns.

WARNING

1. Eye guard:

- **To avoid recoil injury, attach the eye guard before using the sight.**
- **When activated by pressure on the eye guard, the sight emits illumination the enemy can detect. Therefore, avoid touching the eye guard except when you wish to use the sight, and then press your eye area against it to activate it.**

2. Batteries:

- **Dispose of batteries as directed. The contents of the mercury batteries cause extreme irritation to the eyes and to oral and nasal passages.**
- **To prevent explosion, avoid disposing of batteries by burning.**
- **Avoid short circulating batteries.**
- **Avoid recharging batteries.**
- **Remove batteries before storing sight.**
- **Always replace both mercury batteries at the same time.**

c. **Mounting Procedures**. To mount the AN/TVS-5:

(1) Align the scribe line on the sight with the scribe line on the bracket.

(2) Place the sight in the grove of the bracket and tighten the lever screw clockwise.

(3) Secure the lever screw with lacing wire or tape to ensure the sight does not vibrate loose.

(4) Seat the device. After firing the initial burst, retighten the lever screw to ensure the sight is securely mounted. (If you are unable to fire at this time, lightly shake the sight to ensure the sight is mounted correctly.)

d. **Dismounting Procedures.** To dismount the AN/TVS-5:

(1) Remove the lacing wire or tape from the lever screw.

(2) Loosen the lever screw until the sight is free to be lifted off the bracket.

(3) Remove the batteries and place the sight in its carrying case.

e. **Boresighting.** The AN/TVS-5 cannot be boresighted when used with the MK 19. The sight cannot focus on a target closer than 25 meters, and the offset is too great at that distance to boresight accurately.

f. **Zeroing.** To zero the AN/TVS-5 to the MK 19, follow the procedures in TM 11-5855-214-10.

APPENDIX H

CREW DRILLS

The purpose of a crew drill is to build precision, speed, skill, and teamwork in checking equipment, placing the MK 19 into and out of action, and moving the MK 19 to a new position. The main goal is achieved by following prescribed procedures. Precision, speed, skill, and teamwork grow with practice. During the crew drill, all oral and visual signals are repeated. Once crewmembers can perform their duties well, they rotate to allow each member of the crew to learn the duties of others. The ground-mounted MK 19 has three crewmembers: the gunner, the assistant gunner, and the ammunition bearer. The vehicle-mounted MK 19 also has three crewmembers: the squad leader, the gunner, and the driver. Normally the crew is not required to carry the MK 19 and ammunition over long distances. A vehicle usually carries the crew, ammunition, and equipment to the firing position. The terrain and the enemy situation dictate how close the vehicle may be brought to the firing position.

H-1. CREW EQUIPMENT AND DRILL AREA

Dismounted crew drills are conducted with a gunner, assistant gunner, and ammunition bearer. For drills involving a vehicle, the squad leader, gunner, and driver are crewmembers. In addition to individual weapons and equipment, each crewmember for dismounted drills carries specific equipment used with the tripod-mounted MK 19 (Table H-1).

CREW MEMBER	SUGGESTED MINIMUM EQUIPMENT
Gunner	MK 19, binoculars, and compass
Assistant Gunner	M3 tripod, MK 64 gun cradle, with T&E mechanism attached
Ammunition Bearer	One box of ammunition

Table H-1. Equipment.

A crew drill is conducted in an area large enough to allow crewmembers to move freely with equipment and without interfering with the movements of other crewmembers.

WARNING
Each soldier must wear all the protective equipment required by range safety regulations. This may include helmet, earplugs, and protective vest.

H-2. FORMATION FOR CREW DRILL

To form the crew for crew drill, the leader commands FORM FOR CREW DRILL. Upon hearing the leader's command, the crew repeats the command and forms into a column, five paces apart, in the following order: assistant gunner, gunner, and ammunition bearer (Figure H-1).

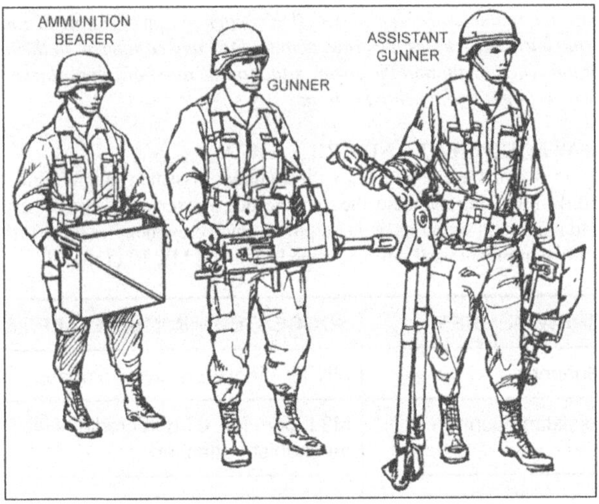

Figure H-1. Three-man MK 19 crew.

Each crewmember carries the following equipment in the specified manner:

- The assistant gunner carries the tripod, with the front leg positioned at a 60-degree angle and the trail legs collapsed, in his right hand; he carries the gun cradle, with the T&E mechanism attached, in his left hand.
- The gunner carries the MK 19, binoculars, and compass.
- The ammunition bearer carries one can of ammunition.

H-3. INSPECTION OF EQUIPMENT

The leader commands CHECK EQUIPMENT. All crewmembers repeat the command and inspect the components. The equipment is inspected and verified by the following crewmembers.

a. The assistant gunner inspects the M3 tripod, gun cradle, and T&E mechanism and ensures:

(1) The indexing levers and clamps on the front and trail legs work, and that the legs are in the "short (low)" position.

(2) The front leg is placed at a 60-degree angle and the trail legs are closely folded and the front leg clamp is tight.

(3) The sleeve lock latch and pintle lock release cam work.

(4) The pintle bushing is clean and has no burrs.

(5) The cradle pintle is clean, and all pins and bolts are seated.

(6) The T&E mechanism is centered.

b. The gunner inspects the MK 19 and ensures:

(1) The weapon is clean.

(2) The bolt is forward, the top cover is closed, and the weapon is on S (SAFE).

(3) The charger handles are forward and locked in the "up" position.

(4) The binocular lenses are clean and the focus mechanism works.

(5) The compass magnetic arrow and bezel ring work freely.

c. The ammunition bearer inspects the ammunition and ensures:

(1) The correct type of ammunition is on hand.

(2) The ammunition can is dent-free.

(3) All ogives are tight and the ammunition is clean.

d. The crewmembers report when the inspections are finished in the following manner:

(1) The ammunition bearer reports "Ammunition correct," or any defects.

(2) The gunner reports "Gun and ammunition correct," or any defects.

(3) The assistant gunner reports, "All correct," or any defects.

H-4. PLACING THE MK 19 INTO ACTION

The leader commands and signals GUN TO BE MOUNTED, and points to the position where the gun is to be mounted; FRONT, and points in the direction of fire; ACTION, and indicates with his fist the MK 19's direction in the chosen position. Crewmembers take the following steps to place the gun into operation.

a. The assistant gunner:

(1) Drags the tripod to the firing position by its front leg with his right hand, and carries the gun cradle with the T&E mechanism attached in his left hand.

(2) When he arrives at the firing position, aligns the front tripod leg for direction and, with one quick motion, snaps the trail legs apart. This action makes the sleeve lock-latch snap in place (Figure H-2).

(3) Stamps each leg into the ground and secures the gun cradle to the tripod by using the pintle lock release cam (Figure H-3).

(4) Secures the T&E mechanism to the tripod, by locking the traversing slide lock lever (Figure H-4).

(5) Helps the gunner mount and load the MK 19.

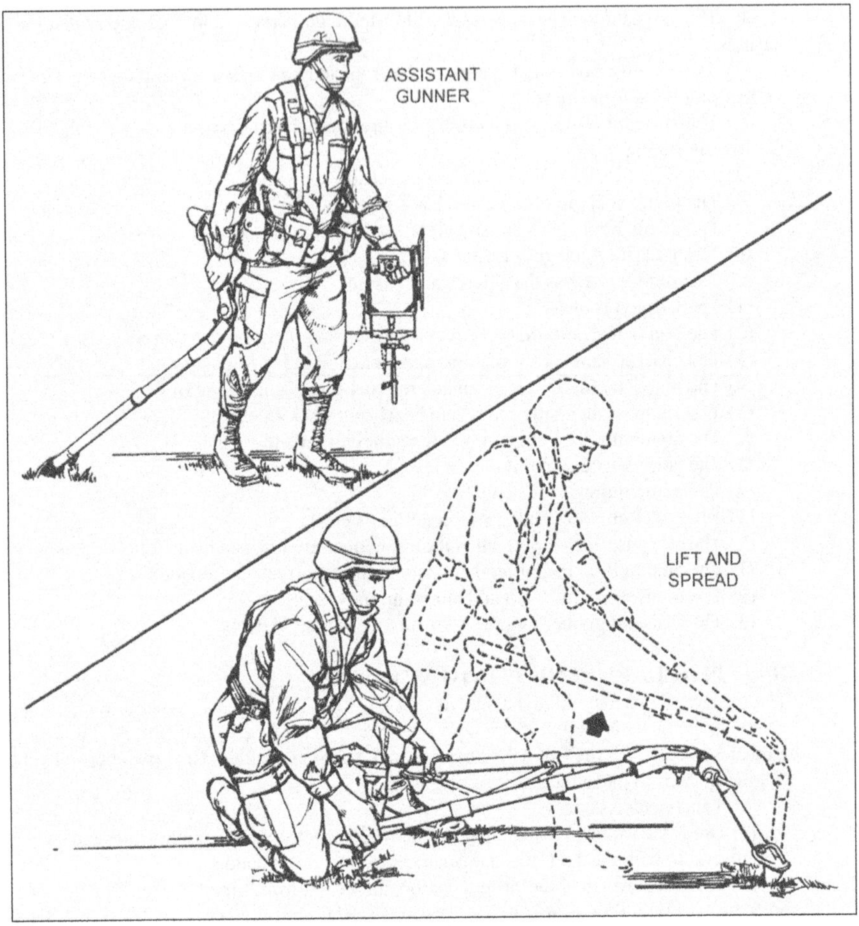

Figure H-2. Positioning the tripod.

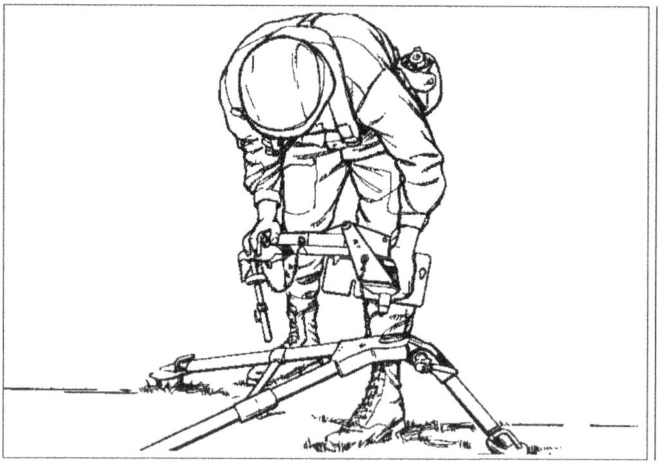

Figure H-3. Mounting the MK 19 mount on the tripod.

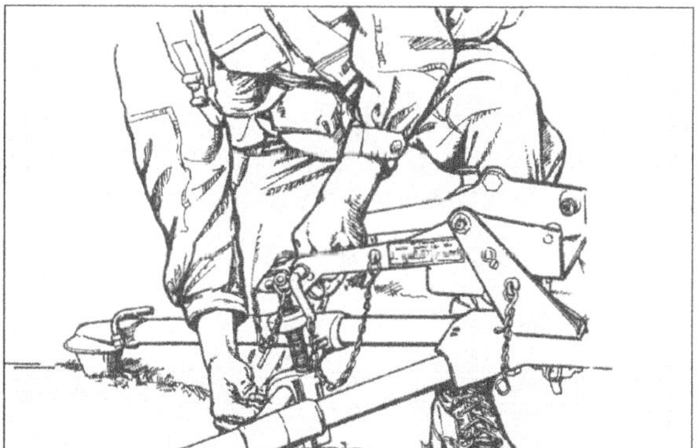

Figure H-4. Securing the T&E mechanism.

b. The gunner:

(1) Carries the MK 19 receiver to the tripod, with the barrel in his left hand, and the control grip in his right hand (Figure H-5).

(2) At the tripod, the assistant gunner grasps the barrel while the gunner holds the grips.

(3) The gunner and assistant gunner align the grooves on the receiver with the lugs in the gun cradle, and slide the receiver forward (Figure H-6).

(4) Secures the rear of the weapon by inserting the retaining pin through the cradle and sear assembly (Figure H-7).

Figure H-5. Gunner moving into position.

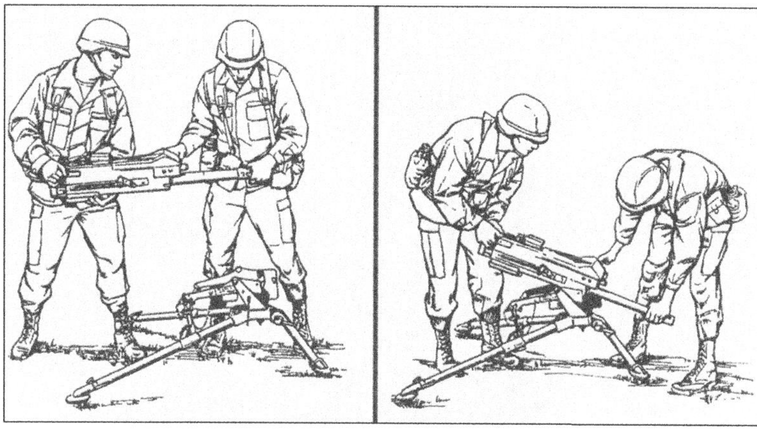

Figure H-6. Placing the MK 19 in the cradle.

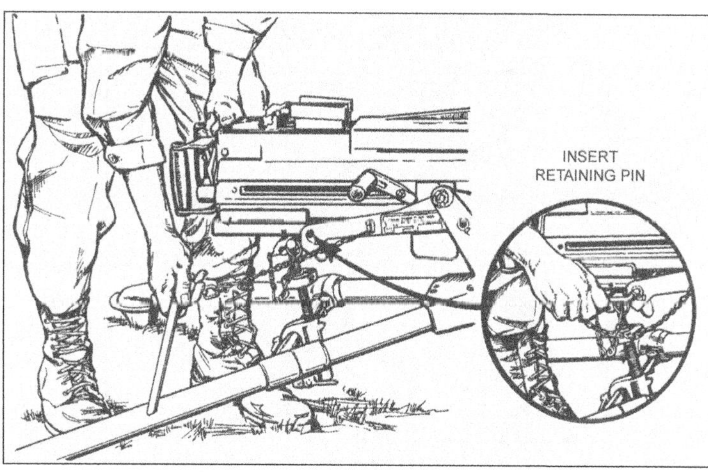

Figure H-7. Securing the MK 19 to the cradle.

c. The ammunition bearer:

(1) Carries the ammunition can to the MK 19 position (Figure H-8).

(2) Takes the lid off and ensures the ogive ends of the rounds are pointed downrange (Figure H-9).

(3) Gives the can to the assistant gunner.

(4) Returns to the ammunition point and stands by to do whatever is needed to complete the mission.

Figure H-8. Ammunition bearer with ammunition can.

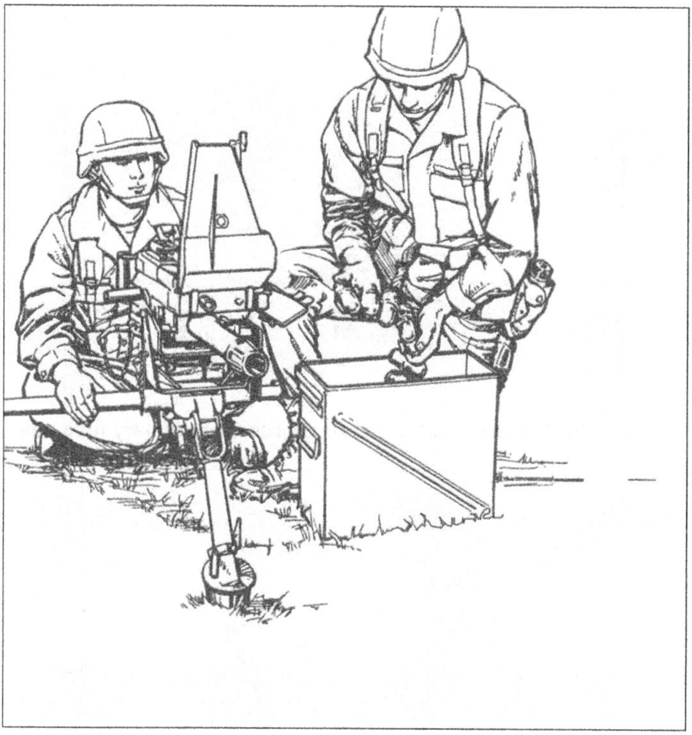

Figure H-9. Alignment of ammunition belt.

d. The gunner opens the top cover. The assistant gunner inserts the first round into the feeder (female link first), pushes the first round across the primary feed pawl (Figure H-10), and moves the feed slide assembly to the left by pushing the secondary drive lever to the right (Figure H-11).

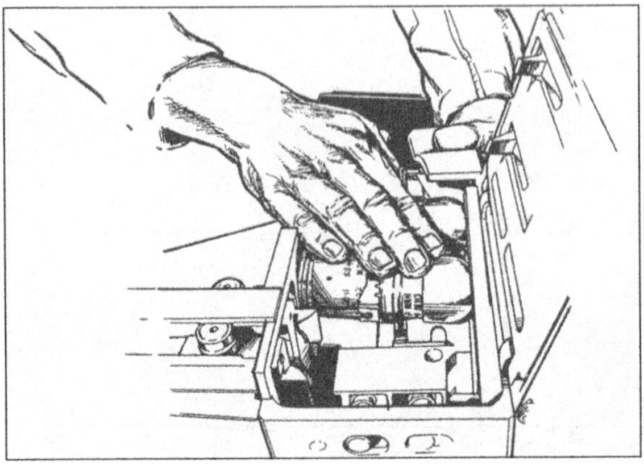

Figure H-10. Placement of the round across the primary feed pawl.

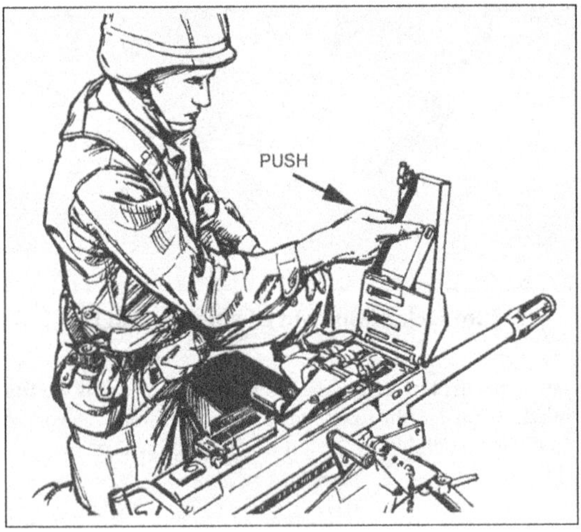

Figure H-11. Feed slide assembly moved to the left.

e. The gunner closes the top cover and charges the weapon (Figure H-12). The gunner presses the trigger, which places the lead round in the bolt extractor. He charges the weapon a second time, forcing the lead round from the bolt extractors into the bolt fingers. The weapon is ready to fire. When the assistant gunner raises the rear sight, the gunner calls "Up" (Figure H-13).

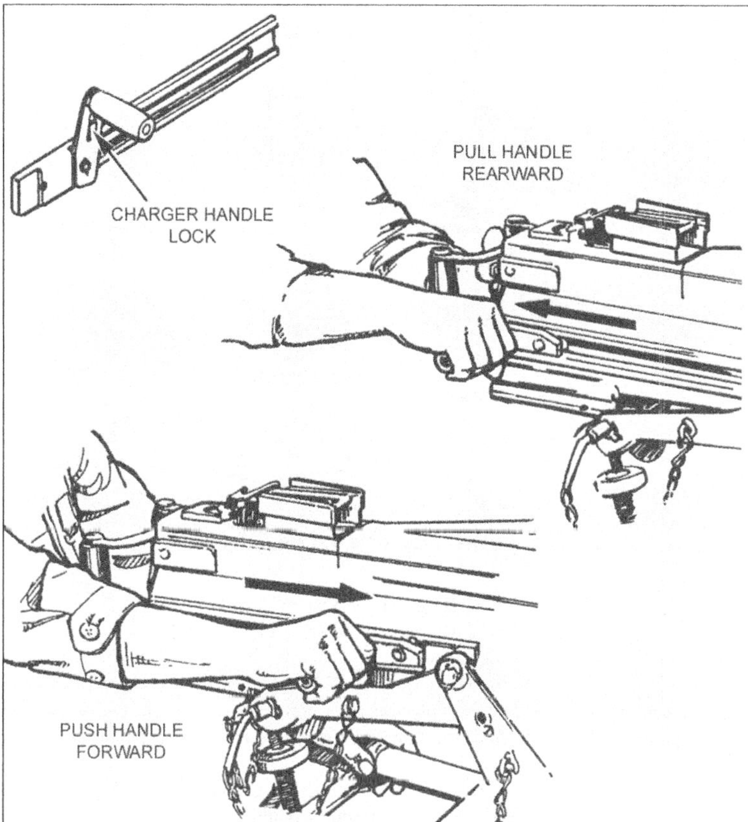

Figure H-12. Gunner charging the MK 19.

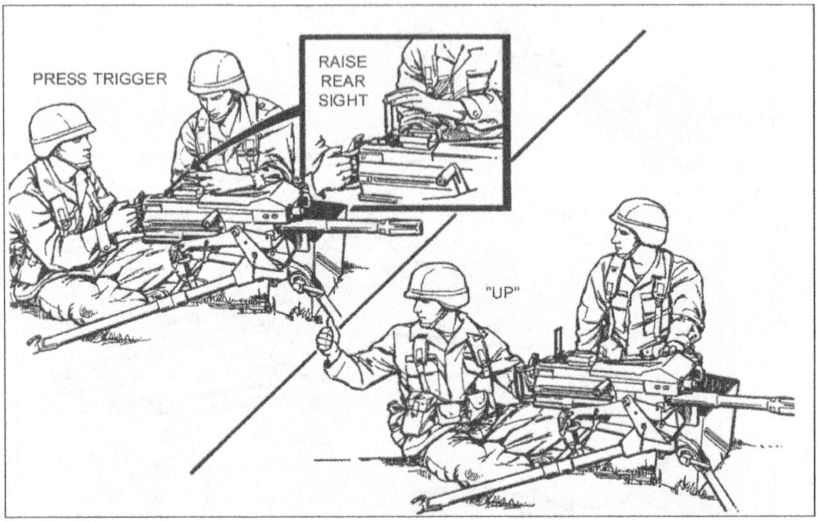

Figure H-13. Raising the rear sight.

H-5. TAKING THE MK 19 OUT OF ACTION

To take the gun out of action, the leader commands OUT OF ACTION. Each crewmember accomplishes the following in sequence.

a. The gunner raises the top cover of the MK 19. The assistant gunner releases the ammunition belt from the feed tray and places it in the ammunition can. The gunner takes the round or spent cartridge case from the bolt fingers, places it in the ammunition can, and closes the top cover.

b. The assistant gunner releases the MK 19 from the gun cradle. The gunner lifts the MK 19 off the mount and moves to a position 10 paces behind the tripod. He places the MK 19 on the ground in front of him with the flash suppressor to the left and with the sights on the top.

c. Once the gunner has moved away from the tripod, the assistant gunner:

(1) Releases the T&E mechanism and takes off the gun cradle/ T&E mechanism.

(2) Releases the tripod sleeve lock latch.

(3) Grasps the front leg and rotates the tripod to a vertical position.

(4) Releases the sleeve latch, and collapses the trail legs.

(5) Loosens the front leg clamp, folds down the front leg to a 60-degree angle, and tightens the clamp.

(6) Returns, with the tripod and gun cradle, to a position five paces in front of the gunner.

d. The ammunition bearer comes forward and gets the ammunition can(s) or any other loose equipment, and returns to a position five paces behind the gunner.

e. The gunner announces "Up."

H-6. TRANSPORT OF THE TRIPOD-MOUNTED MK 19

To move the MK 19 and equipment a short distance, the gunner commands SECURE EQUIPMENT, FOLLOW ME. The MK 19 can be moved using one of three methods:

a. **Dragging.** The gunner grasps the front leg and pulls the MK 19 to the new position, and the assistant gunner secures any loose equipment.

b. **Two-man Carry.** The gunner stands behind the tripod with a trail leg in each hand. The assistant gunner grasps the front leg and avoids touching the barrel if it is hot. The gunner and assistant gunner carry the mounted MK 19. The assistant gunner returns to secure any loose equipment.

c. **Three-man Carry.** The gunner and assistant gunner each holds a trail leg just above the traverse bar. The ammunition bearer grasps the front leg and avoids touching the barrel if it is hot. The crew moves to the new position. The ammunition bearer and the assistant gunner go back and get the ammunition or loose equipment, and return to the new position.

H-7. DISMOUNT THE MK 19

To dismount the MK 19 from the vehicle-mounted or stowed position, each member of the crew performs the following steps in order within four minutes.

a. To dismount the MK 19 from the mounted position the crew performs the following actions.

(1) The squad leader:

(a) Dismounts the vehicle with binoculars and compass.

(b) Obtains the M3 tripod from the stow position.

(c) Takes the M64 gun cradle and T&E mechanism from the gunner and moves to the ground firing position.

(d) At the firing position, the gunner aligns the front tripod leg for direction and snaps the trail legs apart.

(e) Stamps the tripod legs into the ground.

(f) Zero's the T & E mechanism, if it is not zeroed.

(g) Secures the gun cradle to the M3 tripod and the T&E mechanism to the traverse bar and gun cradle.

(h) Assists the gunner in mounting the MK 19.

(2) The gunner:

(a) Raises the top cover, releases the ammunition belt from the feed tray, and places it in the ammunition can.

(b) Takes the round or spent cartridge case from the bolt fingers, places it in the ammunition can, and closes the top cover.

(c) Lifts the MK 19 from the gun cradle after the driver releases it, and hands it to the driver.

(d) Releases the T&E mechanism from the pedestal with the train lock handle and removes the empty case catch bag assembly.

(e) Removes the gun cradle with the T&E mechanism attached and hands it to the squad leader.

(f) Takes the MK 19 from the driver and moves to the firing position.

(g) With the squad leader's assistance, the gunner secures the MK 19 to the gun cradle.

(3) The driver:

(a) Dismounts the vehicle and assists the gunner in clearing the weapon.

(b) Removes the ammunition can and releases the MK 19 and takes it from the gunner.

(c) Gives the MK 19 to the gunner when the gunner is ready.

(d) Stows the empty case catch bag assembly and takes the ammunition can to the firing position, while also removing the lid and ensuring the ogive ends of the rounds are pointed downrange.

(e) Returns to the vehicle and monitors the radio and/or takes more ammunition to the firing point, and maintains rear security.

(4) The squad leader and the gunner load the weapon (as in a crew drill), and place the MK 19 into action.

b. To dismount the MK 19 from the stowed position the crew performs the following actions.

(1) The squad leader:

(a) Secures the binoculars and compass.

(b) Obtains the M3 tripod, the M64 gun cradle, and the T&E mechanism from their stowed position.

(c) Moves to the ground firing position where he aligns the front tripod leg for direction and snaps the trail legs apart.

(d) Stamps the tripod legs into the ground.

(e) If required, zeros' the T&E mechanism to the traverse bar and gun cradle.

(f) Assists the gunner in mounting the MK 19.

(2) The gunner:

(a) Dismounts the vehicle and obtains the MK 19 from its stowed position.

(b) Moves to the firing position with the MK 19.

(c) Secures the MK 19 into the gun cradle with the squad leader's assistance.

(3) The driver:

(a) Dismounts the vehicle and moves to the firing position with a can of ammunition.

(b) Assists in obtaining both the M3 tripod and the MK 19 from the stowed position.

(c) Removes the lid from the ammunition can and ensures the ogive ends of the rounds are pointed downrange.

(d) Returns to the vehicle, monitors the radio and/or takes more ammunition to the firing position, and maintains rear security.

(4) The squad leader and the gunner load the weapon as in the crew drill and place the MK 19 into action.

H-8. REMOUNT THE MK 19

The vehicle crew must be able to mount the MK 19 from the ground-mount to a vehicle-mount or stowed position within four minutes.

a. To mount the MK 19 from the ground to the vehicle the crew performs the following actions.

(1) The squad leader assists the gunner in clearing the weapon by releasing the ammunition belt from the feed tray, and placing it in the ammunition can. He then:

(a) Removes the ammunition can and places it on the ground.

(b) Releases the MK 19 from the gun cradle.

(c) Releases the T&E mechanism from the tripod traverse rod and takes off the gun cradle and T&E mechanism, after the gunner removes the MK 19.

(d) Grasps the front leg of the tripod and rotates the tripod to vertical position, releases the sleeve latch, and collapses the trail legs.

(e) Loosens the front leg clamp, folds down the front leg, and tightens the clamp.

(f) Returns to the vehicle, gives the gun cradle/T&E mechanism to the gunner, and stows the M3 tripod.

(g) Stows the binoculars and compass in the vehicle and resumes his duties.

(2) The gunner raises the top cover of the MK 19 and takes the round or spent cartridge case from the bolt fingers, and places it in the ammunition can, and closes the top cover. He then:

(a) Lifts the MK 19 off the gun cradle when the squad leader releases it and moves to the vehicle.

(b) Gives the MK 19 to the driver and mounts the vehicle.

(c) Takes the gun cradle and T&E mechanism from the squad leader and mounts the gun cradle to the pedestal.

(d) Zeros' the T&E mechanism if needed and reattaches it to the gun cradle.

(e) Attaches the T&E mechanism to the pedestal and clamps it with the train lock handle.

(f) Takes the MK 19 from the driver and installs it on the gun cradle.

(g) Attaches the travel lock assembly if movement is imminent.

(h) Takes the empty case catch bag assembly from the driver and attaches it to the gun cradle.

(3) The driver receives the MK 19 from the gunner and gives it to him when he is ready. He then:

(a) Assists the gunner in installing the MK 19 in the gun cradle.

(b) Moves to the firing position, retrieves the ammunition can and any loose equipment, and returns to the vehicle.

(c) Places the ammunition can on top of the vehicle, with the ogive ends of the rounds pointed down range.

(d) Hands the empty case catch bag assembly to the gunner.

(4) The squad leader and the gunner load the weapon (as in the crew drill) and place the MK 19 into action.

b. To dismount the MK 19 from the ground mount to a stowed position, the crew performs the following actions.

(1) The squad leader:

(a) Assists the gunner in clearing the weapon by releasing the ammunition belt from the feed tray, and places it in the ammunition can.

(b) Removes the ammunition can and places it on the ground. He then releases the MK 19 from the gun cradle.

(c) Releases the T&E mechanism from the tripod traverse rod and takes off the gun cradle/T&E mechanism after the gunner removes the MK 19.

(d) Grasps the front leg of the tripod and rotates the tripod to a vertical position, releases the sleeve latch, collapses the trail legs, loosens the front leg clamp, folds down the front leg, and tightens the clamp.

(e) Returns to the vehicle and stows the M3 tripod, T&E mechanism, binoculars, and compass.

(f) Resumes his duties.

(2) The gunner:

(a) Raises the top cover of the MK 19 and takes the round or spent cartridge case from the bolt fingers.

(b) Places the round in the ammunition can, and closes the top cover.

(c) Lifts the MK 19 off the gun cradle when the squad leader releases it and moves to the vehicle.

(d) Stows the MK 19 in its stowage position.

(3) The driver:

(a) Moves to the firing position.

(b) Collects the ammunition can and any loose equipment.

(c) Stows the ammunition can and assists the squad leader and gunner in stowing any of their equipment.

H-9. CONVERT THE M966 HMMWV TOW MISSILE CARRIER FROM MK 19 MODE TO TOW 2 MODE

To convert the M966 weapon system from the MK 19 mode to the TOW 2 mode, the crew performs the following actions.

a. The squad leader:

(1) Places the traversing unit in the traveling position.

(2) Obtains the nightsight and collimation cases from the stowed position and places them on top of the vehicle near the gunner.

(3) Stows the gun cradle/ammo bracket, T&E mechanism, and empty case catch bag assembly.

(4) Unstraps the missile guidance system (MGS), unlatches its cover, and stows the MGS in the commander's footwell.

(5) Connects the J-1 cable and the battery power conditioner cable to the MGS and MGS battery power conditioner.

(6) Connects the nightsight power conditioner cable to the TOW vehicle power cable (TVPC), uncoils the cable and hands it up to the gunner.

(7) Ensures that the power cable is connected to the TVPC, and that the TVPC switch is on.

(8) Assists the gunner in boresighting the system.

b. The gunner :

(1) Raises the top cover and releases the ammunition belt from the feed tray and places it in the ammunition can.

(2) He takes the round or spent cartridge case from the bolt fingers, places it in the ammunition can, and closes the top cover.

(3) The gunner releases the MK 19 from the gun cradle, lifts the MK 19 out of the gun cradle, and gives it to the driver.

(4) The gunner releases the T&E mechanism from the pedestal with the train lock handle.

(5) Removes the empty case catch bag assembly and gives it to the squad leader. The gunner moves the ammo can and gives it to the driver.

(6) Removes the gun cradle with the T&E mechanism attached and hands it to the squad leader.

(7) Opens the pedestal mount cover and secures it. Lifts the traversing unit with adapter from the stowage base and secures the traversing unit to the pedestal mount with the coupling clamp.

(8) Receives the launch tube from the driver and secures it to the traversing unit.

(9) Receives the optical sight from the driver and installs it on the traversing unit.

(10) Opens the nightsight case, removes nightsight device and installs it on the optical sight and hands the case to the driver.

(11) Takes the TVPC cable from the squad leader and installs the nightsight device.

(12) Opens the collimation case, removes the collimator, attaches to nightsight, and connects collimator power cable.

(13) Assists the squad leader in boresighting.

(14) Collimates sights, removes collimator and power cable, stores collimator and cable in case, and hands case to the driver.

(15) Receives the missile from the driver and loads it in the system.

c. The driver:

(1) Dismounts the vehicle and enters the right rear passenger door and then unlocks and opens the cargo hatch.

(2) Stows the MK 19 when received from the gunner.

(3) Stows the ammunition can when received from the gunner.

(4) Places the optical sight on top of the vehicle for the gunner.

(5) Unstraps a missile, inspects the missile, and gives the missile to the gunner.

(6) Ensures every item in the cargo hatch is secure, closes the cargo hatch, exits the vehicle through the right rear passenger door and takes his place as driver.

H-10. CONVERT THE M966 HMMWV TOW MISSILE CARRIER FROM TOW 2 MODE TO MK 19 MODE

This exercise requires the crew to accomplish its task within three minutes. To convert the M966 weapon system from the TOW 2 mode to the MK 19 mode and put the TOW 2 into its stowed position, the crew performs the following actions.

a. The squad leader:

(1) Disconnects the J-1 cable from the MGS, coils the cable and stows in the traversing unit.

(2) Disconnects the battery power conditioner cable from the MGS battery power cable, and winds it on the stowage handle.

(3) Disconnects the MGS battery cable, recovers the MGS cover, stows the cable in the MGS, replaces the cover on the MGS, and straps the MGS down.

(4) Disconnects the nightsight power cable from the TVPC, and winds and stows the cable.

(5) Receives the nightsight case from the gunner and stows it in the vehicle.

(6) Gives the gunner the gun cradle/ammo bracket, T&E mechanism, and empty case catch bag assembly.

b. The gunner:

(1) Arms the missile (if armed), unlocks the bridge clamp, removes the missile and hands it to the driver.

(2) Disconnects the TVPC power cable from the night sight and drops down the hatch.

(3) Disconnects the nightsight and places it in the case, and hands the case to the squad leader.

(4) Disconnects the optical sight and hands it to the driver.

(5) Disconnects the launch tube, and hands it to driver.

(6) Unclasps the traversing unit, lifts the traversing unit out of the pedestal mount, and takes it through the gunners hatch.

(7) Secures it to the TU stowage base, and closes the pedestal mount access cover.

(8) Takes the gun cradle/ammo bracket and T&E mechanism from the squad leader.

(9) Installs the cradle/ammo bracket on the heavy interdiction missile system (HIMS) mount.

(10) Ensures the T&E mechanism is zeroed, installs the T&E mechanism to the gun cradle and the pedestal mount on the HIMS.

(11) Takes the MK 19 from the driver and installs it on the gun cradle.

(12) He installs the empty case catch bag assembly on the gun cradle.

(13) Receives ammunition can from the driver, while ensuring that the ogive ends are pointing down range, and loads the MK 19.

c. The driver:

(1) Dismounts the vehicle and enters the right rear passenger door, and unlocks and opens the cargo hatch.

(2) Takes the missile from the gunner and stows it in the missile rack.

(3) Unstraps the nightsight case and gives it to the gunner.

(4) Takes the optical sight from the gunner and stows it in the vehicle.

(5) Takes the launch tube from the gunner and stows it in the vehicle.

(6) Unstraps the MK 19 from its stowage, and gives it to the gunner.

(7) Unstraps one can of ammunition and places it on top of the vehicle near the gunner.

(8) Replaces the forward handling ring and dust cover on the missile.

(9) Ensures that every item in the cargo hatch is secure and closes the cargo hatch.

(10) Exits the vehicle through the right rear passenger door and takes his place as the driver.

H-11. ENGAGE TARGETS WITH THE MK 19

The MK 19 squad must be able to successfully engage a target while occupying a firing position. This crew drill is based on the procedures found in Section III of Chapter 5, Application of Fire. The purpose of this drill is to exercise proper fire techniques to successfully engage a target while not hitting friendly troops or attempting to hit a target out of range. The crew occupies a dismounted position when the squad leader announces, "Fire mission."

a. The squad leader then performs the following tasks:

(1) Announces direction using one of the following methods: orally (example:"Right front," or "Left flank"), by a reference point (example:"TRP 1," "Lone tree."), fires tracer from a weapon or another MK 19, or by laying the gun on the target himself.

(2) Announces target description "Troops," "Tank," or if more than one type of target, designated target, "Lead truck," "Right building, far end."

(3) Announces target range, for example, "Four five zero," "One one hundred."

(4) Takes the following corrective actions, as required. If the gunner announces, "Cannot identify," the squad leader attempts to identify and gives further instructions. If the gunner announces "Lost," the squad leader gives additional target location. If the gunner announces "Friendly, the squad leader gives new target description or gives end of mission.

(5) Announces the method of engagement by announcing the class and rate of fire.

(6) Announces method of control.

(7) Issues subsequent fire commands as needed.

(8) Terminates the alert: CEASE FIRE, END OF MISSION.

b. The gunner performs the following tasks:

(1) Attempts to acquire target. If he cannot, the gunner announces "Cannot identify," "Lost," or "Friendly."

(2) When identified, announces "Identified," or "Up."

(3) Makes an adjustment to the lay of the gun as the squad leader gives new data.

c. The driver performs the following tasks:

(1) Provides security for the vehicle or position.

(2) Prepares to get more ammunition to the gun as required.

(3) Prepares to move if the squad leader announces "Cease fire," or "End of mission."

APPENDIX I

GUNNERY EXERCISES

There are seven exercises used to train MK 19 gunners and crews. The first four exercises are individual gunnery firing exercises and allow the gun crew to practice firing prior to the collective gunnery exercises. Commanders dictate the level of MOPP used during MK 19 gunnery.

These seven tables are used for single and multiple gun-crew practice and can be conducted on any MK 19 range. Each table has columns for the task, condition, standard, ammunition required, time required, and a column to record a satisfactory or unsatisfactory (SAT Y/N) score for the task.

I-1. INDIVIDUAL GUNNERY

These first four tables are used for gunnery practice. Each single gun exercise requires the gun crew to mount and dismount the MK 19, and engage point and area targets at various ranges. The two dismounted day exercises require the gun crew to prepare a range card and engage targets using data from the range card (Tables I-1 and Table I-2). The third exercise (Table I-3) is a mounted combat exercise and the fourth exercise (Table I-4) is a night fire exercise.

a. **Instructional Fire Exercise** (Table I-1**).** The instructional firing exercise gives the crew practice firing on a range.

ACTION	CONDITIONS	STANDARDS	AMMO	TIME	SAT Y/N
1. Mount the MK 19 on the M3 tripod.	Given a MK 19, M3 tripod, T&E mechanism, MK 64 gun cradles, and a firing position.	Gun crew must be able to mount the MK 19 in its tripod mode within 1 min.	0	1 min	
2. Prepare a range card.	Given a MK 19 mounted on a M3 tripod, and a firing position with a designated sector of fire and targets.	Gunner must prepare a completed range card within 15 min.	0	15 min	
3. Zero the MK 19 while using a 400-meter BMP target.	Given a tripod-mounted MK 19, 400-meter BMP target, selected firing position, and 4 rds of 40-mm TP.	Gunner must impact on the target with at least 2 of the 4 rds on the BMP target at 400 meters.	4 rds	N/A	
4. Engage a BMP target at 1,100 meters while using range card data (point target).	Given a tripod-mounted MK 19, completed range card, BMP target at 1,100 meters, selected firing position, and 8 rds of 40-mm TP.	Gunner must impact on the 1,100 meter BMP target with at least 2 out of 8 rounds.	8 rds	N/A	
5. Engage a BMP target at 1,500 meters while using range card data.	Given a tripod-mounted MK 19, completed range card, BMP target at 1,500 meters, selected firing position, and 8 rds of 40-mm TP.	Gunner must impact on the 1,500 meter BMP target area with at least 1 out of 8 rds.	8 rds	N/A	
6. Engage dismounted troops at 600 meters while using range card data (area target).	Given a tripod-mounted MK 19, completed range card, dismounted troop targets at 600 meters, selected firing position, and 6 rds of 40-mm TP.	Gunner must impact at least 4 out of 6 rds within 5 meters of a silhouette at a range of 600 meters.	6 rds	N/A	
7. Dismount the MK 19.	Given a MK 19, M3 tripod, and a selected firing position.	Gun crew must dismount the MK 19 from the M3 tripod within 1 min.	0	1 min	
		Total:	**26 rds**		

Exercise Scoring

Fully Trained (T) = Satisfactory on 7 of 7 tasks
Need practice (P) = Satisfactory on 5 or 6 tasks
Untrained (U) = Unsatisfactory on 4 or fewer tasks

Table I-1. Instructional fire exercise.

During scoring procedures, the grader positions himself so that he can observe both the gunner and the target. Once the exercise commences, he:

- Times the tasks having a time standard.
- Observes and informs the gunner of the strike of the rounds.
- Records the results of each task in the right hand column with a "Y" for a satisfactory completion of the task or an "N" for an unsatisfactory completion.
- Sums the results and assigns an overall score in accordance with the scores in the bottom box of the figure above.

b. **Dismounted Range Card Exercise** (Table I-2). The dismounted range card exercise is held after instructional fire and gives the gunner and crew a chance to test their skills against a time-and-hit exercise. It is held as often as the commander feels necessary to maintain crew skills.

(1) Time the gunner and crew while they place the tripod-mounted MK 19 into operation in a selected location.

(2) Time the gunner while he completes a range card with the ranges of all targets within his sector of fire.

(3) Have the gunner suppress or kill single or multiple targets in his sector of fire while using his range card data.

(4) Time the gunner and crew while they take the MK 19 out of operation.

ACTION	CONDITIONS	STANDARDS	AMMO	TIME	SAT Y/N
1. Mount the MK 19 on an M3 tripod.	Given a MK 19, M3 tripod, M64 gun mount, T&E mechanism, and a selected firing position.	Gun crew must mount the MK 19 on the M3 tripod within 1 min.	0	1 min	
2. Prepare a range card.	Given a MK 19 mounted on a M3 tripod, a firing position with an assigned sector of fire, and designated targets.	Gunner must prepare a completed range card with all targets in his sector of fire within 15 min.	0	15 min	
3. Zero the MK 19 while mounted on the M3 tripod at a 400-meter BMP.	Given a tripod-mounted MK 19, selected firing position, 400-meter BMP target, and 4 rds of 40-mm TP.	Gunner must impact at least 2 out of 4 rds on the 400-meter BMP target.	4 rds	N/A	
4. Engage a 600-meter and tripod-mounted BMP target while using range card data (point target).	Given a tripod-mounted MK 19, selected firing position, 600-meter BMP target, and 4 rds of 40-mm TP.	Gunner must impact at least 3 out of 4 rds on the 600-meter BMP target within 1 min.	4rds	1 min	

Table I-2. Dismounted range card exercise.

ACTION	CONDITIONS	STANDARDS	AMMO	TIME	SAT Y/N
5. Engage a 1,100-meter BMP target while using range card data.	Given a tripod-mounted MK 19, selected firing position, BMP target at 1,100 meters, completed range card, and 8 rds of 40-mm TP.	Gunner must impact on BMP target at 1,100 meters with at least 2 out of 8 rds within 2 min.	8 rds	2 min	
6. Engage a 600-meter (area) troop target while using the range card data.	Given a tripod-mounted MK 19, 6 rds of 40-mm TP, selected firing position, and a troop target at 600 meters.	Gunner must impact at least 4 out of 6 rds within 5 meters of a troop silhouette at a range of 600.	6 rds	1.5 min	
7. Engage a BMP target at 1,500 meters while using the range card data.	Given a tripod-mounted MK 19, 8 rds of 40-mm TP, selected firing position, BMP target at 1,500 meters, and a completed range card.	Gunner must impact on the 1,500 meter BMP target with at least 4 out of 8 rds within 2 min.	8 rds	2 min	
8. Engage a BMP target at 1,100 meters and dismounted troops at 600 meters while using range card data.	Given a tripod-mounted MK 19, 12 rds of 40-mm TP, BMP target at 1,100 meters, dismounted troop target at 600 meters, and a completed range card.	Gunner must impact at least 2 rds on the BMP target at 1,100 meters, then shift to a dismounted troop target and place at least 2 rds within 5 meters of a troop silhouette at a range of 600 meters.	12 rds	3 min	
9. Dismount the MK 19 from the M3 tripod.	Given a mounted MK 19.	Gun crew must take the MK 19 out of operation from the tripod mount within 1 min.	0	1 min	
		Total:	**42 rds**		

Exercise Scoring

Fully Trained (T) = Satisfactory on 9 of 9 tasks
Need Practice (P) = Satisfactory on 7 or 8 tasks
Untrained (U) = Unsatisfactory on 6 or fewer tasks

Table I-2. Dismounted range card exercise (continued).

During scoring procedures, the grader positions himself so that he can observe both the gunner and the target. Once the exercise commences, he:

- Times the tasks having a time standard.
- Observes and informs the gunner of the strike of the rounds.
- Records the results of each task in the right hand column with a "Y" for a satisfactory completion of the task or an "N" for an unsatisfactory completion.
- Sums the results and assigns an overall score in accordance with the scores in the bottom box of the figure above.

c. **Mounted Combat Exercise.** The mounted combat exercise gives the gunner and the crew practice fighting from a vehicle. It is held as often as the commander feels necessary to maintain crew skills (Table I-3).

(1) Time the gun crew while they place the MK 19 into operation on its organic carrier. Carriers may be HMMWVs, 2 1/2-ton trucks, M151-series utility trucks, or M113-series APCs.

(2) Using the time-and-hit standards, give the gunner and crew a series of single and multiple targets to suppress or kill in their sector of fire.

(3) Upon completion of mounted firing, time the gunner and crew while they take the MK 19 out of operation.

ACTION	CONDITIONS	STANDARDS	AMMO	TIME	SAT Y/N
1. Mount the MK 19 on its organic carrier.	Given a MK 19, organic carrier, and a selected firing position.	Gun crew must mount the MK 19 on its organic carrier within 2 min.	0	2 min	
2. Engage a BMP target at 400 meters while mounted.	Given a MK 19, organic carrier, 4 rds of 40-mm TP, and a selected firing position.	Gunner must impact on the BMP target with at least 3 out of 4 rds at 400 meters within 1 min.	4 rds	1 min	
3. Engage a BMP target at 800 meters while mounted.	Given a MK 19, organic carrier, 8 rds of 40-mm TP, and a selected firing position.	Gunner must impact a BMP target at 800 meters with at least 3 out of 8 rds within 1.5 min.	8 rds	1.5 min	
4. Engage a 600-meter (area) target.	Given a MK 19, organic carrier, 6 rds of 40-mm TP, and a selected firing position.	Gunner must impact at least 4 out of 6 rds within 5 meters of a troop silhouette at a range of 600 meters.	6 rds	1.5 min	
5. Engage a BMP target at 1,100 meters.	Given a MK 19, organic carrier, 8 rds of 40-mm TP, and a selected firing position.	Gunner must impact the BMP target area at 1,100 meters with at least 2 out of 8 rds within 2 min.	8 rds	2 min	
6. Dismount the MK 19 from its organic carrier.	Given a MK 19 and an organic carrier.	Gun crew must dismount the MK 19 from its organic carrier within 2 min.	0	2 min	
			Total: 26 rds		

Exercise Scoring

Fully Trained (T) = Satisfactory on 6 of 6 tasks
Need Practice (P) = Satisfactory on 5 tasks
Untrained (U) = Unsatisfactory on 4 or fewer tasks

Table I-3. Mounted combat exercise.

During scoring procedures, the grader positions himself so that he can observe both the gunner and the target. Once the exercise commences, he:

- Times the tasks having a time standard.
- Observes and informs the gunner of the strike of the rounds.
- Records the results of each task in the right hand column with a "Y" for satisfactory completion of the task or an "N" for an unsatisfactory completion.
- Sums the results and assigns an overall score in accordance with the scores in the bottom box of the figure above.

d. **Night Fire Exercise** (Table I-4). This LFX develops and improves MK 19 gunnery skills at night or during degraded conditions. This exercise is conducted as often as the commander feels it is necessary to maintain crew skills.

ACTION	CONDITIONS	STANDARDS	AMMO	TIME	SAT Y/N
1. Mount the AN/TVS-5 to the MK 19.	At night, given an AN/TVS-5 complete, a MK 19, M3 tripod, and a T&E mechanism.	Gunner must mount the AN/TVS-5 on the MK 19 within 2 min.	0	2 min	
2. Engage a 400-meter BMP target using a zeroed MK 19 with an AN/TVS-5 mounted for observation.	At night, given an AN/TVS-5 complete, a MK 19, an M3 tripod, 6 rds of 40-mm TP, and a 400-meter BMP target, during the hours of daylight.	Gunner must impact 3 out of 6 rds on a BMP target at 400 meters within 1 min.	6 rds	1 min	
3. Engage a BMP target at 600 meters.	At night, given a MK 19, 8 rds of 40-mm TP, and a target at a distance of 600 meters.	Gunner must impact on a BMP target with at least 4 out of 8 rds at 600 meters within 2 min.	8 rds	2 min	
4. Engage a dismounted troop target at 800 meters.	At night, given a MK 19, 8 rds of 40-mm TP, and a target at a distance of 800 meters.	Gunner must impact within 5 meters of a silhouette target with at least 4 out of 8 rds at 800 meters within 2 min.	8 rds	2 min	
5. Dismount AN/TVS-5 from the MK 19.	At night, given a MK 19 and an AN/TVS-5.	Gun crew must take the AN/TVS-5 out of operation within 2 min.	0	2 min	
			Total: 22 rds		
Exercise Scoring **Fully Trained (T)** = Satisfactory on 5 of 5 tasks **Need Practice (P)** = Satisfactory on 4 tasks **Untrained (U)** = Unsatisfactory on 3 or fewer tasks					

Table I-4. Night fire exercise.

During scoring procedures, the grader positions himself so that he can observe both the gunner and the target. Once the exercise commences, he:

- Times the tasks having a time standard.
- Observes and informs the gunner of the strike of the rounds.
- Records the results of each task in the right hand column with a "Y" for a satisfactory completion of the task or a "N" for an unsatisfactory completion.

- Sums the results and assigns an overall score in accordance with the scores in the bottom box of the figure above.

I-2. COLLECTIVE GUNNERY

These three exercises test the application of fire from two or more guns against linear, deep, linear with depth, and area targets. They test collective individual skills, fire control, leader skills, adjustment of fire, methods of target engagement, and the control of one or more fire units. Tables I-5 through I-7 should be used with unit LFXs. They can be held on any range or piece of terrain that supports the MK 19. Targets may be made up of 55-gallon drums, hulks of old vehicles, silhouettes, chalk, panel markers, or engineer tape.

ACTION	CONDITIONS	STANDARDS	AMMO	TIME	SAT Y/N
1. Engage a linear target with a pair of MK 19s (two guns). (see also Task 2)	Given a linear target 100 meters wide from 800 to 1,200 meters, two MK 19s on tripods with T&E mechanisms, during the hours of daylight or under degraded conditions.	Gun crews must cover 90 percent of the linear target with effective fire while performing the following: 1. The leader lays the guns on his respective flanks. 2. The leader issues the fire command for engaging a linear target. 3. Gunners engage using traversing fire. 4. Observers ensure that fires remain on the target and adjust appropriately.	40 rds	2 min	
2. Engage a linear target with two pairs of MK 19s (four guns).	Given a linear target 200 meters wide from 800 to 1,200 meters, four MK 19s on tripods with T&E mechanisms, during the hours of daylight or under degraded conditions.	Same as Task 1, except the leader may have to subdivide the target.	80 rds	3 min	

Table I-5. Linear and deep targets.

ACTION	CONDITIONS	STANDARDS	AMMO	TIME	SAT Y/N
3. Engage a deep target with a pair of MK 19s (two guns).	Given a deep target 100 meters long with midrange to the target from 800 to 1,200 meters, two MK 19s on tripods with T&E mechanisms, during the hours of daylight or under degraded conditions.	Gun crews must cover 90 percent of the deep target with effective fire while performing the following: 1. The leader lays the guns on his respective ends of the target. 2. The leader issues the fire command for engaging a deep target. 3. Gunners engage using searching fire. 4. Observers ensure that fire remains on the target and adjust appropriately.	40 rds	2 min	
4. Engage a deep target with two pairs of MK 19s (four guns).	Given a deep target 200 meters long with midrange to the target from 800 to 1,200 meters, four MK 19s on tripods with T&E mechanisms, during the hours of daylight or under degraded conditions.	Gun crews should cover 90 percent of the deep target with effective fire while performing the same steps in task 3, except the leader may have to subdivide the target.	80 rds	3 min	
			Total: 240 rdc		

Table I-5. Linear and deep targets (continued).

ACTION	CONDITIONS	STANDARDS	AMMO	TIME	SAT Y/N
1. Engage a linear target with depth using a pair of MK 19s (two guns).	Given a target 100 meters wide and 50 meters long (or 50 meters wide and 100 meters long), from 800 to 1,200 meters, two MK 19s with tripods and T&E mechanisms, during daylight or under degraded conditions.	Gun crews must cover 80 percent of the linear with depth target with effective fire while performing the following: 1. The leader lays the guns on the gun crews' respective flanks (ends) of the target. 2. The leader issues the fire command for engaging a linear target with depth. 3. Gunners engage using traversing and searching fire. 4. Observers ensure that fire remains on the target and adjust appropriately.	40 rds for each target.	3 min	
2. Engage a linear target with depth using two pairs of MK 19s (four guns).	Given a target 200 meters wide and 50 meters long (or 50 meters wide and 200 meters long), from 800 to 1,200 meters, four MK 19s with tripods and T&E mechanisms, during daylight or under degraded conditions.	Same as Task 1, except the leader may want to subdivide the target.	80 rds for each target.	4 min	
			Total: 120 rds		

Table I-6. Linear targets with depth.

ACTION	CONDITIONS	STANDARDS	AMMO	TIME	TPU
1. Engage an area target with two pairs of MK 19s (four guns).	Given an area target from 1,000 to 1,400 meters (midrange) extending over 200 meters in length and width, four MK 19s on tripods with T & E mechanisms, during the hours of daylight or under degraded conditions.	The combined gun crews must cover 70 percent of the area target with suppressive fire while performing the following: 1. The leader determines direction and elevation to the area target (midrange). 2. The leader divides the area target and ensures that each gun is laid on its portion, and assigns TRPs. 3. Each gun registers on its respective TRP. 4. The leader issues the fire command for an area target. 5. All MK 19s engage, gunners use traverse and search manipulations to adjust and distribute fires. 6. Observers ensure the fires remain on target and adjust appropriately.	160 rds	5 min	
2. Engage an area target with three pairs of MK 19s (six guns).	Given an area target from 1,000 to 1,400 meters (midrange), extending over 300 meters in length and 200 meters in width, six MK 19s on tripods with T&E mechanisms, during the hours of daylight or under degraded conditions.	The combined gun crews must cover 70 percent of the area target with suppressive fire while performing the following: (Performance measures are the same as for Task 1.)	240 rds	8 min	
			Total: 400 rds		

Table I-7. Area targets.

GLOSSARY

ACRONYMS AND ABBREVIATIONS

AAR	after-action review
AE	angle of elevation
AI	assistant instructor
AIT	advanced individual training
APC	armored personnel carrier
AS	angle of sight
BMP	fighting vehicle manufactured by the former Soviet Union
BOD	bore obstruction detector
BRDM	Soviet Union vehicle
C4	composition four
CALFEX	combined arms live-fire exercise
CLP	cleaner, lubricant, preservative
cm	centimeter
DA	Department of the Army
deg	degrees
DODIC	Department of Defense Identification Code
EA	engagement area
elev	elevation
EOD	explosive ordinance disposal
FM	field manual
FOV	field of view
fps	feet per second
ft	feet
FTX	field training exercise
GMD	grease, molybdenum disulfide
HE	high explosive
HEDP	high-explosive dual-purpose
HIMS	heavy interdiction missile system
HMMWV	high-mobility multipurpose wheeled vehicle
HWTS	heavy weapon thermal sight
IAW	in accordance with
IR	infrared
JP-4	jet propulsion fuel

LAW	lubricant, automatic weapons, arctic
LBE	load-bearing equipment
LED	light emitting diode
LFX	live-fire exercise
LH	left hand
LSA	lubricant oil, semifluid, automatic weapon
LSAT	lubricant oil, semifluid, automatic weapons, temperate
MACS	multipurpose arcade combat simulator
MGS	missile guidance system
min	minute
mm	millimeter
MOD	model
MOPP	mission-oriented protective posture
MPMG	multipurpose machine gun
MPRC	multipurpose range complex
MTP	mission training plan
MWTS	medium weapon thermal sight
NBC	nuclear, biological, chemical
NCO	noncommissioned officer
NCOES	noncommissioned officers education system
NFOV	narrow field of view
No.	number
NVD	night vision device
PEWS	platoon early warning system
PIBD	point initiating, base detonating
PMCS	preventive maintenance checks and services
QE	quadrant elevation
RBC	rifle bore cleaner
rds	rounds
RH	right hand
RP	reference point
RPG	rocket propelled grenade
sec	seconds
SOP	standing operating procedure
STX	situational training exercise
T&E	traverse and elevating
TESS	tactical engagement simulator system

TM	technical manual
TOE	table of organization and equipment
TP	training practice
TPIAL	target pointer illuminator aiming light
TPT	training practice trainer
TRADOC	U.S. Army Training and Doctrine Command
TRP	target reference point
TVPC	TOW vehicle power cable
TWS	thermal weapon sight
VDT	video disk trainer
W	width
WFOV	wide field of view

REFERENCES

SOURCES USED

These are the sources quoted or paraphrased in this publication.

DA Cir 350-38	Standards in Weapons Training. 3 July 1997.
DA Form 5517-R	Standard Range Cards. February 1986.
DA PAM 385-63	Range Safety. 19 May 2003.
FM 3-22.68	Crew-Served Machine Guns, 5.56-mm and 7.62-mm. 31 January 2003.
FM 3-25.26	Map Reading and Land Navigation. 20 July 2001.
FM 25-2	Unit Training Management. 10 September 1984.
FM 25-3	Training in Units. 12 December 1984.
TC 25-8/ with change	Training Ranges. 16 November 1992.

DOCUMENTS NEEDED

These documents must be available to the intended users of this publication.

* DA Form 5517-R	Standard Range Cards (LRA). February 1986.
DA Form 7518-R	MK 19 Day Practice and Qualification (Hull Targets).
DA Form 7519-R	MK 19 Night Practice and Qualification with Hull Targets and Night Vision Devices.
DA Form 7520-R	MK 19 Day Practice and Qualification (Pop-up Targets).
DA Form 7521-R	MK 19 Night Practice and Qualification with Pop-up Targets or without Night Vision Devices.
* TC 23-13	Crew-Served Weapon Night Vision Sight. 20 January 1967.

* This source was also used to develop this production.

* TM 9-1005-245-13&P/ with 5 changes	Operator's, Unit, and Direct Support Maintenance Manual with Repair Parts and Special Tools List for Machine Gun Mounts and Combinations for Tactical/Armored Vehicles and Ground Mounting M122A1 Machine Gun Tripod Mount, M3 Tripod Mount, M66 Machine Gun Mount Ring, M240E1 Flexible Machine Gun Mount, M142 Machine Gun Mount, .50 Caliber Machine Gun Mount. 22 March 2002.
* TM 9-1010-230-10	Operator's Manual for Machine Gun, 40-mm, MK 19, MOD 3. 30 May 2001.
* TM 9-1010-231-13&P	Operator's, Organizational and Intermediate Maintenance with Repair Parts List and Components List Mount, Machine Gun MK 64: MOD 4, 5, 6, 7, 8 used with Machine Gun, MK 19 MOD 3, 40-mm Machine Gun, M2, Caliber .50 Machine Gun, M60 Machine Gun, 7.62-mm Machine Gun, M60E3, 7.62-mm. 31 October 2000.
* TM 11-5855-214-10	Operator's Manual for Night Vision Sight, Crew-Served Weapon, AN/TVS-5. 15 February 1989.
* TM 11-5855-302-12&P	Operator's and Unit Maintenance Manual, Sight, Thermal, AN/PAS-13 (V2), AN/PAS-13 (V3). 1 January 1997.
* TM 11-5855-308-12&P	Operator's and Unit Maintenance Manual Target Pointer Illuminator/Aiming Light (TPIAL) AN/PEQ-2A. 15 May 2000.
* TM 11-5855-312-10	Operator's Manual, Sight, Thermal AN/PAS- 13B (V2), AN/PAS-13B (V3). 5 January 2001.
* TM 11-5860-201-10	Operator's Manual for Laser Infrared Observation Set, AN/GVS-5. 2 February 1982.

READINGS RECOMMENDED

These readings contain relevant supplemental information.

DoD Reg 5100-76-M	Physical Security of Sensitive Conventional Arms, Ammunition, and Explosives. August 2000.
FM 3-5/ with change	NBC Decontamination. 31 January 2002.
FM 3-21.71	Mechanized Infantry Platoon and Squad (Bradley). 20 August 2002.

* This source was also used to develop this production.

FM 5-250	Explosives and Demolitions. 30 July 1998.

FM 7-8/ with change	Infantry Rifle Platoon and Squad. 22 April 1992.
FM 7-20/ with change	The Infantry Battalion. 29 December 2000.
FM 31-70	Basic Cold Weather Manual. 12 April 1968.

INTERNET WEBSITES

Reimer Digital Library, http://www.adtdl.army.mil.

USAMC Logistics Support Activity, http://www.logsa.army.mil/.

Army Publishing Directorate, http://www.usapa.army.mil.

INDEX

MK 19, 40-mm GRENADE MACHINE GUN, MOD 3 FIRING TABLE I
DAY PRACTICE AND QUALIFICATION WITH HULL TARGETS SCORECARD

For use of this form, see FM 3-22.27; the proponent agency is TRADOC.

PRIVACY ACT STATEMENT

AUTHORITY:	10 USC 3012(g)/Executive order 9397
PRINCIPAL PURPOSE:	To aid individual training on targets at various ranges.
ROUTINE USES:	To evaluate individual proficiency.
DISCLOSURE:	Voluntary. However, mass rating and recording require some tracking method.

1a. LAST NAME	1b. FIRST NAME	1c. MI	2. RANK	3. UNIT

TABLE I (A). DISMOUNTED AND MOUNTED DAY PRACTICE

4. RANGE	5. LANE	6. GRADER	7. DATE

TASK	RANGE (Meters)	AMMO	TIME (Minutes)	ENGAGEMENT STANDARDS	GO	NO GO
ZERO	400	4	NA	2 ROUNDS HIT		
2	1,100 (+/- 200)	8	2.5	2 ROUNDS HIT		
3	1,500 (+/- 200)	10	3.5	2 ROUNDS HIT		
4	600 (+/- 100)	6	2	2 ROUNDS HIT		
5	800 (+/- 100)	6	2	2 ROUNDS HIT		
6	400	4	1.5	2 ROUNDS HIT		
MULTIPLE TARGETS						
7	1,100 (+/- 200)	10	4	1 ROUND HIT		
8	600 (+/- 100)			1 ROUND HIT		
9	800 (+/- 100)	14	4.5	1 ROUND HIT		
10	1,500 (+/- 200)			1 ROUND HIT		
				TOTALS		

8. COMMENTS

9. NUMBER OF ENGAGEMENT MET *(Choose One)*

- ☐ 10 - EXPERT
- ☐ 9 - SHARPSHOOTER
- ☐ 8-7 - MARKSMAN
- ☐ 6 AND BELOW - UNQUALIFIED

10. GUNNER'S SIGNATURE	11. GRADER'S SIGNATURE

TABLE I (B). DISMOUNTED AND MOUNTED DAY QUALIFICATION

12. RANGE	13. LANE	14. GRADER	15. DATE

TASK	RANGE (Meters)	AMMO	TIME (Minutes)	ENGAGEMENT STANDARDS	GO	NO GO
ZERO	400	4	NA	2 ROUNDS HIT		
2	1,100 (+/- 200)	8	2	2 ROUNDS HIT		
3	1,500 (+/- 200)	10	3	2 ROUNDS HIT		
4	600 (+/- 100)	6	1.5	2 ROUNDS HIT		
5	800 (+/- 100)	6	1.5	2 ROUNDS HIT		
6	400	4	1	2 ROUNDS HIT		
MULTIPLE TARGETS						
7	1,100 (+/- 200)	10	3.5	1 ROUND HIT		
8	600 (+/- 100)			1 ROUND HIT		
9	800 (+/- 100)	14	4	1 ROUND HIT		
10	1,500 (+/- 200)			1 ROUND HIT		
				TOTALS		

16. COMMENTS

17. NUMBER OF ENGAGEMENT MET *(Choose One)*

- ☐ 10 - EXPERT
- ☐ 9 - SHARPSHOOTER
- ☐ 8-7 - MARKSMAN
- ☐ 6 AND BELOW - UNQUALIFIED

18. GUNNER'S SIGNATURE	19. GRADER'S SIGNATURE

MK 19, 40-mm GRENADE MACHINE GUN, MOD 3 FIRING TABLE II NIGHT PRACTICE AND QUALIFICATION WITH HULL TARGETS SCORECARD

For use of this form, see FM 3-22.27; the proponent agency is TRADOC.

PRIVACY ACT STATEMENT

AUTHORITY: 10 USC 3012(g)/Executive order 9397
PRINCIPAL PURPOSE: To aid individual training on targets at various ranges.
ROUTINE USES: To evaluate individual proficiency.
DISCLOSURE: Voluntary. However, mass rating and recording require some tracking method.

NOTE: Use these tables if the targets being used are hulls, and any of the following applies:
AN/PEQ-2A mounted on the TWS mounting bracket.
AN/PAS-13 mounted on the TWS mounting bracket.
AN/TVS-5 with the 3d generation tube mounted on the TWS mounting bracket.

1a. LAST NAME	1b. FIRST NAME	1c. MI	2. RANK	3. UNIT

TABLE II (A). DISMOUNTED AND MOUNTED NIGHT PRACTICE

4. RANGE	5. LANE	6. GRADER	7. DATE

TASK	RANGE (Meters)	AMMO	TIME (Minutes)	ENGAGEMENT STANDARDS	GO	NO GO
ZERO	400	4	NA	2 ROUNDS HIT		
2	1,100 (+/- 200)	8	2.5	2 ROUNDS HIT		
3	1,500 (+/- 200)	10	3.5	2 ROUNDS HIT		
4	600 (+/- 100)	6	2	2 ROUNDS HIT		
5	800 (+/- 100)	6	2	2 ROUNDS HIT		
6	400	4	1.5	2 ROUNDS HIT		
MULTIPLE TARGETS						
7	1,100 (+/- 200)	10	4	1 ROUND HIT		
8	600 (+/- 100)			1 ROUND HIT		
9	800 (+/- 100)	14	4.5	1 ROUND HIT		
10	1,500 (+/- 200)			1 ROUND HIT		
				TOTALS		

8. TYPE DEVICE *(Choose One)*
☐ AN/PEQ-2A ☐ AN/PAS-13 ☐ AN/TVS-5

9. NUMBER OF ENGAGEMENT MET *(Choose One)*
☐ 10 - EXPERT ☐ 8-7 - MARKSMAN
☐ 9 - SHARPSHOOTER ☐ 6 AND BELOW - UNQUALIFIED

10. GUNNER'S SIGNATURE	11. GRADER'S SIGNATURE

TABLE II (B). DISMOUNTED AND MOUNTED NIGHT QUALIFICATION

12. RANGE	13. LANE	14. GRADER	15. DATE

TASK	RANGE (Meters)	AMMO	TIME (Minutes)	ENGAGEMENT STANDARDS	GO	NO GO
ZERO	400	4	NA	2 ROUNDS HIT		
2	1,100 (+/- 200)	8	2	2 ROUNDS HIT		
3	1,500 (+/- 200)	10	3	2 ROUNDS HIT		
4	600 (+/- 100)	6	1.5	2 ROUNDS HIT		
5	800 (+/- 100)	6	1.5	2 ROUNDS HIT		
6	400	4	1	2 ROUNDS HIT		
MULTIPLE TARGETS						
7	1,100 (+/- 200)	10	3.5	1 ROUND HIT		
8	600 (+/- 100)			1 ROUND HIT		
9	800 (+/- 100)	14	4	1 ROUND HIT		
10	1,500 (+/- 200)			1 ROUND HIT		
				TOTALS		

16. TYPE DEVICE *(Choose One)*
☐ AN/PEQ-2A ☐ AN/PAS-13 ☐ AN/TVS-5

17. NUMBER OF ENGAGEMENT MET *(Choose One)*
☐ 10 - EXPERT ☐ 8-7 - MARKSMAN
☐ 9 - SHARPSHOOTER ☐ 6 AND BELOW - UNQUALIFIED

18. GUNNER'S SIGNATURE	19. GRADER'S SIGNATURE

MK 19, 40-mm GRENADE MACHINE GUN, MOD 3 FIRING TABLE III
DAY PRACTICE AND QUALIFICATION WITH POP-UP TARGETS SCORECARD

For use of this form, see FM 3-22.27; the proponent agency is TRADOC.

PRIVACY ACT STATEMENT

AUTHORITY: 10 USC 3012(g)/Executive order 9397
PRINCIPAL PURPOSE: To aid individual training on targets at various ranges.
ROUTINE USES: To evaluate individual proficiency.
DISCLOSURE: Voluntary. However, mass rating and recording require some tracking method.

1a. LAST NAME	1b. FIRST NAME	1c. MI	2. RANK	3. UNIT

TABLE III (A). DISMOUNTED AND MOUNTED DAY PRACTICE

4. RANGE	5. LANE	6. GRADER	7. DATE

TASK	RANGE (Meters)	AMMO	TIME (Minutes)	ENGAGEMENT STANDARDS	GO	NO GO
ZERO	400	4	NA	2 ROUNDS HIT		
2	600 (+/- 100)	6	2	1 ROUND HIT		
3	800 (+/- 100)	8	2.5	1 ROUND HIT		
4	400	4	1.5	1 ROUND HIT		
MULTIPLE TARGETS						
5	800 (+/- 100)	12	4	1 ROUND HIT		
6	400			1 ROUND HIT		
7	400	10	3	1 ROUND HIT		
8	600 (+/- 100)			1 ROUND HIT		
				TOTALS		

8. COMMENTS

9. NUMBER OF ENGAGEMENT MET *(Choose One)*

- ☐ 8 - EXPERT
- ☐ 6 - MARKSMAN
- ☐ 7 - SHARPSHOOTER
- ☐ 5 AND BELOW - UNQUALIFIED

10. GUNNER'S SIGNATURE	11. GRADER'S SIGNATURE

TABLE III (B). DISMOUNTED AND MOUNTED DAY QUALIFICATION

12. RANGE	13. LANE	14. GRADER	15. DATE

TASK	RANGE (Meters)	AMMO	TIME (Minutes)	ENGAGEMENT STANDARDS	GO	NO GO
ZERO	400	4	NA	2 ROUNDS HIT		
2	600 (+/- 100)	6	1.5	1 ROUND HIT		
3	800 (+/- 100)	8	2	1 ROUND HIT		
4	400	4	1	1 ROUND HIT		
MULTIPLE TARGETS						
5	800 (+/- 100)	12	3.5	1 ROUND HIT		
6	400			1 ROUND HIT		
7	400	10	2.5	1 ROUND HIT		
8	600 (+/- 100)			1 ROUND HIT		
				TOTALS		

16. COMMENTS

17. NUMBER OF ENGAGEMENT MET *(Choose One)*

- ☐ 8 - EXPERT
- ☐ 6 - MARKSMAN
- ☐ 7 - SHARPSHOOTER
- ☐ 5 AND BELOW - UNQUALIFIED

18. GUNNER'S SIGNATURE	19. GRADER'S SIGNATURE

MK 19, 40-mm GRENADE MACHINE GUN, MOD 3 FIRING TABLE IV NIGHT PRACTICE AND QUALIFICATION WITH POP-UP TARGETS SCORECARD

For use of this form, see FM 3-22.27; the proponent agency is TRADOC.

PRIVACY ACT STATEMENT

AUTHORITY: 10 USC 3012(g)/Executive order 9397
PRINCIPAL PURPOSE: To aid individual training on targets at various ranges.
ROUTINE USES: To evaluate individual proficiency.
DISCLOSURE: Voluntary. However, mass rating and recording require some tracking method.

NOTE: Use this table if you do not have a MK 19/sight combination that applies to Table II, and you are using hull targets OR if targets are pop-up/E type silhouettes.

1a. LAST NAME	1b. FIRST NAME	1c. MI	2. RANK	3. UNIT

TABLE IV (A). DISMOUNTED AND MOUNTED NIGHT PRACTICE

4. RANGE	5. LANE	6. GRADER	7. DATE

TASK	RANGE (Meters)	AMMO	TIME (Minutes)	ENGAGEMENT STANDARDS	GO	NO GO
ZERO	400	4	NA	2 ROUNDS HIT		
2	600 (+/- 100)	6	2	2 ROUNDS HIT		
3	800 (+/- 100)	6	2	2 ROUNDS HIT		
4	400	4	1.5	2 ROUNDS HIT		
MULTIPLE TARGETS						
5	800 (+/- 100)	10	3.5	1 ROUND HIT		
6	400			1 ROUND HIT		
7	400	14	2.5	1 ROUND HIT		
8	600 (+/- 100)			1 ROUND HIT		
				TOTALS		

8. COMMENTS

9. NUMBER OF ENGAGEMENT MET *(Choose One)*

☐ 8 - EXPERT ☐ 6 - MARKSMAN
☐ 7 - SHARPSHOOTER ☐ 5 AND BELOW - UNQUALIFIED

10. GUNNER'S SIGNATURE

11. GRADER'S SIGNATURE

TABLE IV (B). DISMOUNTED AND MOUNTED NIGHT QUALIFICATION

12. RANGE	13. LANE	14. GRADER	15. DATE

TASK	RANGE (Meters)	AMMO	TIME (Minutes)	ENGAGEMENT STANDARDS	GO	NO GO
ZERO	400	4	NA	2 ROUNDS HIT		
2	600 (+/- 100)	6	1.5	2 ROUNDS HIT		
3	800 (+/- 100)	6	1.5	2 ROUNDS HIT		
4	400	4	1	2 ROUNDS HIT		
MULTIPLE TARGETS						
5	800 (+/- 100)	10	3	1 ROUND HIT		
6	400			1 ROUND HIT		
7	400	14	2	1 ROUND HIT		
8	600 (+/- 100)			1 ROUND HIT		
				TOTALS		

16. COMMENTS

17. NUMBER OF ENGAGEMENT MET *(Choose One)*

☐ 8 - EXPERT ☐ 6 - MARKSMAN
☐ 7 - SHARPSHOOTER ☐ 5 AND BELOW - UNQUALIFIED

18. GUNNER'S SIGNATURE

19. GRADER'S SIGNATURE

Made in the USA
Monee, IL
05 November 2026

67811407R00272